Lecture Notes in Artificial Intelligence 4155

Edited by J. G. Carbonell and J. Siekmann

Subseries of Lecture Notes in Computer Science

Oliviero Stock   Marco Schaerf (Eds.)

Reasoning, Action and Interaction in AI Theories and Systems

Essays Dedicated to Luigia Carlucci Aiello

 Springer

Series Editors

Jaime G. Carbonell, Carnegie Mellon University, Pittsburgh, PA, USA
Jörg Siekmann, University of Saarland, Saarbrücken, Germany

Volume Editors

Oliviero Stock
ITC-irst
Via Sommarive 18, 38050 Povo, Trento, Italy
E-mail: stock@itc.it

Marco Schaerf
Università di Roma "La Sapienza"
Dipartimento di Informatica e Sistemistica "Antonio Ruberti"
Via Salaria 113, 00198 Rome, Italy
E-mail: marco.schaerf@dis.uniroma1.it

Library of Congress Control Number: 2006931263

CR Subject Classification (1998): I.2, H.4, F.4.1

LNCS Sublibrary: SL 7 – Artificial Intelligence

ISSN   0302-9743
ISBN-10   3-540-37901-0 Springer Berlin Heidelberg New York
ISBN-13   978-3-540-37901-0 Springer Berlin Heidelberg New York

Springer is a part of Springer Science+Business Media

springer.com

© Springer-Verlag Berlin Heidelberg 2006

Typesetting: Camera-ready by author, data conversion by Scientific Publishing Services, Chennai, India
Printed on acid-free paper   SPIN: 11829263   06/3142   5 4 3 2 1 0

Foreword

Often times, we celebrate people at the end of their scientific career, when we look back at their accomplishments. The occasion for this book is different. It is a pleasure, in this case, to celebrate someone who has pioneered scientific developments in artificial intelligence and has served the scientific community with great energy, and who will certainly remain active in research for many years to come. This book is dedicated to Luigia Carlucci Aiello, better known as Gigina. The Festschrift makes its appearance exactly 50 years after the "official" birth of artificial intelligence (at the historical Dartmouth Conference) at the same time as a similarly round birthday for Gigina.

At Dartmouth College, the initial program for AI was set up by John McCarthy, Marvin Minsky, Herbert Simon and a few others. Alan Turing had been dead only a few years and the brightness of the English mathematician's ideas about computational intelligence was still lingering. The Dartmouth Conference set the scene for exciting research activity in this new field. Even though, at times, the results turned out remarkably different from the initial expectations, the field of AI expanded, adapting to the changing world and continues to flourish today. One of the most emblematic stories of AI is computer chess.

The ambition of being able to compete with the best chess players has been one of the original challenges that have characterized artificial intelligence. At the time of the Dartmouth Conference, in 1956, the belief existed that within some ten years, it would be possible to realize a program that would prevail over the best human players. The truth is that at the beginning of the 1990s, this goal was far from being achieved– the best chess programs were not at the level of the 100th best human chess player.

Later, IBM invested enormously in a special project that led to the development of Deep Thought, first, and then of Deep Blue. Deep Blue was based on special hardware and substantially exploited brute force, with its capability of exploring 200 million moves per second. The system, after a first failed challenge, succeeded in 1997 in overcoming Garry Kasparov, generally considered the best human chess player of all time. Kasparov left the game site furious, claiming that the operators of Deep Blue had cheated, for instance by contravening the agreements about operating the computer between games. In any case, he never accepted the result. Nonetheless, IBM declared victory, and Deep Blue never played again. The results of Deep Blue formed the basis of a number of other practical projects that IBM sustained in the following years. Even though Deep Blue was not known to have incorporated aspects that are typical of human intelligence, the AI community took advantage of the result and declared that one of the best known goals of AI, put forward by the founders of the discipline, had been attained.

1997 marked a milestone, but it was not the end of the story. The following years saw the development of a new generation of programs, running on traditional hardware (powerful PCs), but based on strategies that come closer to those used by humans, including pattern matching, machine learning, opponent modeling, reasoning and planning. In general the new programs, like Deep Junior (world champion in 2006), Shredder, or Zappa, are specifically built for competing at the world championship of computer programs. Yet in 2003, Deep Junior challenged Garry Kasparov, with a tie as the final result. However, this time Kasparov declared himself satisfied with a tie result, and clearly played the final two games on the defensive, impressed by Junior's capability to change strategies in the course of the game. Kasparov seemed to recognize that the future of "chess intelligence" had been declared.

Critics of artificial intelligence have often condemned it for philosophical reasons, ethical reasons and technical reasons. We will not discuss this issue here. Let us just say that in the practical world artificial intelligence has produced many positive results. They may not always appear so clearly because of a paradox: AI is often considered the computer science of the future and, as a consequence, whenever there are practical results, they cannot be attributed to artificial intelligence.

Gigina was born in Fabriano, Marche, a remarkably beautiful region of central Italy, full of historical and artistic monuments that form the casual backdrop for everyday life. Fabriano is famous for its paper industry (there, Gigina and her collaborators must have learned about "paper" production).

After high school she moved to Pisa, where she studied mathematics at the Scuola Normale Superiore with some renowned Italian mathematicians, such as Ennio de Giorgi. At that time, she fell in love doubly: with the theory of computation and with her husband Mario Aiello, an outstanding computer scientist. With him and with other Pisa friends, including Giuseppe Attardi, and Gianfranco Prini she published various brilliant works. At the time, Pisa was the recognized center of computer science research in Italy. In that environment a whole group of young scientists at the Istituto per l'Elaborazione dell'Informazione (Institute for Information Processing) of the National Council for Research began to develop research in artificial intelligence: these include, among others, Ugo Montanari, Giorgio Levi, Franco Sirovich, Alberto Martelli and Franco Turini. Several important papers that appeared in the *Artificial Intelligence Journal* were produced in Pisa at that time. Gigina became a protagonist in the AI scene. In Italy, the scene also included Marco Somalvico and his robotics group at Milan Polytechnic, and, very soon research activity in AI started in Turin and in various major universities. When their son Marco was still a small child, Gigina had to cope with a personal tragedy: the loss of her husband Mario. Subsequently, she decided to move to the United States, to work at Stanford for a few years, in a research environment inspired mainly by John McCarthy. When she came back to Italy she soon took the position of full professor in Ancona, certainly one of the youngest full professors (and female professors!) in Italy. From there she moved to University La Sapienza in Rome, where she taught and established a group including a large number of well-known AI scientists, and where she teaches artificial intelligence now. For two years she also directed IRST in Trento, a major research center in AI, Microsystems and Surfaces.

In the Italian AI research scene in the late 1970s and early 1980s, there were several universities active in AI: Pisa, Rome, Turin, Genova, Milan, Padova, Udine, Brescia, Florence, Bologna, Napoli, Bari, and Palermo. Within the National Council of Research, Istituto per l'Elaborazione dell'Informazione and Istituto di Linguistica Computazionale in Pisa, and Istituto di Psicologia in Rome were also major players. Industrial research was also active, at CSELT in Turin, at Tecsiel and Olivetti, plus at a number of smaller companies. An interest group on AI was established within the Italian Association for Computer science (AICA) in those years. In the mid-1980s, IRST was established in Trento as an important institute devoted mainly to AI. In 1987, thanks to the effort of Marco Somalvico, the major AI conference–IJCAI–was held in Milan. In 1988, the Italian Association for Artificial Intelligence (AI*IA) was established and Gigina was appointed its first President. The Association has brought about a notable set of activities, including a good-quality biennial conference in odd years, a convention in even years, a number of workhops organized by special interest groups and a magazine. The first conference, held in Trento in 1989, had almost 300 participants. AI*IA has become a serious enterprise and AI has received continuous attention in Italy since.

In the meanwhile, the European organizational scene was set between the 1970s and 1980s. Wolfgang Bibel and others worked on establishing a European AI society. At the Amsterdam AISB conference, the European Coordination Committee for Artificial Intelligence was established as an umbrella organization whose members are national AI societies. AICA was among the first, and was later replaced by the novel AI*IA. It was also decided to hold the first European Conference for Artificial Intelligence in Orsay in 1982, the first (even if recorded as "Fifth" to take into account the previous events within AISB) of an ongoing series of important biennial conferences. Bibel was elected the first ECCAI Chair, and further national AI societies including several in the Eastern Block were initiated. ECCAI also established various activities including committees for advising the European Commission. Gigina was among the few key figures in these developments.

Outside Europe, for many years Gigina has been a major actor in the two main international scientific forums of artificial intelligence: IJCAI, as General Chair of IJCAI 1999 in Stockholm, Chair of the Trustees, and on the editorial board of the *Artificial Intelligence Journal*, where she has collaborated to steer the development of the field.

Luigia Carlucci Aiello's research activities have addressed a wide range of topics. Her first interests back in the early 1970s were in the field of pattern recognition. Soon after, she shifted into programming language semantics, program properties and their automatic proofs. She defined a semantics for PASCAL in LCF (Logic for Computable Functions) and machine checked several proofs of (nontrivial) properties. She designed and implemented a theorem prover named PPC (Pisa Proof Checker), published papers describing the (object-oriented) program-

Luigia Carlucci Aiello

ming paradigm used in its development, and its evaluator (an interpreter for the full lambda calculus), and participated and contributed to the development and use of Wehyrauch's FOL system. At the same time, she has investigated the advantages of using meta-level knowledge in AI systems, both to control search and to reason in multi-agent systems. She was involved in the first experiments in Italy on the application of artificial intelligence techniques to the development of expert systems for medical diagnosis and of intelligent tutoring systems. Since at least 1990, she has been active in the field of nonmonotonic reasoning, where her interests have spanned from the characterization of default proofs, to tableau systems for default logic, to modal characterizations of default logic and definability of concepts of natural kinds. Her latest research activity has been in the fields of cognitive robotics, AI techniques applied to security, and AI planning.

In launching this Festschrift initiative, we have involved a number of people who have worked in research with Gigina, or have been her students, who have then embarked on an important path of their own. Others have shared with Gigina a leading role in the strategic service of the AI community, internationally or as presidents of the Italian Association for Artificial Intelligence.

Alberto Martelli, Eugenio Omodeo and Franco Turini are some of the colleagues and friends that worked together with Gigina in that magic period in Pisa.

Gigina with the students of the first AI course she taught in Rome in 1992

Marco Cadoli, Alessandro Micarelli, Daniele Nardi, Fiora Pirri (and Marco Schaerf) are representatives of the large group of AI scientists who have taken the first steps (and often many subsequent steps) with Gigina in Rome.

Pietro Torasso, Roberto Serra, Marco Gori and Marco Schaerf have all followed Gigina as AI*IA Presidents (as has Oliviero Stock, who was also a student of a very young Gigina in Pisa).

Wolfgang Bibel, Alan Bundy, Robert Kowalski, Erik Sandewall, Jrg Siekmann, and Wolfgang Wahlster are certainly among the small group of the most influential scientists in AI. They share with Gigina a passion for reasoning and logic, and have been essential parts of her intellectual milieu. Most of them have worked with her in leading positions in the international AI community, in ECAI, IJCAI Inc. or the AI Journal.

Paolo Traverso worked with Gigina during her period as IRST director. And last but not least, Roberto Cordeschi, a friend of Gigina's and an AI historian, especially in recent times has helped understand the initial years of AI, the period before Gigina's intervention.

This collection is a detailed insight into many aspects of AI and the relationship between theoretical and applied research.

The first set of papers is dedicated to the foundations of AI.

The paper by Roberto Cordeschi, "Searching in a Maze and in Search of Knowledge: Issues in Early Artificial Intelligence" gives an account of the origins of heuristic programming and the shift to knowledge-based or real-life problem solving that followed the initial days of AI.

Wolfgang Bibel in his "Research Perspectives for Logic and Deduction" provides an authoritative manifesto for the role of logic and deduction within AI or, better, within what he has called Intellectics.

Questions and techniques related to computational logic are the themes for various subsequent chapters.

Jörg Siekmann et al.'s paper "Reductio ad Absurdum: Planning Proofs by Contradiction" addresses the questions: how can we proof plan an argument by reduction ad absurdum? When is it useful to do so? What are the methods and decisions involved?

In "Computational Logic in an Object-Oriented World," Bob Kowalski investigates transformations between object-oriented and abductive logic programming systems and argues that ALP multi-agent systems can combine the advantages of logic with the main benefits of object orientation.

In "Best-First Rippling," Alan Bundy and colleagues address the limitations of rewriting systems based on continuous incremental reduction of differences between formulae by introducing a more flexible and efficient best-first technique.

Marco Cadoli and Marco Schaerf's "Partial Solutions with Unique Completion" looks at the computational complexity of several reasoning problems, their formulation by means of quantified Boolean formulae and their solution through an appropriate solver.

In "The Computerized Referee," Eugenio G. Omodeo and colleagues present a system that either certifies a text as constituting a valid sequence of definitions and theorems, or rejects it as defective, and they discuss a series of new enhancements to the system.

Other general themes are the subjects of the subsequent two papers.

In the paper "About Implicit and Explicit Shape Representation" Fiora Pirri addresses a different topic: the analysis of shape and form as the basic features for understanding the relation between images, offering a novel approach to shape approximation and similarity measures and their use.

In "Agents, Equations and All That: On the Role of Agents in Understanding Complex Systems," Roberto Serra and Marco Villani offer a different perspective on a familiar matter; they show how differential equations can describe interactions among agents and point out that the capabilities of the former are broader than is often assumed.

The theme of intelligent robotics is prominent in the set of papers that follow.

In "Coordination of Actions in an Autonomous Robotic System", Erik Sandewall describes the design and formal characterization of a cognitive process, called an action coordinator, that manages restrictions in real-world actions.

Robots in soccer is a very popular theme and could not be passed up in this collection. In "AI and RoboCup", Daniele Nardi and Luca Iocchi provide an AI research perspective on RoboCup, based on the experience accumulated in several years of RoboCup activity.

"Planning Under Uncertainty and Its Applications" by Paolo Traverso is on a theme that bridges between robotics and several other areas. Traverso discusses solutions to the problem of actions that may have different effects that cannot be predicted at planning time and some applications in different domains.

The subsequent set of chapters is dedicated to AI and the Web.

In "Reasoning About Web Services in a Temporal Action Logic" Laura Giordano and Alberto Martelli present an approach to reasoning about Web services, described by specifying their interaction protocols in an action theory based on a dynamic linear time temporal logic.

Alessandro Micarelli and colleagues in "Intelligent Search for the Internet" approach the key theme of personalization and adaptation of human–computer interaction to overcome information overload, by means of machine learning techniques and AI-based information representations.

The history of artificial intelligence has seen many attempts to compete with humans at games.

In "Cracking Crosswords: The Computer Challenge", Marco Gori and colleagues go further and tackle the cracking of crosswords and describe a system which relies strongly on interaction with the Web for clue answering.

The subsequent two papers focus on techniques particularly relevant for specific classes of problems.

Pietro Torasso and Gianluca Torta in their "Model-Based Diagnosis Through OBDD Compilation: A Complexity Analysis" address the problem of evaluating the complexity of diagnostic problem solving, characterized by a potentially exponential size of the search space, often circumvented by compilation of the domain model.

In "Examples of Integration of Induction and Deduction in Knowledge Discovery," Franco Turini and colleagues investigate the use of classification trees in two quite different application areas — business documents and geographic information systems — complementing the use of induction from examples with the exploitation of some form of deductive knowledge.

The final paper in this collection, "SharedLife: Towards Selective Sharing of Augmented Personal Memories" by Wolfgang Wahlster and colleagues, is concerned with a very ambitious topic, namely, the building of an augmented personal memory from the recording of physical and communicative interaction of an individual in an instrumented environment, and its use with the goal of supporting communication between individuals and learning from the experiences of others.

We wish to thank all authors for their enthusiastic participation; Sara Kaufman and Davide Micaletto for their help in producing a polished final version of this book. Some of the best young scientists have grown close to Prof. Luigia Carlucci Aiello and have been inspired by her continuously. May this tribute to her be of inspiration to an even larger, new generation of AI scientists.

Oliviero Stock and Marco Schaerf

Editors and Chapter Authors

Editors

Oliviero Stock is a Senior Fellow at ITC-irst, Trento. He was Director of ITC-irst from 1997 to 2002 and currently coordinates the collaboration project with the University of Haifa, Israel. He is the author of 170 published papers and the author or editor of 11 books. He has been Chairman of the European Coordinating Committee for AI, and President of AI*IA. He is an ECCAI and AAAI Fellow.

    Marco Schaerf is Full Professor of Computer Science at the University of Roma "La Sapienza." He is the current President of the Italian AI Association (AI*IA) and member of the Editorial Board of the *Artificial Intelligence Journal*. His research interests include many areas of artificial intelligence such as knowledge representation, planning, belief revision and computational aspects of AI.

Chapter Authors

Roberto Cordeschi is Professor of Philosophy of Science at the University of Rome "La Sapienza." He is the author of several publications on the history of cybernetics and the epistemological issues of artificial intelligence, including *The Discovery of the Artificial* (Kluwer, 2002).

    Wolfgang Bibel is currently Professor Emeritus of Intellectics at the Department of Computer Science of the Darmstadt University of Technology in Germany. He also maintains an affiliation with the University of British Columbia in Vancouver, Canada, as an Adjunct Professor.

    Jörg Siekmann obtained his PhD at Essex University and returned to Germany in 1976. He is one of the founders of AI in Germany, the first chairperson of the German AI society, fellow of the GI and of ECCAI and now one of the directors of DFKI at Saarbrücken, of which he was among the founders. His research areas are automated reasoning, unification theory, proof planning, and architectures for multiagent systems.

    Erica Melis graduated in mathematics and later focused her research on AI, more specifically on analogical reasoning, proof planning and most recently (at DFKI) on intelligent learning tools. She has pioneered knowledge-based proof planning.

    Robert Kowalski is Professor Emeritus and Senior Research Fellow at the Department of Computing, Imperial College. His early research was in the field of automated deduction. In 1972, he collaborated with A. Colmerauer on the development of logic programming, publishing *Logic for Problem Solving* (1979). His current research focuses on the application of computational logic to cognitive science.

Moa Johansson is a PhD student in the Mathematical Reasoning Group at the University of Edinburgh.

Alan Bundy is Professor of Automated Reasoning in the School of Informatics at the University of Edinburgh. His research is concerned with the interaction and integration of reasoning processes, including inference, search, formalization and learning. He is the winner of the 2007 IJCAI Research Excellence Award.

Lucas Dixon is a Research Associate with the Mathematical Reasoning Group at the University of Edinburgh, where he has developed IsaPlanner, a generic proof planner for the Isabelle Proof Assistant.

Marco Cadoli is Full Professor of Computer Engineering at the University "La Sapienza" of Rome. In 1984, he attended a programming course lectured by Gigina Aiello, and since then has enjoyed a fulfilling and fruitful collaboration with her.

Domenico Cantone earned his PhD in Computer Science from New York University in 1987 under Jacob T. Schwartz. Currently, he is Professor of Computer Science at the University of Catania. His main scientific interests include: computable set theory, automated deduction in various mathematical theories, string matching, and algorithmic engineering.

Eugenio G. Omodeo earned his PhD in Computer Science at New York University in1984 under Martin Davis. Currently, he is Professor of Computer Science at the University of Trieste. His main scientific interests include: computational logic, computable set theory, formal methods in computer science, and logic programming.

Alberto Policriti earned his PhD in Computer Science from New York University in 1990 under Martin Davis. Currently, he is Professor of Computer Science at the University of Udine. His main scientific interests include: computable set theory, model checking and verification, systems biology and algorithms for bioinformatics.

Jacob T. Schwartz, Professor Emeritus of Computer Science and Mathematics at New York University, earned his PhD at Yale University in 1951 and then published the treatise "Linear Operators" with Nelson Dunford. He has worked in programming language design and optimization, computational logic, robotics, parallel computing, and multimedia.

Fiora Pirri received her PhD from the UPMC, Paris VI. She is Professor of Computer Science at the University of Rome "La Sapienza," where she leads the Alcor Laboratory of Cognitive Robotics. She has coauthored more than 70 papers, conference papers, and book chapters on these topics and she is coauthor with Gigina Aiello of *Structures, Logic and Languages*, published by Pearson.

Roberto Serra is Full Professor of Computer Science and Engineering at the University of Modena and Reggio Emilia. He has been President of AI*IA and is the author of four books and about 90 papers in AI and complex systems.

Marco Villani is a tenured Research Associate in Computer Science and Engineering at the University of Modena and Reggio Emilia and the author of several papers dealing with neural networks, genetic algorithms, cellular automata and agent-based models.

Erik Sandewall has been Professor of Computer Science at Linkping University, Sweden, since 1975. His areas of interest in research include reasoning about actions and change, cognitive robotics, systems for human–robot dialogue, and software infrastructure for robotic systems. He is currently Co-editor in Chief of the *Artificial Intelligence Journal.*

Daniele Nardi is Full Professor of Artificial Intelligence at the University of Rome "La Sapienza" in the Department of Computer and System Science. His research interests include various aspects of knowledge representation and reasoning, such as description logics and nonmonotonic reasoning, cognitive robotics, and multiagent and multirobot systems.

Luca Iocchi is Assistant Professor at the University of Rome "La Sapienza," in the Department of Computer and System Science. His main research interests in the area of cognitive robotics and mobile robotics are action planning, self-localization, stereo-vision and coordination among multiple robots.

Paolo Traverso is Head of Division at ITC-IRST, Trento. His main research interests include automated planning, formal verification, and service-oriented applications. He has served on the Board of Directors of the Italian Association for Artificial Intelligence. In 2005, he was elected ECCAI Fellow.

Alberto Martelli is Full Professor of Computer Science at the University of Torino and is an ECCAI Fellow. His main research interests are nonmonotonic reasoning, reasoning about actions and change, and computational logic.

Laura Giordano earned her PhD in Computer Science at the University of Torino in 1993. Since 1998, she has been Associate Professor at the Universit del Piemonte Orientale. Her main research interests are nonmonotonic reasoning, belief revision, reasoning about actions and change, and nonclassical logics.

Alessandro Micarelli is a Full Professor of Artificial Intelligence at the University of "Roma Tre" where he is in charge of the Artificial Intelligence Laboratory of the Department of Computer Science and Automation. His research interests include: adaptive Web-based systems, personalized search, user modeling, and artificial intelligence in education.

Fabio Gasparetti is a Postdoctoral Fellow at the AI Lab of the Department of Computer Science and Automation, University of "Roma Tre". His research interests include: adaptive Web search, user modeling and focused crawling.

Claudio Biancalana is an intern at the AI Lab of the Department of Computer Science and Automation, University of "Roma Tre". His research interests include: personalized search and text categorization.

Marco Gori received his PhD degree in 1990 from the University of Bologna. His main interests are in machine learning, pattern recognition, and game playing. Professor Gori serves as a member of the editorial board of many technical journals. He is the Chairman of the Italian chapter of the IEEE Computational Intelligence Society and he has been the President of AI*IA, and is an IEEE Fellow.

Giovanni Angelini is a PhD student in the Engineering Department at the University of Siena. His main interests are in machine learning, pattern recognition and natural language processing.

Marco Ernandes is a PhD student in Cognitive Science at the University of Siena. He is actively involved in the WebCrow project at the Department of Information Engineering.

Pietro Torasso is Full Professor of Computer Science at the University of Torino, prior to this position he was Associate Professor at Torino, and Full Professor at University of Udine and Campus of Alessandria. He was President of AI*IA from 1995 to 1999 and co-winner of the ECCAI Prize in 1990. Since 2000, he is an ECCAI Fellow.

Gianluca Torta received his PhD in Computer Science at the University of Torino in 2005. Currently, he is a researcher of Computer Science at the University of Turin.

Franco Turini is a Full Professor in the Department of Computer Science of the University of Pisa. His research interests include design, implementation, and formal semantics of programming languages, logic programming and deductive databases, and knowledge discovery. He is a member of the IEEE and ACM.

Miriam Baglioni is a research assistant at the Department of Computer Science of the University of Pisa. Her research interests are in the field of data mining, in particular in the design and implementation of data mining environments.

Barbara Furletti is a PhD student at the Lucca Institute for Advanced Studies. Her research interests are in the field of data mining, particularly the design and implementation of data mining algorithms.

Salvatore Rinzivillo is a research assistant at the Department of Computer Science of the University of Pisa. His research activity includes knowledge discovery from spatial databases.

Wolfgang Wahlster is the Director of the German Research Center for Artificial Intelligence (DFKI) and a Professor of Computer Science at Saarland University. He is an AAAI, ECCAI, and GI Fellow and has received the German Future Prize in 2001 from the President of Germany. He is a Full Member of the German Academy of Sciences and a Foreign Member of the Royal Swedish Academy of Sciences.

Alexander Kröner is a Senior Researcher at DFKI who has written his doctoral thesis under the supervision of Wolfgang Wahlster on the adaptive layout of Web pages.

Dominik Heckmann is a Senior Researcher at DFKI who has written his doctoral thesis under the supervision of Wolfgang Wahlster on ubiquitous user modeling.

Table of Contents

Searching in a Maze, in Search of Knowledge: Issues in Early Artificial Intelligence

Roberto Cordeschi

Università di Roma "La Sapienza"
Dipartimento di Studi Filosofici ed Epistemologici
Via Carlo Fea 2, I-00161 Roma, Italy
cordeschi@caspur.it

"He possesses two of the three qualities for the ideal detective. He has the power of observation and that of deduction. He is wanting only in knowledge..."
Sherlock Holmes, in speaking of Franois de Villard, the French detective

Abstract. Heuristic programming was the first area in which AI methods were tested. The favourite case-studies were fairly simple toy-problems, such as cryptarithmetic, games, such as checker or chess, and formal problems, such as logic or geometry theorem-proving. These problems are well-defined, roughly speaking, at least in comparison to real-life problems, and as such have played the role of *Drosophila* in early AI. In this chapter I will investigate the origins of heuristic programming and the shift to more knowledge-based and real-life problem solving.

1 Introduction

AI has been around for 50 years now:1956 was the year when the Dartmouth meeting officially signalled the birth of AI. An anniversary like this is a fitting occasion to take stock of the events of a half-century worth of research. We can do so in at least two different ways: (i) we could map out possible future scenarios on how AI will evolve, by a look at today's most promising research programs (and those less promising) – an evaluation that not all would agree with; or (ii) we could revisit certain topics and theoretical issues, certain experimental research and methodological controversies that were raised and developed in the era of the Dartmouth pioneers, to think about how the evolution of AI has led us to the point where it stands today.

In this chapter, I have chosen the second option. I aim to discuss some topics, results and controversies which have characterised AI research in the fifteen or so years following Dartmouth (1956-72 ca.). This was, indeed, a lively time in AI – it was then that a large part of its scientific vocabulary came into existence. As we shall see, some topics appear to have been given a systematic form during this time, while others were put aside, to be brought to the fore again later, in more mature contexts.

The topics I shall consider here include heuristic search and heuristic programming (sections 2 and 3), problem representation (section 4), and the early approaches to knowledge representation, regarding both the nature of the problems

O. Stock and M. Schaerf (Eds.): Aiello Festschrift, LNAI 4155, pp. 1–23, 2006.

faced and the different methods employed: toy problems and real-life problems (section 5), and well-structured and ill-structured problems, weak and strong methods (section 6). A conclusion follows in section 7, including a brief look at the most recent developments in some of these areas. My aim is to suggest, through these developments, how AI has not evolved along a linear path, nor can its evolution be described through a succession of "paradigms". All this *may* prove to be instructive when considering the first alternative (i) mentioned above.

2   From a "Mythical Being" to the Actual Decision Maker

Claude Shannon had already started to think about a computer chess program around the mid-1940s. His sketch of the program was based on the idea that the best move could be evaluated using a look-ahead analysis of alternative moves based on the minimax procedure (see [50]). This procedure had been the basis of an earlier chess program hand-simulated by Alan Turing (Hodges [21] 213-214). The origins of the minimax procedure lie in the early formulations of mathematical game theory. The chess player was established as a common metaphor in the analysis of decision making, with its classic formulation in *Theory of Games and Economic Behavior*, published by von Neumann in 1944 together with the economist Oskar Morgenster. In their terminology, chess is a zero-sum game. In theory, this means imagining a perfectly rational player who applies the minimax procedure to every possible move, assigning a value of $+1$ for a victory, 0 for a stalemate and -1 for a loss. In practice, this optimal strategy encounters an insurmountable difficulty owing to the combinatorial explosion of possible moves, which Shannon calculated to be to the order of 10^{120}. The best comment on this situation came from von Neumann and Morgenstern themselves:

> This relative, human difficulty necessitates the use of those incomplete, heuristic methods of playing, which constitute "good" chess, and without this human difficulty there would be no element of "struggle" and "surprise" in this game (von Neumann and Morgenstern, [35] 125).

This is the difficulty that Norbert Wiener seems to have had in mind when in the first edition of *Cybernetics* in 1948, he suggested building a machine which, without playing "an optimum game in the sense of von Neumann", would in any case be able "to [...] offer interesting opposition to a player at some of the many levels at which human chess players find themselves" (see [62] pp. 164-65).

Thus, when Shannon took on the problem of programming a procedure based on the minimax algorithm, he could not avoid the problem of how "to develop a tolerably good strategy for selecting the move to be made" (see [50], p. 260). He suggested the program should have in-built selectivity criteria taken from the Dutch psychologist Adrian de Groot's investigations into the choice processes of chess masters who made their analyses by "thinking-aloud" during the game.

In those years it was the choice-making processes themselves which came under the scrutiny of one pioneer of organisation theory and Operations Research

(OR), Herbert Simon (von Neumann and Morgenstern's quotation above is bor-
rowed from a paper by Simon). Elsewhere I have discussed in some detail how
placing these decision-making processes at the centre means there is a shift of
interest in disciplines dedicated to studying decision-making behaviour – a shift
which includes a renunciation of the *normative* approach of mathematical game
theory, and a move to the study of actual choice processes (see [11]). Briefly,
the *normative* approach consists of an analysis of the choices (or strategies) that
the rational agent should adopt to arrive at the optimal solution to a problem.
Simon shifted his focus to the choice (or strategy) that agents normally make,
since they are conditioned by their (subjective) idea of the task domain. Only
in this case was the analysis of choice processes placed at the forefront, particularly, how they are conditioned by how the agent perceives or represents the
task domain.

In making this non normative point of view his own, Simon concluded a line of
research begun in the 1940s, and contradicted the model of the *Homo oeconomi-
cus*, whose decision-making behaviour is guided by the principle of maximising
utility – which had been the assumption in game theory and OR (more on this
in section 4). Simon studied the agent, or problem solver, not as someone pos-
sessing ideal, omniscient rationality, but in terms of somebody with limited or
"bounded rationality"[1]. The chess player metaphor again came to the fore, but
no longer within the profile of "unlimited intellect" (as Shannon called it), or the
"entirely mythical being" (as Simon would call it) of classic economics. Indeed,
already in 1952 Simon formulated the hypothesis of a chess-game program whose
base was not crucially the minimax algorithm and static evaluation function in
Shannon's sense, but rather the notion of "satisficing" choice (see [51]).

Simon's contribution to AI's birth at the 1956 Dartmouth meeting, as well
as that of other fathers of the new discipline, has been pointed out many times.
Here I will focus on how he and other AI pioneers handled two issues that
proved immediately crucial in building the first intelligent programs: control and
representation. Both involve the question I mentioned above, which comes up
immediately when building a program endowed with the features of the problem
solver as a rational agent – the question of the combinatorial explosion of the
possible alternatives to be evaluated before making a decision. This was the
problem Shannon considered in chess – how to supply the program with *selective*
strategies aimed at freeing chess programming from the impression of its being
based on "brute force" rather than on a fairly expert analysis of the sequence of
moves.

Simon's proposal found a place in the problem of how to control the combinatorial explosion of alternatives. He suggested starting from von Neumann and
Morgenstern's "relative, human difficulty" to study the "heuristic" strategies
which, since elaborated by agents with bounded rationality but notable flexibility (i.e. human beings), use elements of "struggle" and "surprise" rather than

---

[1] See [26] ch. 1, for an introduction,, and see [11] and [12] on the influence of Simon's
theory of decision making on early AI. Kahneman and Frederick [22] discuss the
relationship between Simon's theory and the explanation of "cognitive illusions".

brute force. The aforementioned theory of choice put forward by Simon was influential in the construction of some of the now historic computer programs based on different heuristic strategies, which are among the first examples of what became known in AI as *heuristic programming*: the LT (Logic Theorist), the GPS (General Problem Solver) and some versions of chess programs, all of which came from Simon's work with Allen Newell and Clifford Shaw from the mid-1950s.

These programs "maximally confuse [...] with mutual benefit" the goal of limiting the combinatorial explosion using heuristic strategies and the goal of modelling or simulating by computer the cognitive processes used by human problem solvers (Newell and Simon [39] 279). Nevertheless, cognitive simulation remained the main goal of these programs, and was absent from other programs of the time, where the aim was rather to implement highly *efficient* selective, or heuristic, strategies. The most famous example is Arthur Samuel's checker programs. Starting in 1952, they were (from 1954 on) endowed with learning algorithms which made them milestones in machine learning.

These two aims, the *synthesis* and the *modelling* or *simulation* of human intelligence, immediately defined two distinct areas of research in AI. Minsky pointed this out at MIT in 1961, in the presentation of a version of GPS by Simon. He and Simon agreed on the fact that the aim of the modelling approach was "to understand how humans think", according to Newell, Shaw and Simon's IPP (Information Processing Psychology) program (later merging into Cognitive Science). Meanwhile the other approach (that Minsky adhered to) aimed at obtaining "good problem-solving programs", and according to Minsky, the two approaches were opposed to the neural net and self-organizing system approach. Notice that at the 1958 Teddington Symposium Minsky had already labelled "heuristic programming" as the main task of newly-born AI, openly positioning it against the neural net approach which he judged inefficient[2]. This was a widely held opinion at the time, and Samuel ([48] 207) also distinguished "two general approaches to the problem of machine learning". The first, "which might be called the *Neural-Net Approach*", was considered promising by Samuel, but at the time inefficient and not realizable. The second approach was the computer programming of learning – "much more efficient" and "capable of realization at present"[3].

---

[2] See [10] on both these events.

[3] Samuel's programs were the main attempts to deal with learning during the period I am considering here – to some extent, they were an exception. Learning had been a central topic during Cybernetics, and continued to be investigated in the neural net and self-organising system community (think to James Anderson, Eduardo Caianiello, Teuvo Kohonen and so forth). AI researchers interested in computer programming were more concerned with efficient performance and the mechanisms responsible for this (such as heuristic search and knowledge representation) than with learning. It was during the late 1970s that learning come to the forefront, for example with the AM and EURISKO programs (see section 7). Starting from 1983, the *Machine Learning* collection, edited by Ryszard S. Michalski, Jaime G. Carbonell and others, documented AI investigations on learning.

Heuristic programming then became the basis of the "heuristic search paradigm", to use the expression used in 1968 by Edward Feigenbaum to characterise AI of the previous decade (see [15]). By such an expression Feigenbaum was referring to the earlier AI programs based on heuristic procedures. These were often very different from each other, not only in the specific methods adopted, but also the terminology used – terminology which started developing from circa 1955. Some years after Feigenbaum's review, the book by Nilsson [44] became the first, highly influential systematic treatment of the subject. In the next two sections I shall look briefly at certain topics and controversial issues leading to the formation of a vocabulary of AI in relation to the issues of heuristic search and problem representation.

3 Searching in a Maze

In the pioneering years of heuristics programming the search for a solution to a problem was studied mainly using a representation adaptable to the problems considered by early AI, i.e. puzzles, toy problems and games: the tree-shaped representation. To some extent, the idea behind this type of representation came from game theory and psychology – psychologists had often described the activity of problem solving, both human and animal, as a solution of mazes. In this case, too, the concepts and methods used by early AI overlapped with those of psychology.

"Maze" or "problem maze" was the term used initially (and later abandoned) by the founders of IPP as an equivalent of the game tree. In its turn, "problem maze" was used as the equivalent to "problem space", the term introduced by the founders of IPP between 1956 and 1958 (see [43])[4]. As I said before, a generic definition of "heuristics" was a method or procedure which would reduce the combinatorial explosion of legal moves – moves occurring in the problem space. In the beginning, there was a certain amount of unanimity in placing a "heuristic" in opposition to an "algorithm", which as opposed to a heuristic guarantees the discovery of the solution, if indeed the solution exists. Later this distinction was seen generally not to be as well founded as had been thought back then, e.g., in the case of the resolution method in theorem proving and its various "heuristics"[5].

Coming back to trees, in the case of one-person games or even theorem proving, the search space took on the shape of a game tree. With two-person games like chess, the game tree was different, since it consisted of the moves of two players, and described a search space in which a player uses heuristic strategies with the moves of the opponent in mind. This is an acquired distinction: the first search space can be represented with an AND/OR graph, the second with an OR graph. Yet in the very early stages of heuristic-programming terminology, the

---

[4] But initially "'space' of possible solutions" and "subspace" were used – both identified by a strategy that limits the search (see [41]).

[5] For details see [9]. As Simon put it, "from the beginning there was quite a bit of confusion in the use of the term 'algorithm' and 'heuristics', a confusion to which AI, Cliff and I certainly contributed" (1996, personal communication).

distinction was not always made explicit. The term most used in both cases was "problem space", and sometimes it was used together with "maze", as we have seen. In turn, the problem space was sometimes described explicitly as a directed graph: the graph of the sub-problems in problem decomposition, for example in the GEOMETRY MACHINE, as described by Gelernter and Rochester [20].

What was defined as a problem space in these early years was later defined, on one hand, as "state space" – described as a particular OR graph for one-person games, i.e. a tree – and, on the other hand, as "search space" – generalising to the case of AND/OR graphs, or "game trees" for two-person games (see, e.g., the synthesis in [6]). And problem decomposition has often been considered, at least since Nilsson [44], as a problem representation *distinct* from state space representation, i.e. the "problem reduction representation", that can be described as an AND/OR graph.

Notice, however, that since the beginning Newell and Simon distinguished the problem space from the task environment. Over time, both terms have taken on marked cognitive connotations. This became clearer in the 1960s, with Newell and Simon's research on how human subjects solve different puzzles and toy problems, first and foremost cryptarithmetic ones. The problem space in this framework is something more than just a simple state space. There are differences in the definitions given by the founders of IPP at different times, but they seem to have become clearer in *Human Problem Solving* [40]. Briefly, I think one can say that this is how things stand.

We start with a "laboratory" problem-solving situation. A problem is defined *objectively*, i.e., from the point of view of the observer or experimenter, in a certain task environment, which is the sum of the facts of the world-for example, a cryptarithmetic problem and the rules of the game. Once the problem has been assigned to the subject, the latter constructs his own, *subjective*, inner representation of the problem. The question now is the following: How do we define and study this subject's inner representation? The difficulty is that now, when considering the state space, we have introduced the *psychological* dimension of the problem solver (actually, the *limits* of the problem solver's rationality). So in this case we are interested not in the simple state space, which can be represented as a directed graph, i.e. a space which includes only legal moves – thus alternative paths which include states generated by the application of legal rules. The space we are interested in is the one which should represent the possible alternatives considered by the subject to reach his goal, so not only the effective ones, but also those hypothesised or imagined by the subject while he solves the problem, his wishes, unfeasible trials, errors in applying the rules and so forth.

Strictly speaking all this activity takes place *outside* the simple state space. Thus, if the experimenter intends to examine (e.g., via thinking-aloud protocols) how the subject represents the task environment, the experimenter cannot simply limit himself to examining the alternatives within the state space. This, which we can attribute as an object of study of AI, corresponds to what Newell and Simon defined as a "basic problem space", i.e. the space consisting of the set of states generated by all legal moves (see [54] p. 276, and [40] pp. 665-67, where

Newell and Simon deal with the basic state space in different task environments of logic, of one-person toy problems and two-person games). On the other hand, compared to the basic problem space, the problem space is a notion *augmented* by a consideration of the psychological aspects of the human solver. As such, the problem space is not so much the object of study of AI but rather of IPP, and it is thus introduced and defined:

> We [...] find it necessary to describe not only [a human subject's] actual be-
> haviors, but the set of possible behaviors from which these are drawn; and not
> only his overt behaviors, but also the behaviors he considers in his thinking
> that don't correspond to possible overt behaviors. In sum, we need to describe
> the space in which his problem solving activities take place. We will call it the
> problem space. This is not a space that can yet be pointed to and described as
> an objective fact for a human subject. An attempt at describing it amounts,
> again, to constructing a representation of the task environment – the subject's
> representation in this case. The subject in an experiment is presented with a
> set of instructions and a sequence of stimuli. He must encode these problem
> components – defining goals, rules, and other aspects of the situation – in some
> kind of space that represents the initial situation presented to him, the desired
> goal situation, various intermediate states imagined or experienced, as well as
> any concepts he uses to describe these situations to himself (Newell and Simon
> [40] 59).

The problem space considered in this way is the basis for constructing what Newell and Simon called the "problem behavior graph". This must include the above mentioned idiosyncratic components of a subject, as they are taken from thinking-aloud protocols. Finally, the problem behaviour graph is used to write a simulation program, or computational model of the subject, in the form of a production system. To conclude, it means building a "theory of the subject's thought processes" (p. 60) which is the object of study of IPP. From here stems the insufficiency in limiting ourselves to the simple actual or overt behaviour of the subject and, in computer simulation or modelling, to the state space. I have discussed elsewhere which are the theoretical constructs used in this thought-process theory and the controversial questions it has brought up (see [11]]). Here I would like to look briefly at a notion at the centre of both this process theory and AI: alternative space-representations.

4   Shifting in Problem Representation

The representation of the problem as a state space was most common in early AI, since it was interested in the study of heuristic procedures of problem solving within the "heuristic search paradigm". It would be useful to go back to Newell's words when summing up the common features of early AI programs (these are the ones I mentioned in previous sections, and are to be found along with others, in the well-known collection edited by Feigenbaum and Feldman [17]):

These programs are all rather similar in nature. [...] They all operate on formalized tasks, which although difficult, are not unstructured. All the programs use the same conceptual approach: they interpret the problem as combinatorial, involving the discovery of the right sequence of operations out of a set of possible sequences. All the programs generate some sort of tree of possibilities to gradually explore the possible sequences. The set of all sequences is much too large to generate and examine in toto, so that various devices, called heuristics, are employed to narrow the range of possibilities to a set that can be handled within the available limits of processing effort. Within these bounds there is a good deal of variation among the programs as to the particular heuristic devices used (Newell [36] 393-394).

I shall return later to the "unstructured" structure of tasks as opposed to formalised ones – i.e. those relating to theorem proving, to games like checker or chess, to toy problems such as cryptarithmetic and so forth (see section 6). For now, we can agree that the fundamental ingredient of these earlier AI programs was selective search in the state space, described as a directed graph (or as a tree). As has been seen, the main aim was thus how (i.e. with which methods) to make the search in more or less large state spaces *efficient*. Samuel, for example, started by studying Shannon's static evaluation function. But as has been shown, since the static evaluation function may be wrong, "the minimax procedure no longer serves its original purpose of defining and identifying a move that is theoretically correct. Instead, minimaxing has itself become a heuristic for the choice of move" (see [6] p. 98). Thus Samuel's checker programs, which used this "heuristic", actually supplementing the minimax procedure, analysed hundreds of possible continuations before choosing one. Yet, it was indeed the example of the expert human solver-who usually, in this and other similar cases, analyses only a few continuations[6] – that directly inspired the first attempts at using a different kind of heuristics, for example, those used in certain IPP programs and in the GEOMETRY MACHINE. Some important notions began to emerge with these programs, which broke away from the "efficient search on the basis of analysing many moves" approach – first of all there is the notion of *planning*. Let us look at how.

Planning already appeared in the earlier versions of GPS, presented not under this name, but as a "revised version" of LT, between 1957 and 1958 (see [43]). Planning, as described by the founders of IPP, consists of a procedure which allows the program to formulate a possible solution to a problem in general terms before working out the details. Such a procedure consists of abstracting certain details of the original object and operators, formulating the corresponding problem in this abstract task environment, using the solution to the abstract problem to provide a plan for solving the original problem, and, finally, translating the plan to the original task environment and executing it. So, the plan is used for finding cues in the maze representing the problem space, and should the outcome be successful, the plan produces a "planning space", which is simplified in

---

[6] Remember, for example, de Groot's experiments on chess players I mentioned earlier, and which Newell, Shaw and Simon [43] refer to.

comparison to the original problem space since it is narrower (Newell, Shaw and Simon [42]). The idea of a planning space thus contains a procedure whereby the problem is *abstracted* or *simplified*. The authors of the GEOMETRY MACHINE took up the planning space idea again within their "suggestion space", offered by geometric diagrams, whose cues or "proof indications", as they put it, are transformed into the problem space (see [20] p. 336).

Notice that, at the 1958 Teddington Symposium, Minsky had already dealt with the heuristic value both of planning as a method involving "simplifications [...] of the problem", and of the use of diagrams or figures as "'semantic' models" aiding the search for the problem's solution. In both cases the example was geometry theorem proving (see [32] pp. 14-17)[7]. At the time, experiments on planning were being carried out in logic theorem proving with "revised versions" of LT, and experiments on the use of figures in geometry theorem proving with the GEOMETRY MACHINE. Thus, for different researchers heuristic programming included problem solving methods which could not be reduced to the simple tree-search strategy. This is particularly true concerning the use of different alternative problem spaces.

These pioneering experiments aside, multiple problem-representation and the possibility of shifting from one representation to another became the object of systematic study in AI towards the mid-1960s. The goal remains of reducing the combinatorial explosion in heuristic programming. The fact that for this goal a good representation of the problem is no less important than a good selection strategy is again suggested by observing the human problem solver at work. The human's capacity to reformulate a given problem in *another representation* can drastically reduce the amount of searching needed to find the solution. The classic example, even today (see, e.g., [45]), was formulated by John McCarthy in 1964: the mutilated chess-board.

Saul Amarel first took on a systematic study of multiple problem-representation, and the difficult problems of how to endow a program with the ability to shift from one representation to another. In a great deal of, now classic, research, Amarel showed how the choice of representation influences the efficiency of problem solving. A 1966 meeting on problem representation at Carnegie-Mellon University with the participation of Amarel was a step forward in understanding the question. Newell [37] referred to this meeting in a discussion of different examples of multiple problem representation, including the example discussed by Amarel of three different and differently efficient representations in theorem proving in propositional calculus.

In Amarel's research, probably the most famous example is the missionaries and cannibals toy problem, for which he studied possible representational shifts. He described six increasingly efficient representations of this puzzle, from his verbal version to the one where the state space has a square matrix-like representation, making the solution to the problem immediately visible (see [3]).

---

[7] At the 1956 Dartmouth meeting, Minsky had already described a hand-simulated program that proved the early theorems of Euclidean geometry. He spoke about this at Teddington (see [32] p. 20).

Notice that a form of planning was a fundamental ingredient in this representation. An abstract problem space was formed that enabled a "global view of the situation" for the problem solver, with the result of better discovering the "critical points" in the problem space that allowed an enormous reduction in the search (see [4] p. 230).

This early research in the multiple representation-problem showed how different representations correspond to different ways for the problem solver to consider the problem space – the problem statement, the operators and the goal. With a computer program, a representation that moves efficiently towards the solution corresponds to "having the right point of view", as Amarel put it, or "casting the problem in the appropriate form" – abilities recognised in problem solvers by George Polya (see [2] p. 113).

As seen in the previous section, Nilsson [44] considered planning methods to be the basis of problem solving representations different from state-space representations (through OR graphs). He included planning methods within problem-reduction representations (through AND/OR graphs). In contrast to Nilsson, Simon considered it more useful to have a unified treatment of OR graphs and AND/OR graph within the paradigm of a simple heuristic search through a directed graph. Furthermore, he did not agree that planning methods were methods of problem representation which could be subsumed under the paradigm of a simple heuristic search (see [52] p. 262). For Simon, planning meant a partial departure from such a paradigm, even though the planning and other cases he pointed out – like abstraction (which can be present in the planning method) and semantic models – were experimented on *within* this paradigm from its very beginning. As we have seen, they were in fact the first examples of representations to include the construction of alternative problem spaces and their simultaneous exploration.

For Simon, when we consider a planning space it would be more correct to say that information, rather than a solution to a problem is being sought. Thus, using a multiple representation, such as planning, we have a process of information accumulation or gathering, rather than a process of seeking the solution through the usual "simple tree-search paradigm". "The theory of heuristic search has been usually formulated in terms of a search for a problem solution. It could be generalized to refer to a search for information that will allow the solution to be known" (see [52] p. 268). Important as this distinction may be, for Simon it is an *extension* more than a *replacement* of the simple search paradigm. The re-evaluation of planning notions and the multiple representation in general lead to a better qualification of the role of the heuristic search in problem solving.

5   In Search of Knowledge, I: Toy Problems and Real-Life Problems

Concluding his review of game programming on computers, Samuel observed how it was only a first step towards understanding computer-simulation methods of intelligent behaviour, and that future progress would consist of applying these methods "to real-life situations with increasing frequency", while "the effort devoted to games or other toy-problems will decrease" (see [49] p. 192).

This prediction proved right to a great extent. It is true that many toy problems have continued to play an important role in some AI research and later in Cognitive Science, a role often compared to *Drosophila* in genetics – as Amarel [5] said again in an analysis of another classic AI toy problem, the Hanoi-tower puzzle (by the way, this was a favourite during AI's half century, above all in the experimental research on human subjects by Simon and various co-workers). Furthermore, it is also true that the most famous and acclaimed successes of AI in its first half century have been realised in games, chess in particular[8]. Yet it is also true that, during the AI's second decade, researchers' attention turned to real-life problem solving, with a shift in interest on the role played by knowledge in problem solving. In this way, on one hand, certain research areas came to the fore again (e.g., machine translation) and, on the other, new issues came up, relating to the complexity of real-life problems, which human beings usually confront on the basis of incomplete knowledge, with the aid of background or common-sense knowledge[9]. Finally, more recently real-life problems have been seen in terms of interaction with the real world by "situated" and "embodied" agents, according to the lessons from new robotics (see [10] ch. 6, on all these topics).

In the previous section we saw how the internal limits of the "simple tree-search paradigm" raised the question of its extension to include the representation problem (and the multiple representation-problem in particular). In this and the next section I would like to show how we arrived at knowledge representation: a new and important extension to this "paradigm".

The two issues, (multiple) problem representation and knowledge representation, can be considered distinct, even though they have an intersection (see [15]). Basically, the issue of knowledge representation concerns how to represent knowledge in an explicit form within a program, so that it is easily accessible and modifiable according to circumstances. The issue at hand here is that of the specific structures – data structures – hat the programmer interprets as statements of what the program knows, and which carry out the specific role in *causing* the program's behaviour[10]. In the early AI programs discussed so far, the prior knowledge needed to reduce the search (usually little, given the more or less "toy" nature of the majority of problems considered) was usually represented in an implicit way in the selection rules in heuristic programs (for example, a generator could incorporate information from the task environment, so that only

---

[8] On chess as "the *Drosophila* of AI", see [8].

[9] A well-liked description of this shift in interest is contained in Feigenbaum's account of his lecture on the knowledge-based approach in AI and the DENDRAL program in the early 1970s at Carnegie-Mellon: "'You people are working on toy problems,' he said. 'Chess and logic are toy problems. If you solve them, you'll have solved a toy problem. And that's all you'll have done. Get out into the real world and solve real-world problems"' (see [18] p. 65).

[10] A well-known authoritative statement of this knowledge-based view in AI was later given by Smith [61] in the form of the "knowledge representation hypothesis".

states satisfying given criteria will be generated in the state space – see, e.g., [52] for a discussion).

Notice that the question of how to explicitly represent knowledge in a program was raised, in an embryonic form, at the very beginning of AI. At the 1958 Teddington Symposium, McCarthy [28] described a prototype of a program, the Advice Taker, which was supposed to be able to make plans and to deduce consequences, and so take actions, on the basis of both a set of "advice" coming from the programmer and a body of common-sense knowledge on probability, expectations, etc. for a given context or world. The point at issue here is that this knowledge had to be expressed in an explicit way in the form of sentences of first-order logic, rather than being embodied into the selection rules of the program[11].

Around ten years later, McCarthy and Hayes [31] distinguished the "epistemological" problem from the "heuristic" problem in AI. The former involves how an intelligent agent represents the world via "an adequate model" which makes him able to deduce consequences and carry out actions. The latter concerns the study of specific mechanisms dedicated to this purpose. According to the authors, most of the work in AI up to that time could be regarded as devoted to the heuristic problem (p. 466)[12].

However, perhaps it would be more correct to say that the AI approach preceding the distinction between the epistemological and the heuristic problem often *contaminated* these two problems. Within this framework, Minsky was one of the first to insist explicitly on the role of knowledge representation, presenting a collection of programs in the mid-1960s under the title *Semantic Information Processing* [33]. These were written in LISP and IPL-V, and would have a notable influence on the immediate future of AI: SIR by Bertram Raphael, ANALOGY by Thomas Evans, STUDENT by Daniel Bobrow, SEMANTIC MEMORY PROGRAM by M. Ross Quillian, and QAS by Fisher Black. In opposition to previous programs dealing with "formalized tasks" (games and toy problems) some of these new programs dealt with real-life or every-day tasks, such as understanding natural language and answering questions – thus disproving "one of the most popular misconceptions about artificial intelligence [i.e.] that problem-solving by computers is confined to precisely 'formal' problems", as Minsky put it (see [33] p. 11).

Other programs from around the same time, published in the collection *Representation and Meaning*, edited by Simon and Siklóssy [59], were moving in the same direction, where knowledge and semantics were acquiring a more dominant

---

[11] The main difference between the Advice Taker and the heuristic programs of the time (LT and the GEOMETRY MACHINE) "is that in the previous programs the formal system was the subject matter, but the heuristics were all embodied in the program. In [the Advice Taker] the procedures will be described as much as possible in the language itself and, in particular, the heuristics are all so described" (see [28] p. 77).

[12] But notice: "The use of first order logic in epistemological research is a separate issue from whether first order sentences are appropriate data structures for representing information within a program" (McCarthy [29] 1038).

role than in the past. Simon complained at times about the much lower influence this collection had on the AI community than the one edited by Minsky. With hindsight, this could be explained by the fact that Minsky placed more value on the works he published within a framework that could be perceived as more closely linked to the problem of "representation of knowledge", as he called it. The central role of semantics can be justified in this context, especially in certain real-life tasks like those I have mentioned.

Minsky's starting point is the same as Simon's [52] and Simon and Siklóssy's [59]. In both cases, the attempt was to study some key concepts in early AI (i.e. heuristics, generality and so forth) in light of a parameter centred not on selective search in usually large state spaces "within the older goal-tree search framework", but on "the representation and modification of *plans*", as Minsky put it (see [33] pp. 10 and 9). Here too, the notion of planning was considered important, but the point is that it wasn't only building models or representations which *simplified* the search to present a smaller combinatory branching. The problem raised by Minsky was finding a suitable medium – or a suitable data structure – to represent common sense knowledge, so that when used in the right way by a program, it would display behaviour ascribable to the possession of that knowledge, as I said earlier. Minsky observed how the diversity in the suitability of a medium used to represent knowledge can influence the efficiency of a question-answering system like Raphael's once compared to Black's (p. 17). All of Minsky's discussion revolves around issues of "common sense knowledge" and "specialized knowledge", and on how to manage and retrieve relevant knowledge in the presence of "large bodies of knowledge", with reference also to the Advice Taker (McCarthy's article is reprinted in his collection).

Later Amarel also seemed to distinguish the issue of finding and using new knowledge about the problem by shifting to a better problem representation, from the issue of constructing "a system of concepts and a language which is appropriate for describing knowledge in [a] domain", so that it becomes possible "to express knowledge and store it in the machine in a manner that can be conveniently used for problem solving in [that] domain". In the first case a way of exploiting new knowledge consists of the usual shift to a simplified or abstract problem space. In the second case, the problem at hand is how "to feed background knowledge to a machine in an 'appropriate' form", i.e. in the form of "structured data" (see [4] pp. 227 and 237). Amarel made this distinction on the basis of programs dealing with real-life problems, such as a question-answering program, e.g., Raphael's SIR mentioned earlier, and the DENDRAL program (see section 6): in both these programs the issue of knowledge representation is crucial, albeit to different extents.

Later, when Minsky explicitly took on the problem of knowledge representation, presenting his well-known data structure, the frame, he did so by invoking a "more mature and powerful paradigm" for AI than the "dominant paradigm of the past", i.e. the search paradigm. Notice, however, that the starting point here too was traditional:

> The primary purpose in problem solving should be better to understand the problem space, to find representations within which the problems are easier to solve. The purpose of search is to get information for this reformulation, not – as is usually assumed – to find solutions; once the space is adequately understood, solutions to problems will more easily be found (Minsky [34] 59).

As with Simon, for Minsky too the new approach to "reformulating" the problem space had been "implicit", as he put it, in earlier discussions on planning within the heuristic search paradigm. The limits of search strategies (such as search for differences, hill-climbing, methods of tree-pruning and so forth) that he pointed out are the same as those pointed out by Simon [52]. Yet, while Simon's context is problem representation, Minksy's is frame systems: the possibility of a "reformulation" or "reconfiguration or representation" that Minsky mentions can be found in the alleged flexibility of this type of data-structure system. Frame systems could lead to a way "to improve the strategy of subsequent trials", i.e. to improve the program's future problem-solving performance. This is an issue Minsky had raised previously in the same terms in *Semantic Information Processing*: "What is needed for summarizing a search tree is not a numerical *utility-like* value good only for *comparison* but a *description-like* expression that can be used for *analysis*" (see [33] p. 11). It was hoped that the later role of frames, as "description-like" devices, would make this analysis possible.

In conclusion, the issue of real-life problems has contributed to bringing to the fore the issue of how to represent knowledge in an explicit form, as initially raised by McCarthy in his formulation of the Advice Taker (notwithstanding Minsky's well-known later opposition to McCarthy's "logicism")[13].

6 In Search of Knowledge, II: Well-Structured and Ill-Structured Problems

In the passage quoted at the beginning of section 4, Newell spoke about how earlier AI programs were concerned with "formalized tasks" and not "unstructured" tasks. The distinction between these types of problems or tasks constitutes another way of looking at the role of knowledge in problem solving. This is the last topic mentioned in section 1 to investigate before reaching my conclusions.

Newell and Simon had initially identified "well-structured" problems with problems which can be solved by OR techniques (see section 2 above). In this case, the goals are considered as clearly defined, and the means of problem solving lie in applying algorithms and mainly linear-programming and optimisation techniques:

> Operations research has demonstrated its effectiveness in dealing with the kinds of management problems that we might call "well structured", but it has

---

[13] It is beyond the limits of the present chapter to discuss the different proposals regarding the knowledge representation problem in IA. See [7] for a discussion of the role of knowledge representation in more recent AI research context.

left pretty much untouched the remaining, "ill-structured problems" (Simon and Newell [58] 4).

"Ill-structured" problems are those that are intrinsically complex, that cannot be handled by the traditional methods mentioned above, but by problem solving techniques in the newly-born heuristic programming, i.e. techniques inspired by how human beings actually solve problems, both in "many important situations in everyday life when the objective function, the goal, is vague and nonquantitative" and in situations, like the case of the combinatorial explosion, where "computational algorithms are not available" or cannot be applied in practice (p. 5). For Newell and Simon (writing in 1957) heuristic programming, although in its very early stages, already seemed able to take on not just problems of the latter type, like theorem proving or chess, but also problems closer to real or every-day life, like management and marketing behaviour, albeit at a primitive stage (see Simon [57] for a further analysis of the topic).

An authoritative point of view on the subject was later provided by Reitman [47]. He distinguished problems handled by heuristic problem-solving *procedures* (so by ill-defined procedures, heuristics were still considered in opposition to algorithms) from problems whose formulation and criteria of generation and control of the solution are both well specified and precisely defined. From this point of view, a toy problem or a game, no matter how difficult (take chess for instance), is still an example of a well-structured problem, whereas a real-life problem is still an example of a *really* ill-structured problem (or "ill-defined problem", as Reitman put it).

Actually, Newell later observed that, once one accepts Reitman's definition of ill-structured problems, the programs mentioned in his and Simon's 1958 paper [see 58] were only a demonstration that *some* ill-structured problems were actually being handled by heuristic programming: precisely those problems that "might constitute a small and particularly 'well-formed' subset" of the whole set of ill-structured problems (Newell [38]). In other words, heuristic programs of this special subset had a degree of success by virtue of the fact that certain aspects were well defined: they lacked well-specified methods of solution (in the extent that they used heuristics), but were quite precisely defined regarding the problem statement, goal and solution criteria.

Newell, opposing Reitman, then tried to define ill-structured problems in a more positive way, referring not to the not-well-defined nature of the problem formulation and solution, but to the problem-solving *methods* available. The domain of ill-structured problems is where general or "weak" methods are available: and it is these methods which, according to Newell, characterise AI (examples of such methods are generate-and-test, heuristic search, means-end analysis and so forth – all milestones in heuristic programming). Compared to "strong methods", weak methods are distinguished by their more general nature and less specific knowledge. Newell, though, recognised how the issue of what constitutes an ill structured problem might "remain only half answered" (p. 412), because it remained difficult to capture an aspect which seems crucial: that "vague" or indefinite quality of the information present within ill-structured problems, which seems to continue to distinguish them from well-structured problems. In this "vague" information we can see the non explicit and incomplete knowledge which is typical of common sense when the problem solver is dealing with real-life problems.

There is one other open question. If for simpler toy problems it is possible to ignore, to some extent, the knowledge that characterises the "expert" solution to a problem, this is not possible for more complex toy problems and games (chess again) and real-life problems in general. Such knowledge is present in the form of *less general* methods or heuristics than weak methods. Furthermore, Newell himself pointed out that "there is an inverse relationship between the generality of a method and its power" (p. 372). This conclusion leads to the question: To what extent is a *"general* problem solver" plausible? This is a question that can be asked also for the plausibility of logic as a *general* or uniform representational medium in all problem domains. It is not by chance that the authors of DENDRAL, a program dealing with an ill-structured problem and an ill-defined goal, started precisely from this question:

> The study of generality in problem solving has been dominated by a point of view that calls for the design of "universal" methods and "universal" problem representation. These are the GPS-like and Advice-Taker-like models. This approach to generality has great appeal, but there are difficulties intrinsic to it: the difficulty of translating specific tasks into general representation; and the trade-off between generality and power of the methods (Feigenbaum et al. [19] 187).

With the DENDRAL program (studies starting in 1965 ca.) the focus moved to specialized knowledge, intended not so much as a large data base, but as a body of heuristic rules. In this case, these rules are not only general or weak methods, but are also specialized or task-specific rules (to be more precise, they have stronger constraints imposed by the data). It was with this "shift in paradigm", as Ira Goldstein and Seymour Papert defined it, that expert or knowledge-based systems entered the field of AI[14].

Notice that the problem of ill-structure is not exhausted in the question of specialised knowledge. The problem (as seen at the time of the Advice Taker) concerns common sense knowledge – that sort of background knowledge which is usually not explicit in a problem statement, at least in real-life problems. Compare the missionaries and cannibals problem in its "laboratory" or toy version with its real-life version, where, e.g., the real environment might permit alternative ways of crossing the river – using a bridge, or by an airplane and so

---

[14] Quoted by Feigenbaum [16] 1017. But regarding certain 1956 programs resembling AI expert systems, see Simon [57]. It should be remembered that the concern for GPS-like generality was originally based on the claim of the multipurpose nature of human intelligence. As Slagle (60) [176] observed, "the hope is that once multipurpose programs can be written to solve simple problems [as was the case of GPS at the time], the programs can be extended to solve more difficult problems". A common *Drosophila*-style strategy in AI – although not always as successful as in genetics.

forth (Simon [52] 275). If in the former case the problem, for a person or a computer program, is well-defined, in the latter case we have a constantly changing environment, which makes the problem ill-defined. Now think of chess. In the abstract, it is an example of a well-defined problem, but in reality the player comes up against difficulties and unexpected situations during the game which, *from the player's point of view* (be it a human or machine), render the problem ill-defined, more or less like the missionaries and cannibals problem in its real-life version.

All this leads us to consider different problems as all being ill-structured to a greater or lesser degree for the problem solver, as a bounded-rationality agent (see section 2). This point of view held by Simon detaches the question of ill-structuredness from the nature of the methods employed (weak or strong), and makes such a question relative to the *limits* of the problem solver, establishing a continuum of degree of definiteness between the well structured and ill structured ends of the problem spectrum (see [53] p. 183)[15].

7 Conclusion

In the previous sections, I discussed topics and controversial issues from the fifteen or so years of AI immediately in the wake of Dartmouth. As we have seen, already in this brief initial period the successive approaches to heuristic search, (multiple) problem representation, knowledge representation and knowledge-based systems were presented by different authors as "paradigms" within AI research. Yet, it was the connectionists of the 1980s who explicitly described the history of AI in terms of opposed paradigms – "symbolic" vs. "sub-symbolic". Above I mentioned the early stage in this opposition (which was often experienced in an overly dramatic way in the history of AI), when I spoke about the opposition between early neural nets and newly-born heuristic programming, starting with the 1958 Teddington Symposium (see section 2).

I have shown elsewhere how difficult it is to take the idea seriously that the history of AI consists of a succession of opposed paradigms (up to the most recent, the situatedness paradigm) – at least if one interprets the term paradigm in the same sense as historians of science, i.e. à la Kuhn (see [10] pp. 275-279). Considering the issues discussed in this chapter, the authors mentioned

---

[15] Any consideration of the immediate developments of this discussion would go beyond the limits I set myself for the present chapter. The missionaries and cannibals puzzle was taken up again by McCarthy in the terms above, but with the aim of making stateable in a formal language what is necessary in common-sense knowledge. McCarthy's circumscription rule gave rise to a long series of research on how to formalise and to limit the contexts involved in real-life problem solving. Circumscription assumes the character of a "rule of inference" in toy problems and a "rule for conjecture" in real-life problems: see the paper by McCarthy [30] in the *Artificial Intelligence* special issue on non-monotonic reasoning, which includes classic contributions by D. McDermott, R. Reiter, R.W. Weyerauch and others. A proposal opposed to McCarthy's has been given by Simon [56].

here do not seem to have used the term "paradigm" in this sense. Clearly, the term was used by these authors when they aimed to characterise a research program which brought new issues to the fore. These issues, even though new, were recognised as "implicit" in the previous research program – to such an extent that at times it was also possible to *hybridise* the products of these different research programs.

Thus, it was the limits of the "simple tree-search paradigm" which suggested, or forced, its extension, or integration, to problem representation. It is clear that the latter research project neither replaces nor renders the former superfluous (as should happen if we were to take the term "paradigm" seriously). As we have seen, Minsky recognised how certain issues of his new paradigm were "implicit" in internal discussions in the previous research. Simon stated that his aim was to reconsider the simple tree-search approach – but in order to overcome the "gulf" (as he called it) which separated search from problem representation, he suggested "hybrid systems" (see [52] p. 276): from here comes the need to consider problem solving primarily as information gathering. Feigenbaum, when presenting the first results of his research on DENDRAL, pointed out that his project lay within what he considered "the mainstream of AI research: problem solving using the heuristic search paradigm" (see [15] p. 1016).

A different matter is the case of topics known to be important, but how to tackle them was unknown. This was due to a factor which, as is obvious, has conditioned the history of AI, some of its choices, some of its successes, some of its illusions – advancement in computer technology. Just to stay with one of the issues dealt with in this chapter – the crucial role of knowledge – Daniel Crevier refers to these words by Simon in his 1991 interview with him about early AI's interest in toy problems: "It isn't as though people weren't aware that knowledge was important. They were steering away from tasks which made knowledge the centre of things because we couldn't build large data bases with the computers we had. Our first chess programs and the LT were done on a computer that had a 64- to 100-word core and a scratch drum with 10,000 words of usable space on it..." (see [13] p. 177). So, concentrating initially on prevalently "toy" tasks was to some extent an obligatory choice: with the major exception of chess, those tasks did not need much specialised knowledge. I would like to add that, despite the fact that on various occasions Simon has pointed out – along with others – the *Drosophila*-like role of toy problems in AI, he also warned us to be cautious, particularly in IPP, in "extrapolating what we learn about problem solving in puzzle-like domains to problem solving in information-rich domain" (see [55] p. 142).

In this chapter my aim was not to discuss recent developments in the issues I stated at the beginning (see section 1). Rather, I aimed to show through which clusters of problems these issues were raised and tackled in the pioneering phase of AI. Finally, I do not wish to miss out on a quick look at the recent evolution of one aspect which concerns the heuristic search as much as knowledge representation. Both have competed to raise new research issues and theoretical insights since early times, also by revisiting issues and methods which had previously been if not abandoned, then at least partially set aside, like the issue of general or weak methods. Briefly, this is not an isolated incident in the history of AI (think of very different cases like machine translation or neural nets). Hence, reflecting upon them might suggest caution in imagining future scenarios in the evolution of AI (alternative (i) I referred to above, section 1).

I concluded the previous section with the DENDRAL system. As we have seen, the general or weak methods were judged inefficient, and so had to be "strengthened" by domain-specific knowledge. It was this that essentially assured the system's good performance. Feigenbaum returned to the issue, after the first decade of experimenting on expert systems, by pointing out how their performance was "primarily a consequence of the specialist's knowledge", and "only very secondarily" due to the generality of the methods (see [16] p. 1016). However, the limits of expert systems came out during the 1980s. This is known, and it has been summarised by Lenat and Feigenbuam: "A limitation of past and current expert systems is their brittleness. They operate on a high plateau of knowledge and competence until they reach the extremity of their knowledge; then they fall off precipitously to levels of ultimate incompetence. People suffer the same difficulty, too, but their plateau is much broader and their slope is more gentle. Part of what cushions the fall are layer upon layer of *weaker, more general models that underlie their specific knowledge*" (see [25] p. 196, my italics).

Thus, also experiences as different as Feigenbaum's and those of other researchers in the field of expert systems such as Douglas Lenat's in the 1980s, seemed to converge towards the same conclusion. Lenat's programs in those years – AM and EURISKO – tackled vague and ill-defined problems. For example, AM, though not an expert system, dealt with concept formation in mathematics, with the initial aim of getting close to "the ideal trade-off between generality and power" (see [23] p. 263): a difficult goal, already attempted in AI (see [14] p. 1969). EURISKO was a step forward: it used heuristic rules to produce or *learn* new rules, or metaheuristics[16]. Both of Lenat's programs and the classic expert systems of the time were endowed with domain-specific or specialised knowledge (but the latter with a higher amount than Lenat's programs). They proved, however, drastically lacking in what Lenat and Feigenbaum have called "*consensus reality* knowledge": i.e. knowledge of a *general* and also *analogic* type, based on weak methods, which constitutes the underlying background of every specialised intelligent performance. As the authors concluded for EURISKO, its "ultimate limitation was not what we expected (CPU time), or hoped for (the need to learn new representations of knowledge), but rather something surprising and daunting: the need to have a large fraction of consensus reality already in the machine" (see [25] p. 206). This was a true re-evaluation of general weak meth-

---

[16] As to the use of metaknowledge and metarules in AI systems in the 1970s and 1980s, see [1]. One of the main reasons that weakened AM's performance was its inability to learn new heuristics. It is interesting to note that EURISKO fulfils some of the characteristics Gelernter and Rochester speculated about regarding high forms of learning and above all their "theory machine" (see [20] p. 73 ff.).

ods – actually never totally abandoned, but now placed within the context of the *vagueness* (Newell's term above) that characterises common-sense knowledge underlying domain-specific knowledge.

The rest is known, and it is today's story. As far as Lenat is concerned, there is CYC, a system which could be able to handle consensus reality knowledge in a vast and efficient way: an ambitious and controversial system in progress – presently influential in different research fields on ontology and the web (see, e.g., [24] for a recent statement, and [27], [46]). Its eventual success is part of the possible scenarios of AI's future, but its roots lie in the early stages of AI that I have discussed in the present chapter.

References

1. Aiello, L., Levi, G.: The Uses of Metaknowledge in AI Systems. In: Maes, P., Nardi, D. (eds.): Meta-level Architectures and Reflection. North Holland, Amsterdam (1988)
2. Amarel, S.: On the Mechanization of Creative Processes. IEEE Spectrum 3 (1966) 112-114
3. Amarel, S.: On Representations of Problems of Reasoning About Actions. In: Michie, D. (ed.): Machine Intelligence 3. Edinburgh University Press, Edinburgh (1968) 131-171
4. Amarel, S.: On the Representation of Problems and Goal-Directed Procedures for Computers. In: Banerji, R., Mesarovic, M. (eds.): Theoretical Approaches to Non-Numerical Problem Solving. Springer-Verlag, Berlin Heidelberg (1970) 179-244
5. Amarel, S.: Problems of Representation in Heuristic Problem Solving: Related Issues in the Development of Expert Systems. In: Groner, R., Groner, M., Bischof., W.F. (eds.): Methods of Heuristics. Erlbaum, Hillsdale, NJ (1983) 245-350
6. Barr, A., Feigenbaum, E.F. (eds.): The Handbook of Artificial Intelligence, Vol. 1. HeurisTech and Kaufmann, Stanford and Los Altos (1981)
7. Carlucci Aiello, L., Nardi, D., Pirri, F.: Case Studies in Cognitive Robotics. In: Cantoni, V., Di Ges, V., Setti, A., Tegolo, D. (eds.): Human and Machine Perception 3: Thinking, Deciding, and Acting. Kluwer Academic Publishers, Dordrecht (2001)
8. Coles, S.L.: Computer Chess: The Drosophila of AI. AI Expert 9 (1994) 25-32
9. Cordeschi, R.: The Role of Heuristics in Automated Theorem Proving. J.A. Robinson's Resolution Principle. Mathware and Soft Computing 3 (1996) 281-293
10. Cordeschi, R.: The Discovery of the Artificial. Behavior: Mind and Machines Before and Beyond Cybernetics. Kluwer Academic Publishers, Dordrecht (2002)
11. Cordeschi, R.: IA Turns Fifty: Revisiting its Origins. Applied Artificial Intelligence, 21 (forthcoming 2007)
12. Cordeschi, R.: Steps Towards the Synthetic Method. Symbolic Information Processing and Self-Organizing Systems in Early Artificial Intelligence Modeling. In: Wheeler, M., Husbands, P., Holland, O. (eds.): The Mechanisation of Mind in History. MIT Press, Cambridge, MA (forthcoming 2007)
13. Crevier, D.: AI. The Tumultuous History of the Search for Artificial Intelligence. Basic Books, New York (1993)
14. Ernst, G.W., Newell, A.: GPS: A Case Study in Generality and Problem Solving. Academic Press, New York (1969)

15. Feigenbaum, E.A.: Artificial Intelligence: Themes in the Second Decade. In: Proceedings IFIPS-68. North-Holland, Amsterdam (1969) 1008-1022

16. Feigenbaum, E.A.: The Art of Artificial Intelligence: Themes and Case Studies of Knowledge Engineering. In: Proceedings IJCAI-77. Cambridge, MA (1977) 1014-1029

17. Feigenbaum, E.A., Feldman, J. (eds.): Computers and Thought. McGraw-Hill, New York (1963).

18. Feigenbaum, E.A., McCorduck, P.: The Fifth Generation. Signet, New York (1983)

19. Feigenbaum, E.A., Buchanan, B.G., Lederberg, J.: On Generality and Problem Solving: A Case Study Using DENDRAL Program. In: Meltzer, B., Michie, D. (eds.), Machine Intelligence 6. Edinburgh University Press, Edinburgh (1971) 165-190

20. Gelernter, H.L., Rochester, N.: Intelligent Behavior in Problem-Solving Machines. IBM Journal of Research and Development 2 (1958) 336-345

21. Hodges, A.: Alan Turing: the Enigma. Simon and Schuster, New York (1983)

22. Kahneman, D., Frederick, S.: Attribute Substitution in Intuitive Judgement. In: Augier, M., March, J.G. (eds.), Models of Man. Essays in Memory of Herbert A. Simon. MIT Press, Cambridge, MA (2004)

23. Lenat, D.: On Automated Scientific Theory Formation: A Case Study Using the AM Program. In: Hayes, J.E., Michie, D., Mikulich, L. J. (eds.), Machine Intelligence 9. Halsted, New York (1979) 251-283

24. Lenat, D.: Applied Ontology Issues. Applied Ontology 1 (2005) 9-12

25. Lenat, D., Feigenbaum, E.A.: On the Thresholds of Knowledge. Artificial Intelligence 47 (1991) 185-250

26. Mahoney, J.T.: Economic Foundations of Strategy. Sage, London (2004)

27. Matuszek, C., Witbrock, M., KAHLERT R.C., Cabral, J., Schneider, D., Shah, P., Lenat, D.: Searching for Common Sense: Populating Cyc^{TM} from the Web (2005) *www.cyc.com/doc/white$_p$apers/AAAI051MatuszekC.pdf*

28. McCarthy, J.: Computers with Common Sense. Proceedings of the Symposium on Mechanisation of Thought Processes, Vol. 1. H.M. Stationary Office, Teddington (1959) 77-91

29. McCarthy, J.: Epistemological Problems of Artificial Intelligence. Proceedings IJCAI-77, Cambridge, MA (1977) 1038-1044

30. McCarthy, J. Circumscription-a Form of Non-Monotonic Reasoning. Artificial Intelligence 13 (1980) 27-39

31. McCarthy, J., Hayes, P.: Some Philosophical Problems from the Standpoint of Artificial Intelligence. In: Meltzer, B., Michie, D. (eds.): Machine Intelligence 4. Edinburgh University Press, Edinburgh (1969) 463-502

32. Minsky, M.L.: Some Methods of Heuristic Programming and Artificial Intelligence, Proceedings of the Symposium on Mechanisation of Thought Processes, Vol. 1. H.M. Stationary Office, Teddington (1959) 5-36

33. Minsky, M.L. (ed.): Semantic Information Processing. MIT Press, Cambridge, MA (1968)

34. Minsky, M.L.: A Framework for Representing Knowledge. In: Winston, P. (ed.): The Psychology of Computer Vision. McGraw-Hill, New York (1975) 211-277

35. Neumann, J. von, Morgenstern, O.: Theory of Games and Economic Behaviour. Princeton University Press, Princeton, NJ (1944)

36. Newell, A.: Some Problems of Basic Organization in Problem-Solving Programs. In: Yovits, M.C., Jacobi, G.T., Goldstein, G.D. (eds.): Self-organizing systems. Spartan Books, Washington, D.C. (1962)

37. Newell, A.: On the Representations of Problems. Computer Science Research Review. Carnegie Institute of Technology, Pittsburgh (1966) 19-33
38. Newell, A.: Heuristic Programming: Ill-Structured Problems. In: Aronofsky, J.S. (ed.): Progress in Operations Research, Vol. 3. Wiley, New York (1969)
39. Newell, A., Simon, H.A.: GPS: A Program that Simulates Human Thought. In: Billings, H. (ed.): Lernende Automaten. R. Oldenbourg, Munchen (1961). Reprinted in Feigenbaum, E.A., Feldman, J. (eds.): (1963) 279-293
40. Newell, A., Simon, H.A.: Human Problem Solving. Prentice-Hall, Englewood Cliffs, NJ (1972)
41. Newell, A., Shaw, J.C., Simon, H.A.: Elements of a Theory of Human Problem-Solving. Psychological Review 65 (1958) 151-166
42. Newell, A., Shaw, J.C., Simon, H.A.: Report on a General Problem-Solving Program for a Computer. Proceedings of the International Conference on Information Processing. UNESCO, Paris (1960) 256-264
43. Newell, A., Shaw, J.C., Simon, H.A.: The Processes of Creative Thinking. In: Gruber, H.E., Tirrel, G., Wertheimer, H. (eds.): Contemporary Approaches to Creative Thinking, Atherton, New York (1962) 63-119
44. Nilsson, N.J.: Problem Solving Methods in Artificial Intelligence. McGraw-Hill, New York (1971)
45. Paulson, L.C.: A Simple Formalization and Proof for the Mutilated Chess Board. Logic Journal of the IGPL 9 (2001) 499-509
46. Pirrone, R., Pilato, G., Rizzo, R., Russo., G.: Learning Path Generation By Domain Ontology Transformation. In: Bandini S., Manzoni, S. (eds.): AI*IA 2005: Advances in Artificial Intelligence. Springer-Verlag, Berlin Heidelberg (2005) 359-369
47. Reitman, W.R.: Heuristic Decision Procedures, Open Constraints, and the Structure of Ill-Defined Problems. In: Shelley, M.W., Rryan, G.L. (eds.): Human Judgments and Optimality. Wiley, New York (1964) 282-315
48. Samuel, A.L.: Some Studies in Machine Learning Using the Game of Checkers. IBM Journal of Research and Development 3 (1959). Reprinted in Feigenbaum, E.A., Feldman, J. (eds.): (1963) 71-105
49. Samuel, A.L.: Programming Computers to Play Games. In: Alt, F.L. (ed.): Advances in Computers, Vol. 1. Academic Press, New York (1960) 165-192
50. Shannon, C.E.: Programming a Computer for Playing Chess. Philosophical Magazine 41 (1950) 256-275
51. Simon, H.A.: A Behavioral Model of Rational Choice. Quarterly Journal of Economics 69 (1955) 99-118
52. Simon, H.A.: The Theory of Problem Solving. Proceedings of IFIP Congress (1972) 261-277
53. Simon, H.A.: The Structure of Ill-Structured Problems, Artificial Intelligence 4 (1973) 181-201
54. Simon, H.A.: Information Processing Theory of Human Problem Solving. In: Estes, W.K. (ed.): Handbook of Learning and Cognitive Processes, Vol. 5. Erlbaum, Hillsdale, NJ (1978) 271-295
55. Simon, H.A.: Models of Thought. Yale University Press, New Haven and London (1979)
56. Simon, H.A.: Search and Reasoning in Problem Solving. Artificial Intelligence 21 (1983) 7-29
57. Simon, H.A.: A Very Early Expert System. IEEE Annals of the History of Computing 15 (1993) 64-68
58. Simon, H.A., Newell, A.: Heuristic Problem Solving: The Next Advance in Operations Research. Operations Research 6 (1958) 1-l0

59. Simon, H.A., Siklóssy, L. (eds.): Representation and Meaning. Prentice-Hall, Englewood Cliffs, NJ (1972)
60. Slage, J.R.: Artificial Intelligence: The Heuristic Programming Approach. McGraw-Hill, New York (1971)
61. Smith, B.C.: Prologue to Reflection and Semantics in a Procedural Language. In: Brachman, R.J., Levesque, H.J. (eds.): Readings in Knowledge Representation. Morgan Kaufmann, Los Altos, CA (1985) 31-39
62. Wiener, N.: Cybernetics, or Control and Communication in the Animal and the Machine. 2nd edn. MIT Press, Cambridge, MA (1961)

Research Perspectives for Logic and Deduction

Wolfgang Bibel*

Darmstadt University of Technology
Bibel@gmx.net

Abstract. The article is meant to be kind of the author's manifesto for the role of logic and deduction within Intellectics. Based on a brief analysis of this role the paper presents a number of proposals for future scientific research along the various dimensions in the space of logical explorations. These dimensions include the range of possible applications including modelling intelligent behavior, the grounding of logic in some semantic context, the choice of an appropriate logic from the great variety of alternatives, then the choice of an appropriate formal system for representing the chosen logic, and finally the issue of developing the most efficient search strategies. Among the proposals is a conjecture concerning the treatment of cuts in proof search.

> *Often a key advance is a matter of applying a small change to a single formula.*
>
> Ray Kurzweil [21, p.5]

1 Introduction

Luigia Aiello has made numerous important scientific contributions in many areas of Artificial Intelligence. But it is fair to say that her core interest has always been in a logical approach to Artificial Intelligence (AI) throughout her career. For instance, as early as 1980 her paper [1] appeared in the section on theorem proving at the very first AAAI Conference, noteworthily one out of merely two papers presented by European authors at this legendary conference in the US. Many more papers in a similar vein by herself and her numerous students preceded and followed this particular one.

It is for this reason that I chose to honor her at the occasion of her sixtieth birthday with a perspective contribution to this particular area. I would like to express through it my highest respect for her achievements and my deepest gratitude for the professional and personal friendship and the fruitful cooperation which has lasted for more than a quarter of a century.

The elder AI generation still has vivid recollections of the hot debates of the seventies in the last century within the community concerning the role of logic and deduction in AI. Notwithstanding the GOFAI ("good oldfashioned AI") debate triggered by Rodney Brooks, the central role of logic within many areas of AI and Computer Science (CS) today is undisputed – perhaps even too much so.

---

* Also affiliated with the University of British Columbia.

O. Stock and M. Schaerf (Eds.): Aiello Festschrift, LNAI 4155, pp. 25–43, 2006.

Like in any scientific discipline it is from time to time worthwhile to review the direction of research from a high-level point of view, thereby abstracting from the day-to-day focus on specific research problems, and rather take the entire picture of the discipline into consideration. This is what the current article aims to do. In other words we want to discuss the various dimensions of logic and deduction and their role within AI. Neither the analysis nor the role description can in any way be comprehensive in such a short article; they rather reflect the bias of the author concerning his judgment of particularly important issues. The result of these considerations is a number of concrete proposals for future research which are deemed particularly promising. In short, the text may be regarded as the author's manifesto for an area in which he has worked for nearly four decades.

For completeness the article contains a short summary of the goals of AI – or rather of Intellectics. As we all know those seventies also brought about a schism of our discipline in the way of a separation of the field into the more systems-oriented AI and the Cognitive Science (CogSci) focussing more on the study of natural intelligence and its basis. I am deeply convinced that these two directions have to go together in a synergetic way in order to achieve their mutual and intertwined goals. This deep conviction is the reason for my stubborn adherence to a common name, *Intellectics* [6], proposed in 1980 for the discipline spanning both subareas; in short, Intellectics = AI & CogSci.

As we said the article analyses the nature and role of logic in achieving these general and longterm goals. It begins with a brief view at those goals of Intellectics, thereby pointing out two major subgoals, viz. solving the integration problem and contributing to the solutions of the complex problems with which our societies are currently confronted, whereby logic could play an important role. In addition to the standard applications of logic we then outline as a challenging research line its role in modelling intelligent behavior in a conjunctive way and in compiling from such a descriptive model applicational systems. This development would include a logical modelling language which is not suffering from the limitations experienced with languages like UML.

While logic currently is used exclusively without any kind of grounding its constants, such an association with semantic information could be rather beneficial in terms of efficiency, and is therefore proposed as another challenge. A further section deals with the choice of an appropriate logic in dependence of the intended application and the required features including change, vagueness and uncertainty and proposes research on some measure for a more rational distinction among a variety of logics.

Once we have settled in for a chosen logic for a certain application there is still a wide variety of formal systems to chose from for expressing the logic and support the inferential mechanisms. We once again remind of the important research strategy aiming at a formalism which is as compressed as possible. While remarkable results have been achieved in this line of research such as the ileanCoP system, the approach as a whole is not exhausted at all. Several important longstanding questions have still not been solved and incorporated into actual systems and further ones are raised for future work in this respect.

Among these is the challenge of integrating cuts which lead to shorter proofs. We conjecture that a way to do so is by the use of factoring and engaging a nonclausal form calculus (or a subsequent linear transformation to clausal form).

Finally we discuss that part in proof search on which most efforts were spent in the last decades, which is the development of efficient search strategies. We argue that such strategies would need to be context dependent and that really efficient and specific ones might become too complex to develop by hand. For this reason we propose an automatic design of search strategies based on experimental data, a methodology successfully applied already to solving hard combinatorial problems. We also encourage the community to reconsider the integration of examples into the search for proofs and propose to do so in a preprocessing manner.

2   Main Intellectics Goals

Intellectics aims at a profound understanding of the working of human intelligence in brains (CogSci part) and at mechanizing human-level intelligence (AI part). The fiftieth anniversary of the Dartmouth Conference in 1956 has given rise to numerous reflections on what the field has achieved in the first half century in pursuing these goals and what should be done now in order to progress further. The issue of the *AI magazine* (vol. 26,4) celebrating the 25th anniversary contains numerous statements of this kind. They suggest new challenge problems and research strategies.

But perhaps one should once again take one step further back and ask whether and why we should continue to pursue these grand goals. As far as the CogSci part is concerned the justification is straightforward. Curiosity is inherent in human's nature and we are simply curious what mechanisms make us intelligent. A deeper insight into these mechanisms could have numerous beneficial implications including cues how to educate humans more effectively, how to communicate more smoothly among ourselves, how to improve our problem solving capabilities, and so forth.

The justification of the AI part is not as obvious. Why should the more than six billions of humans on earth strive for a new breed of intelligent agents? Well, first of all Intellecticians are convinced that the CogSci part of the goals cannot be achieved without actually realizing human intelligence in an artificial way. In other words, AI is a prerequisite for CogSci in this sense so that AI inherits its justification already from CogSci. Reversely, AI also needs CogSci insights for inspirations and new ideas how to proceed. In fact one approach to AI consists in *reverse engineering* of the brain [21, Ch.4] which can only be achieved on the basis of CogSci input (not least the one from the neurosciences). This mutual dependency and the common goals are good reasons for regarding AI and CogSci as a single discipline. Apart from the basic justification of AI just mentioned, the short history of AI research has produced plenty of evidence that it generates techniques which have become extremely useful in numerous applications, even so many that any short list of examples would leave us with too a distorting picture.

I myself have always regarded the ultimate AI goal as a rather distant one providing the Intellectics community with a socially uniting umbrella, but too distant to influence our daily work in concrete terms. In this vein I continue to think that our next subgoals should rather be guided by responsibilities to the society at large, although in such a way that they are deemed compatible with, and their achievements steps towards, AI's grand goal. In this sense I see two such subgoals as of paramount importance, one basic the other applicational. The first, basic subgoal consists in solving the urgent issue of integration, the second in attacking fundamental problems in our societies whose solution could be achieved by AI technologies. Both subgoals will be expanded further in the following.

Intellectics in general and AI in particular today is rather fragmented. The vision community within AI has little or no interaction with the knowledge representation community, to give one out of many possible examples. As a result we have vision systems such as those built into the autonomous vehicle Stanley which triumphally won the DARPA contest in 2005; but Stanley "knows" literally nothing at all about the world it sees. How could we integrate into such vision systems knowledge systems without reprogramming everything? How could we then extend the resulting system in the same vein by integrating speech and natural language understanding systems, planning systems and a variety of systems with further functions including those beyond AI in a way so that the final integrated system features a truly intelligent behavior? These questions refer to what we call the *integration problem*. We believe that logic holds the key for solving this problem as discussed in the subsequent section.

Humanity faces dramatically complex problems to be solved in a relatively short time, foremost the problems caused by the world climate change due to the man-made increase of greenhouse gas like carbon dioxide and methane in the atmosphere the consequences of which can be traced in many global phenomena like the glacier retreats, the warming of oceans leading to a dramatic reduction of life in it and to numerous other frightening consequences, the disappearance of virgin forrests, the spreading of deserts, to mention just a few [25]. Despite the exponential growth of technological advances we have to make sure that enough time is left for humanity to be able to harvest the fruits of these advances. Namely, the intrusion of these advances into the social mechanisms seems to take place at a much slower pace, as some of those (like politics, law, social struggles etc.) have not changed much since ancient Greek and Roman times. Hence, it is currently undecided where this brinkmanship of meddling into the global mechanisms of nature will lead us. Therefore scientists have the reponsibility to contribute to the solutions of these fundamental problems rather than pursuing prestigeous goals for the goals sake.

AI technology can contribute substantially in this endeavor in many ways as has been described in great detail in my recent book [8]. I see for instance a key role for knowledge and problem solving systems in a more rational approach towards solving societal and global problems such as the one just mentioned (viz. the world climate). Problems of this complexity cannot be coped with by the locally oriented problem solving attitude of humans but only by a truly global consideration of all aspects involved of the kind as realized through objectively accumulated knowledge bases and general problem solving mechanisms. In [10] this potential has been outlined for the complex area of law. Progress in such domains which are fundamental for the prosperity of societies – and there are many more than just the legal domain – would have an even far greater impact also for the standing of AI as a discipline than, say, a program which beats the worldmaster in chess, notwithstanding the fact that this is indeed a truly impressive achievement. Another vital domain of application of this kind of AI technology is science itself as has recently been pointed out in several foresighting reports [11,15] followed by the 23 March 2006 special issue of *Nature*. Again we believe that logic is substantial for these kinds of contributions.

The emphasis on these two selected subgoals is not meant to diminuish all the fascinating work currently going on in all other subareas of Intellectics. Rather we want to point out that these particular ones deserve at least the same level of attention. We do sense an imbalance in this respect which to some extent may be due to the schism between AI and CogSci.

3 Why Logic?

Logic is often paraphrased as the language of thought. Because thinking is a crucial component of intelligence, logic on this account will most likely play an important role in an artificially intelligent agent at some level of abstraction. Aspects of this role can already be observed in knowledge systems, in automated theorem proving (ATP), logic programming, problem solving, and so forth.

Given these successful applications of logic we feel that no further justification for the relevance of logic is needed. Nevertheless there is a fundamental criticism of a logical approach to achieve artificial intelligence. According to this argument our brains function in a rather different way. For instance, catching a ball does not involve solving diffential equations but a direct transformation of the observed movements of the ball into an appropriate movement of the player's arms and legs. Similarly, it is supposed that reasoning as well is realized in the brain by analogue direct transformations rather than by logical deductions. While this may well be the case it is still important to understand the underlying mechanism in terms of the higher level of abstraction of logic, as it is important to understand a ball's movements in terms of differential equations. How we eventually will realize such behavior in artificial systems is quite an independent question.

Besides this role of logic as the language of thought there might be another similarly prominent role for logic in Intellectics. As we said in the preceding section understanding intelligent, cognitive behavior and making machines exhibit such behavior is the goal of Intellectics. In order to characterize this additional role of logic we begin with mentioning that there are at least three different viewpoints from which this goal can be approached. The first is the viewpoint of observing and analyzing intelligent behavior in existing creatures, foremost in humans. The second takes the perspective of future artificially intelligent systems or robots and their potential architectural design without much ado about how to realize it with present technology. And the third focusses on concrete steps towards realizing selected intelligent functions with present technology which eventually might be part of a future intelligent agent.

Each of the viewpoints has its merits and each is necessary for a future overall success. Depending on which of these viewpoints one takes rather differing standpoints and preferences may be chosen. In the past these differences were, as already indicated in the Introduction, the cause for schism and hot debates in the community. For instance, CogSci – rooted deeply in the first and analytic viewpoint – separated from AI in the late seventies (see [16, pp. 33 f] for some background information) mainly because the AI community to a large extent became obsessed with quick commercial successes based more or less exclusively on the third viewpoint which is synthetic and bottom-up. How about the second viewpoint?

It seems unlikely to me that the current bottom-up and patchwork-like approach in AI will ever lead to a truly intelligent agent. Ultimately this goal will not be reached but in a top-down fashion starting from the insights gained by CogSci, Neuroscience and by introspection (cf. [21, p.168] for a similar argument). For that purpose these insights need to be accumulated in a computational model reflecting the many facets of human intelligence. The generation of such a model is a Herculean task given the complexity of intelligent behavior.

The only way I could think of mastering it would be a *conjunctive* one in the logical sense. That is, if we have two independently generated parts M_1, M_2 of the model then these can be combined by a simple conjunctive (or additive) operation like logical conjunction $M_1 \wedge M_2$. The reason for this requirement lies in our human way of insight. We are bound to understand just small fragments of the entire workings of intelligence at any given time. So in addition to forming each single fragment in some representation there must be some operation which makes a coherent mosaic out of the myriads of represented fragments. The operation must be simple in order to cope with the shere amount of pieces.

In other words I am pleading here for the accumulation of a coherent, formalized and implemented model of intelligence in the sense of the second viewpoint. This task would involve many scientists, even generations of scientists over a long period of time. Could and should we afford such a grand endeavor? Yes, we could because the project would necessarily employ an economic "anytime" procedure of the following kind.

Such a model would have to be in some way *descriptive* in order to comply with the requirement of conjunctiveness, whereby descriptive (or declarative) is meant in a rather broad sense possibly even including natural language descriptions, pictures and scenarios, the simulation of dynamic processes, etc. At any given time the model accumulated upto a given point could be synthesized to a working agent featuring all the aspects accumulated in the model. This means that the model could at the same time serve a variety of many practical purposes in the sense of the third viewpoint. Namely, for each particular application one selects the desired parts in the model existing upto that point in time and synthesizes from there the applicational system. In other words the second and the third viewpoint could from there on be pursued in a synergetic manner.

This vision assumes two major prerequisites in order to be realizable. The first is the existence of a conjunctive formalism which is descriptive in the sense just indicated. Since logic is both descriptive and conjunctive it is exactly here where we see a central role for logic, possibly an extended logic with many more features than currently familiar. The second prerequisite is a mechanism which synthesizes working systems out of such a formal model. This is less illusionary than one might think at first sight. Just think of current practice in systems engineering which often generates a model in some language such as UML (ie. universal modelling language) and extracts from it the systems code, to some degree in an automated way.

The analogy with UML demonstrates that we are proposing here a rather realistic and fruitful research project with two major research lines. One is the development of a language like UML but without UML's severe limitations which at present are painfully felt in many applications. Most likely such a language would be more logic-like. The other line consists of a further automation of the synthesis of systems code out of a formalized model. Both goals are of utmost importance for current software practice. It is these two goals which drive the Mercury project [2] (see also below Section 6) for exactly the reasons I have given here. So the grand endeavor of accumulating a model of intelligence could in fact be pursued in parallel with and on the basis of research on very practical tasks. In addition, the methodology promoted here for evolving a model of intelligence would of course be useful for any area which strives for understanding complex structures, eg. those present in social systems.

As we just touched upon current software practice, it is interesting to take a retrospective view and note how little progress the software community has achieved in the last thirty years in terms of a truly user-oriented software approach of the kind which the present author described more than thirty years ago (see eg. [3] and the references given therein). In essence that approach (originally termed *predicative programming*) shares the methodology with what is described here for developing a model of intelligence. So in a nutshell I have here just reiterated for the development of intelligent agents what I proposed three decades ago as a better way of producing software. In this context it is encouraging that the recent years have indeed seen remarkable steps into this direction also within the software community. Namely, computation independent models (CIM), model-driven architectures, post requirement specification traceability, and several further terms of this kind are now the catch-words of the day circumscribing an approach to software generation of the kind envisioned with predicative programming.

4   On the Issue of Grounding

Logic employs a formal language without any grounding. A constant symbol such as *a* or *table* has no semantics at all. A logical statement like *On*(*glass, table*) may be subject to many different interpretations, not the one intended by the choice of the names of the constants. Human knowledge seems to operate in a completely different way (and the specific discipline of semiotics studies the meanings of human symbols) – or does it? Well, we do not really understand the working of the brain wrt. its knowledge processing, so at the time being we cannot be sure.

Nevertheless anyone can experience by introspection that many seemingly logical conclusions in everyday life are drawn by inspection of a mental model rather than by deductive inference [19]. For instance if someone tells me that a book lies on the table with a glass standing on the book, I "see" that in this imagined scenario the glass is above the table without regress to formal rules concerning the transitivity of the "on"-relation and its connection with the "above"-relation.

If, on the basis of such experience, we take for granted that the brain realizes logical conclusions partially by way of inspecting mental models the question naturally arises whether an analogue mechanism might not be similarly useful in AI systems. It is therefore our proposal to investigate this possibility in future research. Here are a few ideas how this might be achieved.

In the brain the word, say, *table* is associated with the sensory information deriving from a number of concrete tables experienced out there in the world. Already the issue of this association raises a number of questions. Does the mental model of a table refer to the sensory data of a particular, selected table? Or does the brain use some mechanism of abstraction to generate some mental model of table with the common characteristics of all experienced tables? Anyway, an artificial agent could similarly associate with a name such as *table*, used in its logic, corresponding sensory data and classify it as a table, ie. as a particular unary predicate. Similarly with any other constant, function or predicate. There might be many ways to exploit such an association in the reasoning processes carried out within a logical system. Such an association would be a first step towards involving a true semantics in machines and would have great relevance for many applications where knowledge plays a key role. We will come back to one of these applications in Section 7 where we discuss examples supporting the inferential process which in AI was the first and sofar only attempt to integrate this kind of information.

5   Which Logic?

The history of logic has produced a great variety of logics. One of the reasons for this variety is the choice one has between axiomatization and logical structure. In other words, some more advanced features of natural expression can be characterized in an axiomatic way or, alternatively, be built into the logic itself.

An example is how actions and change are formalized. Logic has tradition-ally been used to model static logical reasoning while changes in the modelled world have been considered as quite different a matter. Modelling applications like planning and computation however have forced us to integrate change into logical frameworks. How this best is done is still a matter of debate. The situa-tion calculus was one of the very first attempts of this kind which uses first-order logic to characterize the features needed to model actions and change in an ax-iomatic way. In a series of papers eventually resulting in [9] the present author has developed a transition logic as an alternative. The idea is to regard changes or transitions as first class citizens (in the form of transitional rules) within the formal framework and otherwise keep the logical part more or less like in first-order logic. A related approach focussing on an integration of concurrency is described in [20]. Modal logic, linear logic as well as the more recent com-putability logic [18] in contrast build changes into the logical connectives inside the logic.

There is an even larger area open for research in this topic of modelling changes. Namely, the bread and butter of actual modelling and simulation sys-tems (eg. for modelling the climate, ocean currents, physiology, etc.) are differ-ential equations. But in addition we would urgently need knowledge of a logical kind to be integrated into such systems. However the author is not aware of any practical formalism which could provide the theoretical basis for such an integration.

Another example of modelling features in logic is vagueness and uncertainty where the same kinds of alternatives have been developed. Namely, there is the basically first-order treatment extended by probabilistic features as in [28] and, alternatively, there are logics which express vagueness and uncertainty by means of the logical operators like fuzzy logic, nonmonotonic logic and so forth.

Upto this day there is no systematic study on which a rational choice of an appropriate logic in dependence of the intended application and the required features could be based. Research rather proceeds in trying out many possibil-ities at the same time and rather independently. This unsystematic strategy is one reason for a waste of resources in the community which should be taken notice of.

This is not to say that we are in lack of any comparative arguments concern-ing the various logics. An important one is the complexity of the proof search. For instance in description logics we have a detailed classification in this re-spect which of course is extremely helpful. However, there seems to be no way around involving logics which – assuming $P \neq NP$ – are computationally non-polynomial. Among them there is still a great variety of logics waiting for a distinction through some measure. Apart from the ones mentioned to handle changes and uncertainty there are many more where such a distinction would be rather helpful.

So here we have come across another proposal for future research in our do-main namely to work out comparative arguments or measures distinguishing different logics in view of intended applications. While I refrain from stating any preferences in this respect which would be based merely on personal prejudice, my experience still tells me that classical first-order logic might in some form or another continue to play a major role in future formal logical reasoning as would higher-order logic whose potential seems still underestimated.

6 Formal Systems

There are numerous formal systems (or calculi) which encode in one way or another a given logic. Let us take first-order logic as our paradigm example because of its wide dissemination although any other logic could as well have been selected to make the point.

Chapter 4 in [7, pp. 97ff] demonstrates that such formal systems may differ in the degree of their compression. Thereby we consider a formal system S_1 to be more compressed than another S_2 if any proof in S_1 is shorter (in terms of the number of symbols required) than the corresponding one in S_2. For example, Gentzen's formal system of natural deduction NK is less compressed in this sense than the tableau calculus.

Occasionally compression may result in a substantial change in the complexity. For instance, the elimination of the cut rule leads to a less compressed calculus and to possibly exponentially longer proofs. In most cases however, compression has less dramatic effects. For instance the connection calculus is more compressed than the tableau calculus although the proof lengths differ only by a polynomial factor. This however does not mean that such a compression is worthless. On the contrary, our experience shows that the performance may increase dramatically as we demonstrated through a comparison of leanTAP with leanCoP in [27]. In fact, the intuitionistic version of leanCoP, called ileanCoP, is now by a wide margin the fastest theorem prover in existence for intuitionistic first-order logic [26].

In the preceding discussion we measured compression in terms of lengths of proofs. It is interesting to have a look also at the length of the program underlying the theorem prover as a measure. leanCoP needs 333 bytes (!) for representing the program. The smallest version of ileanCoP derives from leanCoP by adding 191 bytes so that altogether it needs exactly 524 bytes without the approximately 30 lines required for prefix unification. Figure 1 shows the three clauses of the long version of the source code in a way such that both, leanCoP (everything except the underlined symbols) and ileanCoP can be seen. Note thereby that not a single symbol had to be changed in leanCoP but only additional information had to be included in the form of new terms and literals. The way how this extension can be achieved comes close to the kind of predicative programming the author had in mind 30 years ago (and already mentioned further above). In contrast provers like Otter need hundreds of thousand times longer code. Modifying such a monster is simply impossible for any person (except for the system's author within a period of a few years after completion).

```
(1)     prove(Mat,PathLim) :-
(2)         append(MatA,[FV:Cla|MatB],Mat), \+ member(-(_):_,Cla),
(3)         append(MatA,MatB,Mat1),
(4)         prove([!:[]],[FV:[-(!):(-[])|Cla]|Mat1],[],PathLim,[PreSet,FreeV]),
(5)         check_addco(FreeV), prefix_unify(PreSet).
(6)     prove(Mat,PathLim) :-
(7)         \+ ground(Mat), PathLim1 is PathLim+1, prove(Mat,PathLim1).

(8)     prove([],_,_,_,[[],[]]).
(9)     prove([Lit:Pre|Cla],Mat,Path,PathLim,[PreSet,FreeV]) :-
(10)        (-NegLit=Lit;-Lit=NegLit) ->
(11)          ( member(NegL:PreN,Path), unify_with_occurs_check(NegL,NegLit),
(12)            \+ \+ prefix_unify([Pre=PreN]), PreSet1=[], FreeV3=[]
(13)            ;
(14)            append(MatA,[Cla1|MatB],Mat), copy_term(Cla1,FV:Cla2),
(15)            append(ClaA,[NegL:PreN|ClaB],Cla2),
(16)            unify_with_occurs_check(NegL,NegLit),
(17)            \+ \+ prefix_unify([Pre=PreN]),
(18)            append(ClaA,ClaB,Cla3),
(19)            ( Cla1==FV:Cla2 ->
(20)                  append(MatB,MatA,Mat1)
(21)              ;
(22)                  length(Path,K), K<PathLim,
(23)                  append(MatB,[Cla1|MatA],Mat1)
(24)            ),
(25)            prove(Cla3,Mat1,[Lit:Pre|Path],PathLim,[PreSet1,FreeV1]),
(26)            append(FreeV1,FV,FreeV3)
(27)          ),
(28)          prove(Cla,Mat,Path,PathLim,[PreSet2,FreeV2]),
(29)          append([Pre=PreN|PreSet1],PreSet2,PreSet),
(30)          append(FreeV2,FreeV3,FreeV).
```

Fig. 1. Main part of the ileanCoP source code

The way compression is achieved by the connection method (CM) for various logics has been described many times (eg. see [7] and the references given therein) so that we can – and for reasons of space must – restrict ourselves to stating the most important of its essential features. The CM analyzes the structure of a given formula F without changing F whatsoever which has a particularly beneficial effect on the length of proofs. It focusses on establishing a spanning set of connections which characterizes the formula's validity. Thereby the procedure is connection- and goal-oriented, and unification is employed preferably relying on a particular partial-ordering on the set of terms rather than on Skolemization [4, Sect. IV.8]. In the case of non-classical logics the unificational part is extended, eg. with prefix unification in the case of intuitionistic and modal logics. Complementary to the main top-down procedure bottom-up preprocessing steps may reduce the proof problem substantially. A CM proof in some connection calculus represents many different proofs in say a Gentzen-type formal system, ie. it identifies them by disregarding and abstracting from irrelevant differences. It is obvious that this dispensing with irrelevant burden has a beneficial effect on the efficiency of the resulting systems. Sofar the CM's features.

The research program underlying the CM approach is not exhausted at all. For instance, the transformation to normal form, even if done wisely, still introduces a lot of redundancy which distracts the proof search. A leanCoP for nonclausal-form formulas along the lines of [4, p.150ff] is therefore highly desired but requires a mind with the unique talents like that of Jens Otten nurtured by an appropriate research climate. Further compressions like those described in Chapter 4 in [7, pp. 97ff] (such as equality handling etc.) need then to be integrated into such a system. And last not least a further boost for the performance would come from a compilation of the various leanCoPs into a low-level programming language preferably by some automatic mechanism. The way how this can be achieved is shown by the remarkable Mercury project whose goal is to combine the virtues of declarative programming with features from current software practice, especially providing for (separate) compilation of declarative code among many other attractive features [2].

6.1   A Conjecture Concerning the Cut

We have just pointed out the importance of compression and its role in the CM. In recent years there have been complementary attempts towards the compression of logical calculi. A prominent one is Guglielmi's calculus of structures [13]. It overcomes the restriction in Gentzen's calculi that inference rules can only be applied to surface (or main) connectives. Rather it allows inference rules to be applicable at any time to any logical connective anywhere inside the formula, a technique termed *deep inference*. Due to this enhanced flexibility the cut formula can be restricted to atoms with predicate symbols exclusively from the conclusion of the cut inference. This leads to a finitely generating system even if the cut rule is included. Recall that the inclusion of the cut rule leads to a potentially exponential compression as already mentioned at the beginning of this section to see the relevance of this achievement. In Section 4.2 of [14] a similar result is presented using meta-variables on arbitrary cut-formulas which insofar is to be considered as a less compressed approach. No clue is given in either approach when to apply the cut rule during the search for a proof as it could be applied at any point. A strategy in this regard is then a major challenge for future research in this area.

   A first step into such a direction has been made in [23] with the so-called folding up technique. It derives lemmata in a bottom-up manner during the main top-down procedure which is derived from an amalgamation of the connection and the tableau calculus underlying the high-performance proof system Setheo [24]. The paper shows that folding up can be viewed as a controlled integration of the cut rule.

   Folding up builds into the connection tableau calculus a technique whose effect can alternatively be achieved with the form of factoring reduction which has been termed FACTOR in [7, pp. 56]. In order to illustrate this relationship let us consider the matrix $\{\{p,t\},\{\neg p,q,s\},\{\neg t,p,s\},\{\neg q,r\},\{\neg s,r\},\{\neg r,\neg p\}\}$ which is the prime example for illustrating folding up in [23]. FACTOR applied to this matrix twice leads to the nonclausal form (NCF) matrix $\{\{p,\{\{t\},\{\neg t,s\}\}\},\{\neg p,q,s\},\{\{\{\neg q\},\{\neg s\}\},r\},\{\neg r,\neg p\}\}$ by factoring the p in the first and third clause and the r in the fourth and fifth clause of the matrix.[1] A connection calculus like the extension procedure for NCF formulas or matrices in [4, p.150ff] would then behave exactly like folding up and establish the partial proof of the matrix as in [23, Figure 8] with 5 extension and 1 reduction steps, ie. altogether 6 connections. In other words the effect of folding up and hence of the corresponding use of cut could equally be achieved with this kind of factoring.

Note that this matrix is in fact not complementary, but could be made so eg. by adding the clause $\neg s$ to it. Also note that FACTOR could be applied yet another time to the resulting NCF matrix by factoring additionally $\neg p$ which yields the matrix $\{\{p, \{\{t\}, \{\neg t, s\}\}\}, \{\neg p, \{\{q, s\}, \{\neg r\}\}\}, \{\{\{\neg q\}, \{\neg s\}\}, r\}\}$. Now the extension procedure would require only 5 extension steps (or connections) which illustrates that even in the special case of factoring just literals this kind of factoring is more powerful than folding-up. As a final remark we mention that FACTOR could alternatively have been applied to the original matrix by factoring the s in the second and third clause and the r in the fourth and fifth clause of the matrix resulting in $\{\{p, t\}, \{\{\{\neg p, q\}, \{\neg t, p\}\}, s\}, \{\{\{\neg q\}, \{\neg s\}\}, r\}, k\{\neg r, \neg p\}\}$. A different (partial) extension proof would then be found.

This result established by Letz concerning the relationship of folding up and the cut along with the rather obvious fact just illustrated that in general folding up can always be viewed as applying FACTOR a finite number of times and then use a connection calculus for the resulting NCF matrix suggests an even more general conjecture. Namely, we *conjecture that the general cut can linearly be simulated by FACTOR applied to arbitrary submatrices (ie. not only atomic ones) a finite number of times and then a connection calculus applied to the resulting NCF matrix.* I pose this conjecture as a research challenge.

There is more evidence than the one just mentioned supporting the conjecture. It is known that the pigeonhole formulas are hard for resolution requiring exponential proof lengths [17]. In contrast they can be established with polynomial proofs both in a Gentzen system with cut (or a Frege system) and in a connection calculus using among others FACTOR as a preprocessing rule [5].[2] The general reason for this advantage could be the following.

---

[1] For readers unfamiliar with this kind of matrix notation for formulas we mention that (in the positive interpretation) such a (clause-form) matrix, ie. a set of clauses, can be read as a disjunction of its clauses (ie. sets of literals) which may in turn be interpreted as conjunctions of literals. In an NCF matrix the elements of the clauses may in turn be matrices, ie. disjunctions of clauses, rather than just literals, and so forth until any nesting depth. In the negative interpretation (commonly used eg. in the resolution literature) the role of disjunction and conjunction are interchanged, ie. a matrix is interpreted as a conjunction (rather than disjunction) of clauses, and so forth. The formal details may be found in a standard textbook like [7].

[2] Other reduction rules used are PURE, UNIT and Prawitz' matrix reduction. Renaming, a rather strong rule, is not required for proving the formulas, but is just used in the paper (on the metalevel) to be able to apply the induction hypothesis.

The cut enables a compression of a proof in a Gentzen system.[3] Resolution, in contrast, although superficially of the form of a cut, does not feature the full power of the cut rule since otherwise there would be polynomial resolution proofs for the pigeonhole formulas. It is unknown how much of the power of cut is inherent in resolution. It is conjectured that there is some of it and that this part makes resolution occasionally more efficient than CM-type proof systems for clausal logic (without FACTOR).

Since there is a close relationship between the formula to be proved and its Gentzen-type proof, an elimination of the redundancy in the formula by compression also decreases the potential for further compression of the proof by the cut rule. FACTOR enables the elimination of redundancy without loss of information, ie. the formula's "entropy" (in analogy with Shannon's information theoretical concept) increases by its application. When it reaches its maximal value, there is no room anymore left for compression of the corresponding proof by way of the cut rule.[4] That basically is another way of putting the conjecture above. Once this potential is exhausted by future CM-systems they should uniformly outperform standard resolution systems since then the only remaining advantage of resolution will also be incorporated into them, and otherwise their connection and goal-oriented behavior endows them with a clear advantage over standard resolution systems.

It is generally believed that cut formulas have to be invented creatively without much clue given by the conclusion formula which seems to speak against our conjecture. However, Guglielmi's result just mentioned might be seen as an indication that all propositional information about the cut formula might indeed be contained in the conclusion while its first-order features, ie. the terms, could anyway be determined by unificational mechanisms in the usual way. Some

---

[3] As pointed out by Alessio Guglielmi (private communication) the cut may also play the role of enabling case analysis. For example, the cut $A \rightarrow B, \neg A \rightarrow B \vdash B$ features a way of trying to prove B in the two mutually exclusive cases in which the hypotheses A and $\neg A$ are assumed. This aspect of case analysis has been studied extensively in deduction either explicitly (like for instance in Plaisted's work) or implicitly (like in the connection calculus).

[4] In letters dated 4 August 1980 to both, Georg Kreisel and Dag Prawitz, the author already pointed out the conjecture that the shortening effect of cuts is mainly due to the redundancy (or "bureaucracy") in derivations and would disappear when the author's derivational skeletons [4, p.190] were used instead. This conjecture raised the curiosity and interest of Kreisel which he expressed in his response letter.

In a nutshell this early conjecture can be phrased as follows. Let P_1, P_2 denote the premises of a cut with conclusion C. Let s_1, s_2 denote the skeletons of cutfree derivations of P_1, P_2, resp., and s the skeleton of the derivation of C obtained from the derivations of P_1, P_2 by the well-known process of cut-elimination. Then the length of s is a polynomial function of that of s_1, s_2, ie. not an exponential function as in the case of Gentzen-type derivations (which carry all that redundancy).

The conjecture stated in the present paper is even stronger than that earlier one and, in the present terms, states that this function is even a linear one for some C' obtained from C by applications of the (first-order version of the) FACTOR operation to C, and even if Gentzen-type derivations would be taken instead of skeletons.

"creativity" would of course still be needed as is already illustrated by our example above where different sequences of applying FACTOR have led to different proofs, so that the mechanism would have to explore the finite space of different such sequences for the most suitable one (in the sense of the maximal entropy value). A parallel approach to this exploration might eventually be taken into account (similarly as in reality where more than one mathematicians are trying to solve mathematical problems).

If this conjecture could be proved correct then an algorithmic realization of this idea would still be complicated, especially when first-order unificational mechanisms have to be integrated (into FACTOR etc.), thus posing a further research challenge in this case. Even if the conjecture would turn out not to be generally valid FACTOR would still remain an attractive reduction operation which has been neglected in current systems. Note thereby that, even if a connection calculus for NCF would not be available or undesirable, one could still apply FACTOR, then apply to the result a linear or quadratic transformation to clause form [7], and finally apply any proof method to the result, which occasionally would be a more compressed formula than the original one.

Guglielmi's calculus has motivated the question for the *essence* of proofs after eliminating all bureaucracy caused by individual formal systems. The answer given in [22] naturally is closely related to that given by the CM, namely that this essence is basically given by the connection structure underlying a proof. The paper in addition clarifies the effect of the cut rule in terms of a composition operation on such structures (without addressing the questions underlying our conjecture). This result might be helpful as well in the context of incorporating the cut into proof search one way or another. But it does not yet attack the first-order features as has been done within the CM with its skeleton concept pointed out in footnote 3.

Another step towards compressing Gentzen's sequent calculus has recently been taken in [18] which introduces the so-called cirquent calculus. "Roughly speaking, the difference between the two is that, while in Gentzen-style proof trees sibling (or cousin, etc.) sequents are disjoint and independent sequences of formulas, in cirquent calculus they are permitted to share elements." In other words, a proof is no more a tree of sequents but becomes a compressed tree-like structure whereby different branches share joint parts. Such a calculus can be sensible to resources and in fact it has evolved in the context of attempts to develop a computational logic of the linear-logic kind. Whether or not computation and planning will be modelled in such a purely logical way in the future or rather in the transitional way described in the previous section is independent of the interesting compressional idea behind the cirquent calculus.

7   Search Strategies

Whatever formal system we might have chosen proofs in it cannot be found without search. Any such proof search has two quite different parts. One consists of a mechanism which is not really search in the sense of the word. Rather it combines a number of operations which are needed to test one single alternative for success or failure. In contrast to this the other part consists of choosing true alternatives where the wrong choice might well lead into a blind alley or at least into a superfluous detour (in a confluent system). For instance if we think of a tableau system in propositional logic for simplicity, it is the alternative branching points which give rise to search in the true sense while the remaining steps are rather straightforward. Let us refer to the sequential and the choice part to distinguish the two in the following.

A lot of efforts in a variety of directions have been invested into dealing with the choice part. One direction has been to deal with the alternatives in a parallel way. Since always only a limited number of different processors are available parallelization has the potential to provide some improvement but not a cure to the underlying exponential explosion. Since the single processor machines became so much more efficient they outraced the advantage of multiprocessor machines in this application and will do so for some time to come. An exception might be pursuing a finite (and small) number of alternatives in parallel like those in the application of FACTOR described in the previous section.

A second direction of research tried to take advantage of the information gathered in one alternative to be used also in another one in order at least not to waste redundant efforts completely. Intelligent backtracking was a popular technique in this direction. To some extent the same effect can be achieved with compression as discussed above.

A third direction tried to enhance the chances for selecting the right alternative. Many attempts have been made in this regard, including some rather naïve ones as seen from hindsight. Just think of the many so-called refinements of resolution some of which were mere adhoc attempts based on evidence of a few selected examples. Others were indeed based on solid theoretical arguments. For instance, Setheo [24] featured a preference in its selection strategy based on basic probabilistic arguments. However there could be many more preferences of this sort but they are difficult to develop and integrate into the overall strategy. It seems therefore that the ideal preference measure might be too complex to develop by hand.

Faced with all these difficulties some people nurtured the hope that human ingenuity might interact with systems in cases of difficult choices. I continue to regard this as a vage hope. Human ingenuity fails to blossom in the complex technical contexts and at states of our systems when these would need advice most. Human advice should therefore be integrated into the way the problem to be solved is stated upfront rather than investing in interactive systems and proof planning approaches which require an interaction on a rather deep technical level.

For all these experiences the most promising perspectives are in an automatic design of the search strategy based on experimental data. The technique for this direction has been formally developed in [12] in the context of metaheuristics. What is needed then is the integration of this technique into the framework of theorem proving. Basically this amounts to learning the search strategy from the data of successful proof search. It is to be expected that the resulting strategy depends upon the theory within which the proofs are to be searched for. That is the strategy in a purely logical framework without special axioms will presumably be different from one in, say, linear algebra.

I also believe that this way analogical reasoning will eventually be made possible in a practical sense. That is if the learned strategy used data from successful proofs, it will succeed in finding analogical proofs by way of the learned strategy; in other words, the analogy is coded into the strategy rather than in some logical form. This fits well with the observation that strategies used by humans typically are fuzzy. For instance, in chess many such fuzzy rules can be learned from textbooks. Similarly, in law such rules or strategic principles are common place known under the term *topoi*. Although sometimes seemingly contradictory, they are extremely helpful in human problem solving in chess, law, mathematics and many other disciplines. Capturing them in precise rules seems nearly impossible while computationally learning such strategic principles appears a promising perspective.

In human theorem proving examples play a prominent role. For instance, in [29] the attention is focussing on mathematical proofs including psychological phenomena like gazing at some structures or immediacy in recognizing truth. It is therefore surprising that in current systems such kind of feature is hardly ever present. It is well-known how examples can guide the proof search and avoid blind alleys [7, pp. 143ff] but the technique has not found its way into applications. The reference just given mentions as a possible reason for this fact the problem of how to generate examples or counterexamples automatically for a given theory (and gives references to respective approaches). Possibly it has been overlooked that the technique could be used in a preprocessing manner rather than by interrupting the proof search at certain choice points and query the available examples for guidance. While such an interrupt does not fit into the fast processing of modern proof systems, a preprocessing of this kind could indeed easily and elegantly be integrated and this way approximate the human way of mathematical proofs more closely.

Namely, as explained in the given reference examples require the open subgoals during a proof search to be satisfiable for the interpretations given by the examples. The respective information could be collected prior to the proof search for a number of examples and for each potential literal in the clauses of the theory and stored along with the literals by way of an appropriate data structure. During the actual proof search this information could then easily be checked and the choice made appropriately. I consider the ignorance of this possibility to be a major oversight on the side of the ATP community and its realization a project of high priority.

8   Conclusions

This paper has explored the most important dimensions of the space of logical research. In each of these dimensions we have pointed out opportunities for future research which are deemed of great relevance for the success of the logical approach towards the grand goals of Intellectics. A list of these research proposals has been given at the end of the Introduction.

The author shares the confidence with Luigia Aiello that the succeeding generation of researchers will be picking up these challenges and pursue their solutions with the same enthusiasm as we did during the hey-days of our careers to the benefit of humankind. I combine this confidence with the hope that society will appreciate this work more than it did sofar and provide the logic talents with a research environment appropriate for their work (eg. with a research institute of the kind of a Max-Planck-Institute) which, like in Mathematics, requires an extreme amount of concentration, certainly more so than in "softer" disciplines.

Acknowledgments. I greatly appreciate discussions on the topics of the paper with Kai Brünnler, Uwe Egly, Alessio Guglielmi, Reinhold Letz and last not least Jens Otten who also provided the leanCoP figure. I also thank two anonymous referees for their comments. All these interactions have led to substantial improvements of the text. For any remaining errors I am of course fully responsible.

References

1. L. Aiello. Automatic generation of semantic attachments in FOL. In Robert Balzer, editor, *Proceedings of AAAI-80*, Menlo Park CA, 1980. AAAI.
2. R. Becket, Maria Garcia de la Banda, Kim Marriott, Zoltan Somogyi, Peter J. Stuckey, and Mark Wallace. Adding constraint solving to mercury. In *Proceedings of the Eighth International Symposium on Practical Aspects of Declarative languages*, Charleston, South Carolina, January 2006. Springer Verlag.
3. W. Bibel. Syntax–directed, semantics–supported program synthesis. *Artificial Intelligence*, 14:243–261, 1980.
4. W. Bibel. *Automated Theorem Proving*. Vieweg Verlag, Braunschweig, second edition, 1987.
5. W. Bibel. Short proofs of the pigeonhole formulas based on the connection method. *Journal of Automated Reasoning*, 1990.
6. W. Bibel. Intellectics. In S. C. Shapiro, editor, *Encyclopedia of Artificial Intelligence*, pages 705–706. John Wiley, New York, 1992.
7. W. Bibel. *Deduction: Automated Logic*. Academic Press, London, 1993.
8. W. Bibel. *Lehren vom Leben – Essays über Mensch und Gesellschaft*. Sozialwissenschaft. Deutscher Universitäts-Verlag, Wiesbaden, 2003.
9. W. Bibel. Transition logic revisited. 2004. Submitted.
10. W. Bibel. AI and the conquest of complexity in law. *Artificial Intelligence and Law Journal*, 12:159–180, 2005.
11. W. Bibel. Information technology. Technical report, European Commission, 2005. ftp://ftp.cordis.lu/pub/foresight/docs/kte_informationtech.pdf.
12. M. Birattari. *The Problem of Tuning Metaheuristics as seen from a machine learning perspective*, volume 292 of *DISKI*. Akademische Verlagsgesellschaft Aka, Berlin, 2005.
13. K. Brünnler and A. Guglielmi. A first order system with finite choice of premises. In Vincent Hendricks, Fabian Neuhaus, Stig Andur Pedersen, Uwe Scheffler, and Heinrich Wansing, editors, *First-Order Logic Revisited*, Logische Philosophie, pages 59–74, Berlin, 2004. Logos Verlag.

14. E. Eder. The cut rule in theorem proving. In Steffen Hölldobler, editor, *Intellectics and Computational Logic*, volume 19 of *Applied Logic Series*, pages 101–123. Kluwer Academic Publishers, Dordrecht, 2000.

15. S. Emmott. Towards 2020 science. Technical report, Microsoft Research Cambridge, 2006. http://research.microsoft.com/towards2020science/.

16. E. A. Feigenbaum. Stories of AAAI – Before the beginning and after – A love letter. *AI Magazine*, 26(4):30–35, 2005.

17. A. Haken. The intractability of resolution. *Theor. Comput. Sci.*, 39:297–308, 1985.

18. G. Japaridze. Introduction to cirquent calculus and abstract resource semantics. *Journal of Logic and Computation*, 2005. To appear, see http://arxiv.org/abs/math.LO/0506553.

19. P. N. Johnson-Laird and Ruth M. J. Byrne. *Deduction*. Lawrence Erlbaum Associates, Hove and London (UK), 1991.

20. O. Kahramanoğulları. Towards planning as concurrency. In M.H. Hamza, editor, *Proceedings of the IASTED International Conference on Artificial Intelligence and Applications, AIA 2005, February 14-16*, pages 387–394, Innsbruck, Austria, 2005. Acta Press.

21. R. Kurzweil. *The Singularity Is Near – When Humans Transcend Biology*. Viking, Penguin Group, New York NY, 2005.

22. F. Lamarche and L. Straßburger. Naming proofs in classical propositional logic. In Paweł Urzyczyn, editor, *Typed Lambda Calculi and Applications, TLCA 2005*, volume 3461 of *Lecture Notes in Computer Science*, pages 246–261, Berlin, 2005. Springer Verlag.

23. R. Letz, K. Mayr, and C. Goller. Controlled integration of the cut rule into connection tableaux calculi. *Journal of Automated Reasoning*, 13:297–337, 1994.

24. R. Letz, J. Schumann, S. Bayerl, and W. Bibel. SETHEO — A high-performance theorem prover for first-order logic. *Journal of Automated Reasoning*, 8(2):183–212, 1992.

25. J. Lovelock. *The Revenge of Gaia: Why the Earth Is Fighting Back – and How We Can Still Save Humanity*. Allen Lane, 2006.

26. J. Otten. Clausal connection-based theorem proving in intuitionistic first-order logic. In *TABLEAUX 2005, International Conference on Automated Reasoning with Analytic Tableaux and Related Methods*, volume 3702 of *Lecture Notes in Computer Science*, pages 245–261, Berlin, 2005. Springer.

27. J. Otten and W. Bibel. leanCoP: Lean connection-based theorem proving. *Journal of Symbolic Computation*, 36:139–161, 2003.

28. D. Poole. First-order probabilistic inference. In *Proceedings of the 8th International Joint Conference on Artificial Intelligence*, pages 985–991, 2003.

29. J. A. Robinson. Proof = guarantee + explanation. In Steffen Hölldobler, editor, *Intellectics and Computational Logic*, volume 19 of *Applied Logic Series*, pages 277–294. Kluwer Academic Publishers, Dordrecht, 2000.

Reductio ad Absurdum:
Planning Proofs by Contradiction

Erica Melis, Martin Pollet, and Jörg Siekmann

Universität des Saarlandes and
German Research Center for Artificial Intelligence (DFKI)
66123 Saarbrücken, Germany

Abstract. Sometimes it is pragmatically useful to prove a theorem by contradiction rather than finding a direct proof. Some reductio ad absurdum arguments have made mathematical history and the general issue if and how a proof by contradiction can be replaced by a direct proof touches upon deep foundational issues such as the legitimacy of tertium non datur arguments in classical vs. intuitionistic foundations.

In this paper we are interested in the pragmatic issue when and how to use this proof strategy in everyday mathematics in general and in particular in automated proof planning. Proof planning is a general technique in automated theorem proving that captures and makes explicit proof patterns and mathematical search control. So, how can we proof plan an argument by reductio ad absurdum and when is it useful to do so? What are the methods and decision involved?

1 Introduction

Heuristic guidance plays a major role in mathematical problem solving. This has been an issue in human search behavior for a mathematical proof, see, e.g., [17,20], as well as in artificial intelligence (see [16]) and in automated theorem proving. Newell argued that Polya's heuristics are beyond the current state of the art in artificial intelligence, which today is no longer the case in general (see [12]). However, it turned out that each of Polya's heuristics actually represents a whole class of related more specific heuristics.

Heuristics that are mainly *domain-independent* have been incorporated into some early AI-systems for mathematical problem solving, most notably into Gelernter's geometry prover [8], Lenat's AM system [9], and in Woody Bledsoe's work. Gelernter's system used a given diagram to check the satisfiability of sub-goals and assigned priorities to goals dependent on the expected length of their solution. AM was based on general heuristics to search for new concepts and conjectures.

The still dominating (purely logic-based) automated theorem proving paradigm – mostly based on the resolution principle [18] – hardly uses any mathematical knowledge or mathematically inspired heuristics and makes up for this deficiency by its general refinements and ultra-fast search based on well-engineered representational techniques (see [19], vol.II, chapter 26). Of course, a resolution-based

O. Stock and M. Schaerf (Eds.): Aiello Festschrift, LNAI 4155, pp. 45–58, 2006.

system always searches for a refutation, i.e, using the insight from Herbrandt's Theorem, the theorem to be shown is negated and there is a proof once the system derives the empty clause as the final contradiction.

The situation is different for proof planning systems [5], where the proof steps are more general and more human-like holding the promise that traversing the search spaces can be based on (human) mathematical principles. Therefore, knowledge-based proof planning [10, 14, 15] uses extensive means for heuristic guidance and mathematical control knowledge, which has to be acquired.

In this paper, we propose some manually acquired control knowledge typical for theorems in an undergraduate textbook such as R.G. Bartle and D.R. Sherbert's 'Introduction to Real Analysis' [1] from which our examples are taken. Our focus is on proofs by contradiction and this exercise serves the purpose to see what knowledge is available, how it can be expressed, and, generally, to shed more light on the use of domain-dependent mathematical control knowledge in automated theorem proving based on proof planning.

Why is the control knowledge interesting that helps to select the proof by contradiction strategy? A first answer is that although proofs by contradiction are relatively frequent in mathematics, mathematicians make this choice usually 'instinctively' and do not reason about it explicitly. That is, we have to bring this implicit knowledge to the surface and make it visible – for machines and students alike. Secondly, since this is obviously a proof strategy that humans use to their advantage, we like to give it to a machine as well.

2   Reductio ad Absurdum

Proofs by contradiction have a long history in mathematics and they were used to advantage already in Greek mathematics. Well known is Euclid's Theorem, which states that there are infinitely many prime numbers. The proof assumes [1] that this is not the case, i.e., the number of primes is finite. Let $p_1, p_2, \ldots p_n$ be all these finitely many primes and consider the number $p = p_1 \cdot p_2 \cdot \ldots \cdot p_n$. Now take the number $p + 1$. If this is a prime number, then it is greater than any of the p_i, which contradicts our assumption since it would be a prime not yet among the assumed ones. Or else, $p + 1$ has a prime divisor, say q, then q would have to be one of the p_i and consequently q divides $p + 1$ and p, which leads to a contradiction since q would also have to divide the difference $p + 1 - p$, which is 1.

Another famous proof, due to Hippasus from Metapontum, a student of Pythagoras and member of the Pythagorean School shows that $\sqrt{2}$ is not rational. Because of the geometrical interpretation of $\sqrt{2}$ (i.e., the diagonal in a square of length 1) this problem had already puzzled Indian mathematicians two millenia b.c. and the Babylonian estimated $\sqrt{2} = 1 \cdot 60^0 + 24 \cdot 60^{-1} + 51 \cdot 60^{-2} + 10 \cdot 60^{-3}$ (which is about correct for the first five decimals) as recorded in a cuneiform script from about 1800 b.c. We recapitulate the theorem and a proof.

---

[1] Lets take this formulation for the purpose of explanation: as Michael Beeson has pointed out, the proof is actually not necessarily by contradiction [2,3].

Theorem. $\sqrt{2}$ is irrational.

Proof (Hippasus, 500 b.c.). Assume $\sqrt{2}$ is rational, i.e., there exist natural numbers m and n with no common divisor such that $\sqrt{2} = \frac{m}{n}$. Then $n \cdot \sqrt{2} = m$ and, thus $2n^2 = m^2$. Hence, m^2 is even and since odd numbers square to odds, m is even; say $m = 2k$. Then $2n^2 = (2k)^2 = 4k^2$, that is, $n^2 = 2k^2$. Thus, n^2 is even too, and so is n. That means that both n and m are even, which contradicts the fact that they do not have a common divisor.

This is a particularly interesting theorem not only because the Pythagorean's – believing in a rational world order – drowned Hippasus as a punishment for such offensive thinking but it was also posed as a challenge to 'the seventeen provers of the world': the results of this contest have been published in the Springer lecture notes [24] to mark the 50th anniversary of the first theorem ever, that was proven by a computer. [2] Our system participated in this contest as well (see [24]) but a fully automated proof planning of the proof, remarkably similar to the above proof of the Pythagorean School, was established a little later by Omega and has now been published in [23].

Proofs by contradiction have become an issue in the foundational discussion between classical versus intuitionistic mathematics. The above reasoning can be formulated as

$$\text{if } Ax \cup \{A\} \vdash F \text{ and } Ax \cup \{A\} \vdash \neg F \text{ then } Ax \vdash \neg A \qquad (1)$$

or alternatively as

$$\text{if } Ax \cup \{\neg A\} \vdash F \text{ and } Ax \cup \{\neg A\} \vdash \neg F \text{ then } Ax \vdash A \qquad (2)$$

The difference is that in (1) we conclude from the axioms and A as well as the contradiction $\{F, \neg F\}$ that $\neg A$ holds. In contrast, the second formulation states that we are allowed to conclude A from the axioms and $\neg A$ as well as the contradiction. Using the classical law:

$$\text{if } Ax \vdash \neg\neg A \text{ then } Ax \vdash A \qquad (3)$$

the above (1) and (2) collapse into the same kind of reasoning. This holds, however, only if we accept the tertium non datur postulate $F \vee \neg F$ from which (3) follows. Intuitionism rejects this postulate and hence, this is not a valid form of reasoning in intuitionism (see, e.g., [7]).

3   Knowledge-Based Proof Planning

Proof planning is a technique for theorem proving in which proofs are planned at a higher level, where individual choices can be mathematically motivated by the semantics of the domain. In particular, proof planning tackles theorems not only

---

[2] Martin Davis' program based on the decidable fragment of first order logic called Presburger Arithmetic, showed the remarkable theorem that the sum of two even numbers is again even.

with logical operators but also by using domain knowledge and explicitly encoded control [10]. However, a monolithic proof planner, in which the order of problem solving operations is pre-defined does not take full advantage of the runtime knowledge that is available from the mathematical domain. For instance, failure analysis is a natural and important ingredient of mathematical proof construction.

Our experiments with proof planning in the past decade indicate, that the search process would benefit from more flexibility of choice [14] and more and better control knowledge for specific domains and mathematical techniques – such as proofs by contradiction in real analysis – which is the subject of this paper.

The ΩMEGA project, which is essentially based on proof planning represents one of the major attempts to build an all encompassing assistant tool for the working mathematician, which combines interactive and automated proof construction for domains with rich and well-structured mathematical knowledge. The inference mechanism at the lowest level is an interactive theorem prover based on a higher order natural deduction (ND) variant of a soft-sorted version of Church's simply typed λ-calculus [6]. While this represents the "machine code" of the system, the user will seldom want to see, the search for a proof is conducted at a higher level by a proof planning process.

Proof planning differs from traditional search-based techniques in automated theorem proving not least with respect to its level of abstraction: the proof of a theorem is planned at an abstract level where an outline of the proof is found first. This outline, that is, the abstract proof plan, can be recursively expanded with operators and tactics eventually down to a proof within the logical calculus. The plan operators represent mathematical techniques familiar to a working mathematician.

Knowledge-based proof planning [10,13] employs even more techniques from artificial intelligence such as hierarchical planning, constraint solving and control rules for meta-level reasoning. While the knowledge of a mathematical domain represented by operators (called *methods*) and control rules is specific to the mathematical field, the representational techniques and reasoning procedures are general-purpose.

The methods (partially) describe changes of proof states by pre- and postconditions which are called *premises* and *conclusions* in the following. The premises and conclusions of a method are formulae (more precisely, sequents) in a higher-order language and the conclusions are considered as logically inferable from the premises.

Hence, a mathematical theorem proving problem is expressed as a planning problem whose initial state consists of the proof assumptions and whose goal description consists of the conjecture. Proof planning searches for a sequence (or a hierarchy) of instantiated methods, i.e. a *solution plan*, which transforms the initial state with assumptions into a state containing the conjecture.

Methods, Control Rules, and Strategies in a Context. In order to make the ingredients of proof planning more explicit, let us repeat what methods and control rules contribute to proof planning and then extend the discussion to strategies and contexts.

Methods have been perceived by Alan Bundy as tactics augmented with preconditions and effects, called *premises* and *conclusions*, respectively. A method represents the inference of the conclusion from the premises. Backward methods reduce a goal (the conclusion) to new goals (the premises). Forward methods, in contrast, derive new conclusions from given premises.

For example, the following method `ComplexEstimate` is an essential ingredient in epsilon-delta proofs of limit theorems.

method: `ComplexEstimate`(a, b, e_1, ϵ)		
premises	(0), $\oplus(1)$,$\oplus(2)$, $\oplus(3)$	
conclusions	$\ominus$ L12	
appl.cond	$\exists \sigma(subst(a,b) = \sigma)\,\&$ $\exists k, l(casextract(a_\sigma, b) = (k,l))\ \&\ b = k * a_\sigma + l$	
proof schema	$(0).\varDelta$ $\vdash\ \|a\| < e_1$	()
	$(1).$ $\vdash\ \|a_\sigma\| < \epsilon/(2 * \mathbf{V})$	(OPEN)
	$(2).\varDelta$ $\vdash\ \|k\| \leq \mathbf{V}$	(OPEN)
	$(3).$ $\vdash\ 0 < \mathbf{V}$	(OPEN)
	$(4).\varDelta$ $\vdash\ \|l\| < \epsilon/2$	(OPEN)
	L0. $\vdash\ b = b$	(Ax)
	L1. $\vdash\ b = k * a_\sigma + l$	(CAS;L0)
	. $\vdash\ \ldots$	(...)
	L12$\varDelta$ $\vdash\ \|b\| < \epsilon$	(schema;L1,(3), (0),(1),(2),(4))

This frame-like data structure should be read as follows: the method's name is ComplexEstimate and it has the parameters a, b, e_1, ϵ. The premises are the lines (0), (1), (2), and (3) which are schematically detailed in the proof schema slot. The $\oplus$ in the premises slot indicates these are added as subgoals by the application of the method. The conclusions is a goal schematically detailed in line L12 of the proof schema and the $\ominus$ indicates that it is removed by the application of the method. The application condition is formulated in a meta-language and for ComplexEstimate it expresses that a and b have to be unifiable by a substitution σ and b can be decomposed into a linear combination of the substituted a (which is tested by a computer alegbra system). The slot proof schema contains a sequence of (proof) lines that are introduced with the expansion of the method and used in the final proof. $\mathbf{V}$ is a newly introduced (auxiliary) variable.

Control rules represent mathematical knowledge about how to proceed in a particular mathematical situation, and they guide the proof planning process. They can influence the planner's behavior at choice points (e.g., which goal to tackle next or which method to apply next) by preferring members of the list of possible goals or of the list of possible methods. This way promising search paths are preferred and the general search space can be pruned.

Methods and control rules are the main ingredients of current proof planning systems, however, they do not always provide enough structure and flexibility for the problem solving process as the past decade of experimentation revealed.

First, there is a problem with the planning algorithm itself, which cannot be decomposed into its main components nor can new techniques easily be added.

Secondly, the proof planning process is too uniform, irrespective of the current context and independent of the kind of theorem to be shown.

For instance, if we want to prove a continuity theorem in the theory of analysis, it makes a difference whether we prove it via limit theorems, via an epsilon-delta technique, via converging sequences, or by contradiction which is the focus of this paper. Every mathematician has a variety of different strategies at her disposal to tackle such specific problems.

A *Strategy* as it is now used in our multi-strategy proof planning system employs a specific subset of the search algorithms, methods and control rules that are typical for the particular proof technique we want to simulate. For instance, one strategy may use an external system for some computation, another one may attack the problem with a completely different set of methods such as the epsilon-delta techniques, the methods and control rules typically used for a proof by induction or the methods and control rules for a proof by contradiction. It may cooperate with another strategy from a different theory, or it may use a different backtracking technique.

Meta-reasoning as to which strategy to employ on a problem introduces an additional explicit choice point and, thus, the system searches at the level of strategies as well.

All of this, however, is still in stark contrast to the situation of a mathematician who operates in a specific *context* which includes the current theory under development, preferences, knowledge representation techniques tailored to the specific context, definitions that are formulated in a way to serve the current purpose, typical techniques to prove this particular theorem as well as tricks of the trade typical for the field within which the theorem is stated. For example, a theorem about a continuous function requires certain methods, control rules and strategies to tackle its proof, which a student learns in the calculus courses, whereas a theorem, say in group theory, requires a very different set of methods and control, usually taught in an algebra class.The choice of either of them *is prior and above* a strategy and determined by the current context. The notion of a context provides a powerful structuring technique, in particular, for large-scale applications of mathematical assistant systems; this is ongoing work in the Omega group.

4   The Extended Limit Domain

The 'Limit Domain' is a well-known set of theorems to be shown by epsilon-delta-proofs. This domain typically comprises limit theorems and theorems about continuity. For example, the theorem that f is continuous at a, formulated as

$$\lim_{x \to 0} (f(a + x) - f(a) = 0 \to continuous(f, a).$$

has been shown by our system (see [15]).

Another well-known example is the LIM$^+$ theorem which was posed by Woody Bledsoe as a challenge to (classical) automated theorem proving systems. It states that the limit of the sum of two functions f and g equals the sum of their limits. Hence there are two assumptions which define the limit of f and g.

$$\forall \epsilon_1 (0 < \epsilon_1 \Rightarrow \exists \delta_1 (0 < \delta_1 \wedge \forall x_1 (|x_1 - a| > 0 \wedge |x_1 - a| < \delta_1 \Rightarrow |f(x_1) - l_1| < \epsilon_1)))$$
(4)

and

$$\forall \epsilon_2 (0 < \epsilon_2 \Rightarrow \exists \delta_2 (0 < \delta_2 \wedge \forall x_2 (|x_2 - a| > 0 \wedge |x_2 - a| < \delta_2 \Rightarrow |g(x_2) - l_2| < \epsilon_2))).$$
(5)

and the theorem is:

$$\forall \epsilon (0 < \epsilon \Rightarrow \exists \delta (0 < \delta \wedge \forall x (|x-a| > 0 \wedge |x-a| < \delta \Rightarrow |(f(x)+g(x))-(l_1+l_2)| < \epsilon)))$$
(6)

This and many more open challenge problems in this domain have been solved now with proof planning. Here is the proof of LIM$^+$, first provided in [11]:

The system first decomposes the conjecture and the assumptions. Among others, this yields the new assumptions [3] $|f(v_{x_1}) - l_1| < v_{\epsilon_1}$ and $|g(v_{x_2}) - l_2| < v_{\epsilon_2}$ and the two new goals $0 < v_\delta$ and $|(f(c_x) + g(c_x)) - (l_1 + l_2)| < c_\epsilon$.[4] The first goal, $0 < v_\delta$, is closed by the method TellCS which closes the goal and adds it to the constraint store of the constraint solver CoSIE [25]. The second goal $|(f(c_x) + g(c_x)) - (l_1 + l_2)| < c_\epsilon$ requires further decomposition, which is done by ComplexEstimate [11]

In the concrete example LIM$^+$, ComplexEstimate employs the new assumption $|f(v_{x_1}) - l_1| < v_{\epsilon_1}$ and yields four new goals:

$$\epsilon_1 < \frac{c_\epsilon}{2 * v} \qquad (7)$$
$$|1| \leq v \qquad (8) \qquad |g(c_x) - l_2| < \frac{c_\epsilon}{2} \qquad (10)$$
$$0 < v \qquad (9)$$

(7), (8), (9) can be closed by TellCS. Goal (10) is reduced by a method called Solve using the derived assumption $|g(v_{x_2}) - l_2| < v_{\epsilon_2}$ and yields the subgoals $v_{\epsilon_2} \leq \frac{c_\epsilon}{2}$ and $v_{x_2} = c_x$ which can be closed by the method TellCS.

When all goals are closed, the constraint solver CoSIE computes appropriate instances for variables that are consistent with the collected constraints. In this case, it generates the following instantiation:

---

[3] Notation: Proof planning replaces quantified variables either by constants or placeholder variables. The placeholder variable substituted for a quantified variable x is denoted by v_x. The constant substituted for a quantified variable x is denoted by c_x.

[4] During the decomposition of the assumptions further goals are created and the decomposition of the conjecture yields further assumptions are derived. However, in order to illustrate the basic proof planning approach we ignore these details.

$v_\delta \mapsto min(c_{\delta_1}, c_{\delta_2})$, $v_{\epsilon_1} \mapsto \frac{c_\epsilon}{2}$, $v_{\epsilon_2} \mapsto \frac{c_\epsilon}{2}$. Note, that these happen to be the same values that are used in a typical human proof of LIM$^+$, say, in a standard textbook such as [1]). (end of proof)

In the meantime, we investigated and solved many more problems from the Extended Limit Domain whose problems are all the theorems, examples, and exercises of two chapters of the introductory textbook for Real Analysis [1] that deal with limits of sequences and limits of functions. For many of these theorems there are several ways to prove them, e.g., epsilon-delta-proofs, proofs using other limit theorems, and proofs involving the estimation of an upper bound. The latter involves the use of the Dominance method. Dominance(a_n), where the parameter (a_n) denotes a sequence converging to zero, reduces a goal $\lim(x_n) = l$ to the goal $\exists k \forall n(k \in \mathbf{N} \land (n \in \mathbf{N} \to (n > k \to \frac{|x_n - l|}{|a_n|} < c)))$ for a constant c. This is justified by the theorem that says

if there is a sequence (a_n) converging to zero, a constant number c, and a natural number k such that for all $n \geq k$ holds $|x_n - l| < c \cdot a_n$, then $\lim(x_n) = l$.

Not without surprise there are also many theorems in Bartle and Sherbert that are shown by contradiction, although in principle they could be proven directly. But the authors found it simpler and esthetically more pleasing to do otherwise.

5   Planning Proofs by Contradiction

In order to show the conjecture T under the assumptions $A_1, \ldots A_n$, a *proof by contradiction* assumes $\neg T$ and the proof assumptions $A_1, \ldots, A_n$ and tries to prove a contradiction $F \land \neg F$ from these assumptions for some formula F.

In common undergraduate textbooks such as Bartle and Sherbert's introduction to real analysis, this principle is used often whenever it is convenient. In the past we looked at this textbook quite frequently for inspiration to proof planning and in the meantime we have solved (almost) all of these problems with our system.

But there remained several proofs by contradiction: so the first question to be answered is, under which conditions is a proof by contradiction appropriate. And secondly, the difficulty and the 'creative trick' in human proofs by contradiction (unlike say in a refutation by resolution) is to find an appropriate fact F which can be contradicted.

In order to solve the problems in Bartle and Sherbert we experimented with control knowledge on *when* to prove a conjecture by contradiction and *which* formula F to refute. In all the examples cited below we used the heuristic that F is one of the assumptions (A_k). Now this is not much of a restriction, if we include anything among the assumptions A_i. So, what we have in mind is the common situation in a textbook, where the assumptions $A_1, \ldots, A_n$ are immediately given either in the theorem itself as its hypotheses or more or less immediately in the chapter prior to the theorem, i.e., in the *current context*.

Proofs by contradiction of this nature are now realized in our system by the method $\mathtt{Contradict}(A_k)$, which has the parameter A_k.

	Method : $\mathtt{Contradict}$ (A_k)	
Premises	$L_0, \oplus L_3$	
Conclusions	$\ominus L_4$	
Parameter	L_0	
Proof Schema	$(L_0) \qquad\quad \vdash A_k$	$()$
	$(L_1)\ \Delta \quad \vdash \neg T$	(Hyp)
	$(L_2)\ \Delta, L_1 \vdash \neg A_k$	(OPEN)
	$(L_3)\ \Delta, L_1 \vdash A_k \wedge \neg A_k$	$(\wedge\mathrm{I};\ L_0 L_2)$
	$(L_4)\ \Delta \qquad \vdash T$	$(\neg\mathrm{E};\ L_1 L_3)$

The $\ominus L_4$ conclusion of the method indicates that the sequent in line L_4, T, is removed as a goal from the state when $\mathtt{Contradict}(A_k)$ has been applied. The L_0 premise which is not annotated indicates that the sequent of L_0 has to be an assumption in the state before $\mathtt{Contradict}(A_k)$ is applied. The $\oplus L_3$ premise indicates that the sequent $\Delta, L_1 \vdash A_k \wedge \neg A_k$ of line L_3 in the Proof Schema is added as a new subgoal to the state when $\mathtt{Contradict}(A_k)$ has been applied, where Δ, L_1 is the union of $A_1, \ldots A_n$ and $\neg T$. The Proof Schema contains the lines/nodes that are inserted into the partial proof plan when $\mathtt{Contradict}(A_k)$ is expanded.

Now, when should this method be used in the proof planning process? Some very general guidelines for choosing a proof by contradiction are given in [21] for proofs in the natural deduction calculus of Wilfried Sieg. These guidelines say: if you cannot derive a goal forwardly from the proof assumptions and if the goal is a negation, disjunction, or existentially quantified formula, then try a proof by contradiction. This means that first all possible forward proofs have to be attempted and only when this fails, search for a proof by contradiction. This control is, of course, not particularly efficient and in many cases it may even lead to infinitely many attempts. However, setting a heuristic bound on these attempts it works quite well in practice.

Many mathematical domains have domain-specific knowledge, which can be exploited to control the application of the method $\mathtt{Contradict}$. For instance, in the Extended Limit Domain one would *not* try to prove a goal $\neg(a < b)$ by contradiction although it is a negation but rather rewrite this goal to $b \leq a$ using knowledge that is specific for real numbers, or more generally for totally ordered structures.

For the Extended Limit domain we define: an (in)equality is called *simple*, if its lhs and rhs terms are constant expressions. 'Constant expressions' contain only variables (constants as opposed to placeholder variables) that do not depend on the instantiations of other variables (i.e., variables introduced through $\forall$-elimiination in goals or $\exists$-elimination in assumptions). For instance, the expression c_ϵ replacing the $\forall$-quantified variable ϵ in the goal (6) is a constant, whereas the expression replacing $\exists$-quantified δ is not. Vice versa, the expression introduced for ϵ_1 in (4) is a placeholder variable whereas the expression c_{δ_1} introduced for δ_1 in (4) is constant.

For proof planning in the Extended Limit domain we found the following two heuristics sufficient to trigger a proof by contradiction

- If the goal is a simple equation or inequality, in which (preferably) at least one constant represents a limit and if the goal is not entailed by (in)equality assumptions, then prefer proof by contradiction.
  The mathematical insight for this heuristic is that simple (in)equalities can be satisfied by (finite) forward entailment (realized by the method AskCS, which uses the constraint solver of the system to compute entailment). In case the simple (in)equation is not finitely entailed, a proof by contradiction provides an alternative. Mathematically, such an alternative is needed, when a property of the elements of a sequence cannot be finitely transferred to the corresponding property of the limit since this transfer requires an infinitesimal process.
- Contradict will be applied at most once in the same branch of a proof plan.

These two heuristics are expressed in the following control rule.

```
IF goal-simple-ineq
   AND not-yet-applied (Contradict)
   AND all-ineqAssumptions-known
THEN prefer (AskCS
             Contradict)
```

This control rule encodes the heuristic that, if the goal is a simple equation or inequality and Contradict has not been applied in the current branch of the proof plan already, and all entailment information is available, then the planner should first try to check the entailment of the goal using the method AskCS and if this fails, a proof by contradiction should be preferred.

We could test the entailment via a meta-predicate in the preconditions of a control rule but instead this test is done in the application condition of the method AskCS. Thereby we avoid duplication of work in case a proof by contradiction is unnecessary.

6   Results

We have collected all the theorems shown by contradiction in the textbook on Real Analysis of Bartle and Sherbert and surprisingly the rather simple heuristic above worked in most cases. Omega's proof planner finds proofs by contradiction inter alia for the following theorems:[5]

Theorem 3.1.5
A sequence of real numbers can have at most one limit.

This 'uniqueness of limit' theorem is shown in [1] with the following proof. Suppose on the contrary that x' and x'' are both limits of $X = (x_n)$ and that $x' \neq x''$.

---

[5] The numbering is the same as that of the theorems in the textbook [1].

We choose $\epsilon > 0$ such that the ϵ-neighborhoods $V_\epsilon(x')$ and $V_\epsilon(x'')$ are disjoint, that is, such that $\epsilon < \frac{1}{2}|x' - x''|$). Now let K' and K'' be natural numbers such that if $n > K'$ then $x_n \in V_\epsilon(x')$, and if $n > K''$ then $x_n \in V_\epsilon(x'')$. However, this contradicts the assumption that these ϵ-neighborhoods are disjoint. (Why?) Consequently, we must have $x' = x''$.

The answer to the corresponding exercise question (Why) as well as the whole proof is found by Omega in a similar way. The formula F that is refuted in the proof of Bartle and Sherbert is 'the ϵ-neighborhoods are disjoint'.

The next theorem 3.2.4 is related to the dominance property mentioned in section 4 and is used in [1] as the prerequisite for theorem 3.2.5. and theorem 3.2.6. all of which have been shown by contradiction without this prerequisite with the proof planner.

Theorem 3.2.4
If (x_n) is a convergent sequence and if for all n $x_n \geq 0$, then $\lim(x_n) \geq 0$.
    The formula F that gives the contradiction is $F := x_n \geq 0$

Theorem 3.2.5
If (x_n) and (y_n) are convergent sequences of real numbers and if $x_n \leq y_n$, then $\lim(x_n) \leq \lim(y_n)$.

Theorem 3.2.6
If (x_n) is a convergent sequence and if $a \leq x_n \leq b$, then $a \leq \lim(x_n) \leq b$.

The next theorem 'assures us that the value L of the limit is uniquely determined, when it exists. As this uniqueness is not part of the definition of limit, it must be deduced.' ([1], p.113).

Theorem 4.1.5
If $f : A \mapsto R$ and if c is a cluster point of A, then f can have only one limit at c.
    Here the formula F that gives the contradiction is obtained from 'the ϵ-neigborhoods of L and L' are disjoint (an assumption of a subproof).

The following is taken from the exercises in [1], chapter 4.1 aimed at the expertise of a freshman.

Theorem 4.1.(3)
Let $f : \mathbb{R} \mapsto \mathbb{R}$ and let $c \in \mathbb{R}$. If $\lim_{x \to 0} f(x + c) = l_1$ and $\lim_{x \to c} f(x) = l_2$, then $l_1 = l_2$.

The above sample is taken from many more theorems for which the proof planner found a proof with the fixed set of methods and control rules we have collected for this branch of mathematics (and reported elsewhere [11,14,?] except that the above control rule capturing the heuristic for proof by contradiction has been added. As a result, the theorems in the two chapters of the textbook [1] that require the use the of method **Contradict** are recognized and handled properly.

7   Discussion

The restriction to experiments with theorems from analysis is certainly not a principal one as these results apply in other areas of mathematics as well, where proofs by contradiction are abundant. The Extended Limit Domain is already a very rich one, in particular, when it comes to search spaces and the use of domain-dependent knowledge and this 'small world' encapsulates many of the problems from mathematical problem solving in general.

In particular, the search for the formula F that is finally used in the contradiction is an 'AI-complete' problem and many proofs by contradiction have in fact become famous, because of an ingenious choice of F. An example is the proof that Euler's number e is not rational, which was shown by Leonhard Euler in 1737. In fact, e is not only irrational but a transcendental number, which was shown 150 years later by Charles Hermite. Euler's proof reasons by contradiction and assumes that e is a rational number and hence, can be represented as $e = \frac{p}{q}$ for some integers p and q. The contradiction follows from the fact that $q! \cdot e$ is an integer, whereas its expansion

$$q! + \frac{q!}{1} + \frac{q!}{2} + \frac{q!}{3} + \ldots + \frac{q!}{q} + \frac{q!}{(q+1)!} + \frac{q!}{(q+2)!} + \ldots \tag{11}$$

is not an integer, hence, e is irrational. So the ingenuity of this proof is to construct the appropriate formula F that states that $q! \cdot e$ is an integer and show that the expansion of $q! \cdot e$ is not an integer.

However, as it is always possible to tune and set the dials of a system to prove a theorem – no matter how famous and difficult – for which a proof is known in the literature, the goal of our game is to take Woody Bledsoe's criticism of automated theorem proving seriously:

> *Automated theorem proving is not the beautiful process we know as mathematics. This is 'cover your eyes with blinders and hunt through a cornfield for a diamond-shaped grain of corn... Mathematicians have given us a great deal of direction over the last three millenia. Let us pay attention to it.* (Woody Bledsoe, 1986)

'Nothing can be explained to a stone' (McCarthy 1967) and in particular 'Nothing mathematically interesting can be told to a (resolution-based) automated theorem proving system' is our mantra capturing this general issue. The Omega system has been 'told' one extra control rule and to our own suprise this turned out to be sufficient to find many proofs by contradiction in [1]. But no doubt things will not stay this way and we have to uncover stronger means to the end of Reductio ad Absurdum arguments, in particular, more ingenious ways to determine the formula F. This is the subject of ongoing research in the Omega group.

Acknowledgement. The reported work was funded by the DFG-project MIPPA (Me 1136/2-1) and the collaborative research center 378 of the German National Science Foundation.

References

1. R.G. Bartle and D.R. Sherbert. *Introduction to Real Analysis.* John Wiley& Sons, New York, 1982.
2. M.J. Beeson. Automatic generation of epsilon-delta proofs of continuity *Artificial Intelligence and Symbolic Computation,* J. Calmet and J. Plaza (eds), pages 67-83. LNAI 1476, Springer Verlag, 1998.
3. M.J. Beeson. Automatic generation of a proof of the irrationality of e. *Journal of Symbolic Computation,* 32(4): 333-349, 2001.
4. W.W. Bledsoe. Non-resolution theorem proving. *Artificial Intelligence,* 9:1–35, 1977.
5. A. Bundy. The use of explicit plans to guide inductive proofs. In E. Lusk and R. Overbeek, editors, *Proc. 9th International Conference on Automated Deduction (CADE-9),* volume 310 of *Lecture Notes in Computer Science,* pages 111–120, Argonne, 1988. Springer.
6. A. Church. A formulation of the simple theory of types. *Journal of Symbolic Logic,* 5:56–68, 1940.
7. M. Dummett. Elements of Intuitionism. Oxford, 2nd edition, 2000.
8. H. Gelernter. Realization of a geometry theorem-proving machine. In *Proceedings of the International Conference on Information Processing, UNESCO,* 1959.
9. D.B. Lenat. AM: an AI approach to discovery in mathematics. In R. Davis and D.B. Lenat, editors, *Knowledge-Based Systems in Artificial Intelligence.* Mc-Graw Hill, New York, 1981.
10. E. Melis. AI-techniques in proof planning. In *European Conference on Artificial Intelligence,* pages 494–498, Brighton, 1998. Kluwer.
11. E. Melis. The "limit" domain. In R. Simmons, M. Veloso, and S. Smith, editors, *Proceedings of the Fourth International Conference on Artificial Intelligence in Planning Systems,* pages 199–206, 1998.
12. E. Melis and A. Meier. Proof planning with multiple strategies. In J. Loyd, V. Dahl, U. Furbach, M. Kerber, K. Lau, C. Palamidessi, L.M. Pereira, and Y. Sagivand P. Stuckey, editors, *First International Conference on Computational Logic,* volume 1861 of *Lecture Notes on Artificial Intelligence,* pages 644–659. Springer-Verlag, 2000.
13. E. Melis and J.H. Siekmann. Knowledge-based proof planning. *Artificial Intelligence,* 115(1):65–105, November 1999.
14. E. Melis, A. Meier, and J. Siekmann. Proof planning with multiple strategies. *Artificial Intelligence,* submitted 2006.
15. A. Meier and E. Melis. Proof Planning Limit Problems with Multiple Strategies. SEKI Technical Report SR-2004-04, FR Informatik, Universitaet des Saarlandes, 2004.
16. A. Newell. The Heuristic of George Polya and its Relation to Artificial Intelligence. Technical Report CMU-CS-81-133, Carnegie-Mellon-University, Dept. of Computer Science, Pittsburgh, Pennsylvania, U.S.A., 1981.
17. G. Polya. *How to Solve it.* Princeton University Press, Princeton, 1945.
18. J.A. Robinson. A machine-oriented logic based on the resolution principle. *JACM,* 12, 1965.
19. A. Robinson and A. Voronkov. *Handbook of Automated Reasoning,* vol. 1 and 2. Elsevier, 2001.
20. A.H. Schoenfeld. *Mathematical Problem Solving.* Academic Press, New York, 1985.

21. W. Sieg and J. Byrnes. Normal natural deduction proofs (in classical logic). *Studia Logica*, 60:67–106, 1998.
22. J. Siekmann, C. Benzmüller, and S. Autexier. Computer supported mathematics with Omega. *Journal of Applied Logic*, 2006. in press.
23. J. Siekmann, C. Benzmüller, A. Fiedler, A. Meier, I. Normann, M. Pollet, Proof Development in OMEGA: The Irrationality of Square Root of 2. In: Kamareddine, F. (Ed.), Thirty Five Years of Automating Mathematics. Kluwer Applied Logic series. Kluwer Academic Publishers. 2003.
24. F. Wiedijk (ed.). The Seventeen Provers of the World. LNAI vol. 3600 (the first volume in the AI-Systems subseries), Springer-Verlag, 2006.
25. J. Zimmer and E. Melis. Constraint solving for proof planning. *Journal of Automated Reasoning*, 33(1):51–88, July 2004.

Computational Logic in an Object-Oriented World

Bob Kowalski

Imperial College London
rak@doc.ic.ac.uk

Abstract. Logic and object-orientation (OO) are competing ways of looking at the world. Both view the world in terms of individuals. But logic focuses on the relationships between individuals, and OO focuses on the use of hierarchical classes of individuals to structure information and procedures. In this paper, I investigate the similarities and differences between OO and abductive logic programming multi-agent systems (ALP systems) and argue that ALP systems can combine the advantages of logic with the main benefits of OO. In ALP systems, relationships between individuals are contained in a shared semantic structure and agents interact both with one another and with the environment by performing observations and actions. In OO systems, on the other hand, relationships are associated with objects and are represented by attribute-value pairs. Interaction between objects is performed by sending and receiving messages. I argue that logic can be reconciled with OO by combining the hierarchical, modular structuring of information and procedures by means of objects/agents, with a shared semantic structure, to store relationships among objects/individuals, accessed by observations and actions instead of by message passing.

Keywords: object-orientation, logic programming, Linda.

1 Introduction

There was a time in the 1980s when it seemed that Computational Logic (CL) might become the dominant paradigm in Computing. By combining the declarative semantics of logic with the computational interpretation of its proof procedures, it could be applied to virtually all areas of Computing, including program specification, programming, databases, and knowledge representation in Artificial Intelligence.

But today it is Object-Orientation (OO), not Logic, that dominates every aspect of Computing from modelling the system environment, through specifying system requirements, to designing and implementing the software and hardware. Like CL, OO owes much of its attraction, not only to its computational properties, but also to its way of thinking about the world. If these attractions have real substance, then they potentially undermine not only CL's place inside Computing, but also its way of modelling and reasoning about the world outside Computing.

O. Stock and M. Schaerf (Eds.): Aiello Festschrift, LNAI 4155, pp. 59–82, 2006.
© Springer-Verlag Berlin Heidelberg 2006

The aim of this paper is to try to understand what makes OO so attractive and to determine whether these attractions can be reconciled with CL, both in Computing and in the wider world. I will argue that logic-based multi-agent systems can combine the advantages of CL with the main benefits of OO.

I will illustrate my argument by using abductive logic programming (ALP) multi-agent systems [14]. However, most of the argument applies to more general logic-based multi-agent systems, and even to heterogeneous systems that use different programming languages, provided their external interfaces can be viewed in logical terms.

ALP multi-agent systems (ALP systems, in short) are semantic structures, consisting of individuals and relationships, as in the conventional semantics of classical logic. However, in ALP systems, these structures can change state, in the same way that the real world changes state, destructively, without remembering its past. Some individuals in the structure are agents, which interact with the world, by observing the world and by performing actions on the world. Other individuals passively undergo changes performed by agents, and still other individuals, like numbers, are immutable and timeless.

ALP agents, which are individuals in the ALP semantic structure, also have an internal, syntactic structure, consisting of goals and beliefs, which they use to interact with the world. Their beliefs are represented by logic programs, and their goals are represented by integrity constraints. Their observations and actions are represented by abducible (undefined) predicates.

I argue that such logic-based multi-agent systems share many of the attractions of OO systems. In particular, they share with objects the view that the world consists of individuals, some of which (objects or agents) interact with other individuals and change the state of the world. However, whereas in OO systems relationships among individuals are associated with objects and are represented as attribute-value pairs, in ALP systems relationships belong to the semantic structure of the world.

Both agents in ALP systems and objects in OO systems encapsulate their methods for interacting with the world, hiding their implementation details from other agents and objects. Both agents and objects can inherit their methods from more general classes of agents or objects. Whereas objects use methods implemented in conventional, imperative programming languages, ALP agents use methods implemented by means of goals and beliefs in logical form. The methods used by ALP agents have both a procedural behaviour, as well as a declarative semantics. In the declarative semantics, the goals and beliefs of an agent have a truth value in the semantic structure that is the ALP system as a whole. Normally, beliefs that are true and goals that can be made true are more useful to an agent than ones that are false[1].

Both ALP systems and OO systems share a local notion of change, in which changes can take place in different parts of the world locally, concurrently and independently. This local notion of change contrasts with the global notion that

---

[1] A false belief can be more useful than a true belief, if the truth is too complicated to use in practice.

is prevalent in most logical treatments, including the possible world semantics of modal logic and the situation calculus. The global notion of change is useful for theoretical purposes, but the local notion is more useful both as a model of the real world and as a model for constructing artificial worlds.

ALP agent systems differ from OO systems in one other important respect: Whereas objects interact by sending and receiving messages, agents interact by observing and performing actions on the shared semantic structure. This semantic structure acts as a shared environment, similar to the blackboard in a blackboard system [9] and to the tuple-space in a Linda programming environment [10]. In the same way that Linda processes can be implemented in different and heterogeneous programming languages, the methods used by ALP agents can also be implemented in other programming languages, provided their externally observed behaviour can be viewed in logical terms.

In the remainder of the paper, I will first introduce ALP systems in greater detail and then distinguish between the semantic and syntactic views of OO systems. I will then compare OO systems and ALP systems by investigating how each kind of system can be simulated by the other. The directness of these simulations is the basis for the comparison of the two approaches. The simulations are informal and should be viewed more as illustrations than as outlines of formal theorems.

2   The Logical Way of Looking at the World

In logic there is a clear distinction between syntax and semantics. Syntax is concerned with the grammatical form of sentences and with the inference rules that derive conclusion sentences from assumption sentences. Semantics is concerned

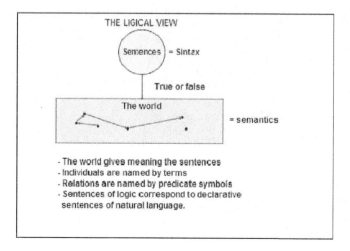

Fig. 1. The logical view

with the individuals and relationships that give sentences their meaning. The relationship between syntax and semantics is pictured roughly in figure 1.

The distinction between atomic sentences and the semantic relationships to which they refer is normally formalised by defining an interpretation function, which interprets constant symbols as naming individuals and predicate symbols as naming relations. However, it is often convenient to blur the distinction by restricting attention to *Herbrand interpretations*, in which the semantic structure is identified with the set of all atomic sentences that are true in the structure. However, the use of Herbrand interpretations can sometimes lead to confusion, as in the case where a set of atomic sentences can be considered both semantically as a Herbrand interpretation and syntactically as a set of sentences. Sometimes, to avoid confusion, atomic sentences understood as semantically as relationships are also called *facts*.

For notational convenience, we shall restrict our attention to Herbrand interpretations in the remainder of the paper. However, note that, even viewing Herbrand interpretations as syntactic representations, there is an important sense in which they differ from other syntactic representations. Other syntactic representations can employ quantifiers and logical connectives, which generalize, abstract and compress many atomic sentences into a smaller number of sentences, from which other sentences, including the atomic sentences, can be derived.

2.1 ALP Agents

Traditional logic is often accused by its critics of being too concerned with static states of affairs and of being closed to changes in the world. The first of these criticisms has been addressed in various ways, either by making the semantics more dynamic, as in the possible worlds semantics of modal logic, or by making

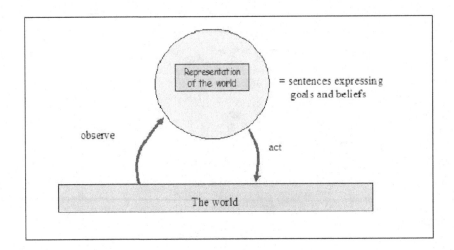

Fig. 2. Observation-thought-decision-action cycle in ALP agents

the syntax more expressive by reifying situations or events, as in the situation or event calculus.

The second of these criticisms has been addressed by embedding logic in the thinking component of the observation-thought-decision-action cycle of an intelligent agent, for example as in ALP agents, pictured in figure 2.

In ALP agents, beliefs are represented by logic programs and goals are represented by integrity constraints. Integrity constraints are used to represent a variety of kinds of goals, including maintenance goals, prohibitions, and condition-action rules. Abducible predicates, which are not defined by logic programs, but are restricted by the integrity constraints, are used to represent observations and actions.

ALP agents implement *reactive* behaviour, initiated by the agent's observations, using forward reasoning to trigger maintenance goals and to derive achievement goals. They also implement *proactive* behaviour, initiated by achievement goals, using backward reasoning to reduce goals to sub-goals and to derive action sub-goals. In addition to reactive and proactive thinking, ALP agents can also perform *pre-active* thinking [?], using forward reasoning to simulate candidate actions, to derive their likely consequences, to help in choosing between them.

2.2   An ALP Agent on the London Underground

Passengers on the London underground have a variety of goals - getting to work, getting back home, going out shopping or visiting the tourist attractions. In addition, except for renegade terrorists, everyone is also concerned about safety. This can be represented by goals in logical form, which might include the (simplified) goal:

If there is an emergency then I get help.

To recognize when there is an emergency and to find a way to get help, a passenger can use beliefs[2] in logic programming form:

I get help if I alert the driver.
I alert the driver if I press the alarm signal button.

There is an emergency if there is a fire.
There is an emergency if one person attacks another.
There is an emergency if someone becomes seriously ill.
There is an emergency if there is an accident.

The beliefs about getting help are declarative sentences, which may be true or false about the effect of actions on the state of the world. The beliefs about emergencies are also declarative sentences, but they are simply true by definition, because the concept of emergency is an abstraction without a direct interpretation in concrete experience.

In ALP, beliefs can be used to reason forwards or backwards. Forward reasoning is useful for deriving consequences of observations and candidate actions.

---

[2] For simplicity, this representation of goal and beliefs ignores the element of time.

Backward reasoning is useful for reducing goals to sub-goals. A combination of forward and backward reasoning in the London underground example is illustrated in figure 3.

The mental activity of an ALP agent is encapsulated in the agent, hidden from an observer, who can see only the agent's input-output behaviour. In the case of the London underground passenger, this behaviour has the logical form:

If there is a fire, then the passenger presses the alarm signal button.

As far as the observer is concerned, this externally visible, logical form of the passenger's behaviour could be implemented in any other mental representation or programming language.

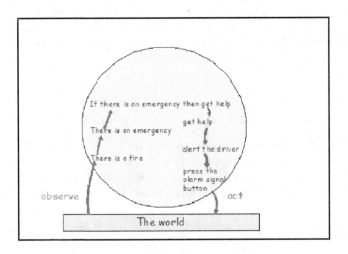

Fig. 3. ALP combines forward and backward reasoning

2.3   ALP Agent Systems

Similarly to the way that agents interact with other individuals in real life, ALP agents interact with other individuals embedded in a shared environment. This environment is a semantic structure consisting of individuals and relationships. Some of the individuals in the environment are agents of change, while others undergo changes only passively. To a first approximation, we can think of an ALP environment as a relational database, which changes destructively as the result of agents' actions.

The environment that ALP agents share is a dynamic structure, in which relationships come and go as the result of actions, which occur locally, concurrently and independently of other actions. Because this environment is a semantic structure, relationships can appear and disappear destructively, without the environment having to remember the past. In the London underground example, the observations and actions of the passenger, train driver and fire department agents are illustrated in figure ??. Instead of standing apart and, as it were, above the world, as pictured in figures 1-3, the agents are embodied within it. Their actions change the shared environment by adding and deleting facts (or relationships):

The passenger's action of pressing the alarm signal button
    deletes the fact that the alarm is off and
    adds the fact that the alarm is on.
The driver's action of calling the fire department
    adds the fact that the fire department has been called.
The fire department's action of putting out the fire
    deletes the fact that there is a fire in the train.

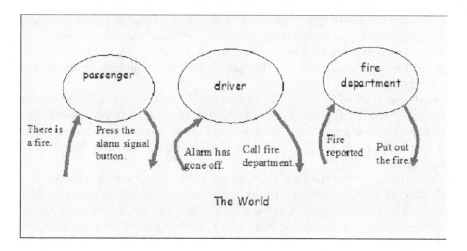

Fig. 4. An ALP system

The driver's action of calling the fire department can be viewed as sending the fire department a message, in the form of a fact that is stored in the shared environment. The fire department observes the message and, if it chooses, may delete it from the environment. Other agents may be able to observe the message, as long as it remains in the environment, provided they can access that part of the environment.

Notice that, just as in the case of a single agent, an observer can see only the agents' external behaviour. In this case also, that behaviour has a logical form:

If there is a fire, then the passenger presses the alarm signal button.
If the alarm has gone off, then the driver calls the fire department.
If a fire is reported, then the fire department puts out the fire.

These implications can be combined with sentences describing the effect of the agents' actions on the world:

If a person presses the alarm signal button, then the alarm goes off.

If a person calls the fire department, then a fire is reported.

to derive, as a logical consequence, the input-output behaviour of the combined system as a whole:

If there is a fire, then the fire department puts out the fire.

3   Object-Oriented Systems

Despite the dominant position of OO in Computing, there seems to be no clear definition or consensus about its fundamental concepts. One recent attempt to do so [3] identifies inheritance, object, class, encapsulation, method, message passing, polymorphism, and abstraction, in that order, as its most frequently cited features. However, the relative importance of these concepts and their precise meaning differs significantly from one OO language to another. This makes comparison with logic very difficult and prone to error. Therefore, the claims and comparisons made in this paper need to be judged and qualified accordingly.

Nonetheless, viewed in terms of the concepts identified in [3], the argument of this paper can be simply stated as claiming that all of these concepts are either already a feature of ALP systems (and other, similar logic-based multi-agent systems) or can readily be incorporated in them, with the exception of *message-passing*.

OO shares with logic the view that the world consists of individuals, some of which (objects or agents) interact with other individuals and change the state of the world. In OO, objects interact with one another by sending and receiving messages, using encapsulated methods, which are hidden from external observers, and which are acquired from more general classes of objects, organised in hierarchies.

Whereas ALP agents use goals and beliefs to regulate their behaviour, objects use methods that are typically implemented by means of imperative programming language constructs. An object-oriented system corresponding to the multi-agent system of figure 4 is pictured in figure 5. Both ALP systems and OO systems can be viewed as semantic structures, in which the world is composed of individuals that interact with one another and change state. However, there are important differences between them:

1. *The treatment of individuals.* In ALP systems, individuals are distinguished between active individuals, which are agents, and other individuals, which are passive. However, in OO systems, both kinds of individuals are treated equally as objects.
2. *The treatment of attributes and relationships.* In the semantic structures of logic, individuals have externally visible attributes and relationships with other individuals. Attributes of individuals are treated technically as a special case of relationships.
   In OO systems, attributes of objects are internalised. Moreover, relationships between objects are treated as attributes and similarly internalised. Either

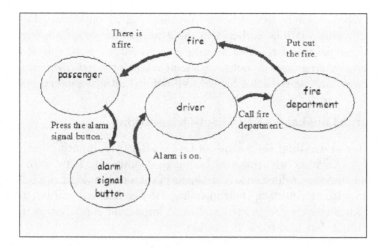

Fig. 5. An object-oriented system corresponding to the multi-agent system of figure 4

one of the objects in a relationship has to be treated as its "owner". Or the relationship needs to be represented redundantly among several "owners".

3. *The way of interacting with the world.* ALP agents interact with the world by observing the current state of the world and by performing actions to change it. A relationship between several individuals can be accessed in a single observation; and a single action can change the states of several relationships.

Objects in OO systems, on the other hand, interact with one another rather than with a separate semantic world structure. They do so by sending and receiving messages. But the concept of "message" is not defined. In many cases, messages are used to send information from one object to another. In other cases, they are used to request other objects for help in solving sub-goals. But in the general case, messages can be used for any, arbitrary purpose.

3.1   Object-Oriented Systems as Syntactic Structures

The relationship between logic and objects can be viewed in both semantic and syntactic terms. However, it is the syntactic structuring of information and methods into encapsulated hierarchies of classes of objects that is perhaps the most important reason for the practical success of OO in Computing.

In ALP systems, information and methods are syntactically formulated by means of goals and beliefs in logical form. In OO systems, methods are typically implemented in an imperative language. In both cases, internal processing is encapsulated, hidden from other agents or objects, and performed by manipulating sentences in a formal language.

In logic, there is a well understood relationship between syntax and semantics, in which declarative sentences are either true or false. In ALP agents, declarative sentences representing an agent's goals and beliefs are similarly true or false in the semantic structure in which the agent is embedded.

In OO systems, where methods are implemented in imperative languages, there is no obvious relationship between the syntax of an object's methods and the semantic structure of the OO system as a whole. In part, this is because purely imperative languages do not have a simple truth-theoretic semantics; and, in part, because messages do not have a well defined intuitive interpretation.

3.2   Natural Language and Object-Orientation

We can better understand the nature of OO syntax by comparing both logical syntax and OO syntax with natural language. Comparing logic with natural language, one important difference is that sets of sentences in logic can be written in any order, without affecting their meaning. However, in natural language the order in which sentences are written makes an important contribution, not only to their meaning, but also to their intelligibility.

In contrast with logic, but similarly to natural language, OO has a major concern with the way that sentences are structured. OO associates methods and attribute-value pairs with the objects they concern. Natural languages, like English, employ a similar form of object-orientation by using grammatical structures in which the beginning of a sentence indicates a *topic* and the following part of the sentence expresses a *comment* about the topic. This kind of structure often coincides with, but is not limited to, the grammatical structuring of sentences into *subjects*[3] and *predicates.*

Consider, for example, the pair of English sentences [6, p. 130]:

The prime minister stepped off the plane.
Journalists immediately surrounded her.

Both sentences are formulated in the active voice, which conforms to the guidelines for good writing style advocated in most manuals of English.

The two sentences refer to three individuals/objects, the prime minister (referred to as "her" in the second sentence), journalists and the plane. The prime minister is the only object in common between the two sentences. So, the prime minister is the object that groups the two sentences together. However, the topic changes from the prime minister in the first sentence to the journalists in the second.

Now consider the following logically equivalent pair of sentences:

The prime minister stepped off the plane.
She was immediately surrounded by journalists.

Here the two sentences have the same topic, which is the individual/object they have in common. However, the second sentence is now expressed in the passive voice.

Despite the fact that using the passive voice goes against the standard guidelines of good writing style, most people find the second pair sentences easier

---

[3] In this analogy between objects and topics, objects are more like the grammatical subjects of sentences than the grammatical objects of sentences.

to understand. This seems to suggest that people have a preference for organising their thoughts in object-oriented form, which is even stronger than their preference for the active over the passive voice.

However, OO is not the only way of structuring sentences. Both linguists and proponents of good writing style have discovered a more general way, which includes OO as a special case. As Joseph Williams [22] argues:

> *Whenever possible, express at the beginning of a sentence ideas already stated, referred to, implied, safely assumed, familiar, predictable, less important, readily accessible.*
>
> *Express at the end of a sentence the least predictable. The newest, the most important, the most significant information, the information you almost certainly want to emphasize.*

This more general way of structuring sentences also includes the use of logical form to make sets of sentences easier to understand. For example:

A if B.
B if C.
C if D.
D.

Or:

D.
If D then C.
If C then B.
If B then A.

3.3   Classes in Object-Oriented Systems and Sorts in Logic

Perhaps the most important practical feature of OO systems is the way in which objects acquire their methods from more general classes of objects. For example, an individual passenger on the underground can obtain its methods for dealing with fires from the more general class of all humans, and still other methods from the class of all animals.

Thus, classes can be organised in taxonomic hierarchies. Objects acquire their methods from the classes of which they are instances. Similarly sub-classes can inherit their methods from super-classes higher in the hierarchy, possibly adding methods of their own.

However, classes and class hierarchies are neither unique nor original features of OO systems. Classes correspond to types or sorts in many-sorted logics, and hierarchies of classes correspond to hierarchies of sorts in order-sorted logics.

Sorts and hierarchies of sorts can be (and have been) incorporated into logic programming in many different ways. Perhaps the simplest and most obvious way is by employing explicit sort predicates, such as Passenger(X) or Human(X), in the conditions of clauses, together with clauses defining sort hierarchies and instances, such as:

Human(X) if Passenger(X)
Passenger(john).

However, even unsorted logic programs already have a weak, implicit hierarchical sort structure in the structure of terms. A term $f(X)$, where f is a function symbol, can be regarded as having sort $f()$. The term $f(g(X))$, which has sort $f(g())$ is a sub-sort of $f()$, and the term $f(g(a))$, where a is a constant symbol is an instance of sort $f(g())$. Two terms that differ only in the names of variables, such as $f(g(X))$ and $f(g(Y))$ have the same sort. Simple variables, such as X and Y, have the universal sort. Both explicitly and implicitly sorted logic programs enjoy the benefits of inheritance[4].

Although sorts and inheritance are already features of many systems of logic, OO goes further by grouping sentences into classes. Sentences that are about several classes, such as the methods for humans dealing with fire, have to be associated either with only one of the classes or they have to be associated with several classes redundantly[5].

3.4   Object-Oriented Logic Programming

We can better understand the relationship between OO systems and ALP systems if we see what is needed to transform one kind of system into the other. First, we will show how, under certain restrictions, logic programs can be transformed into OO systems. Later, we will show how to extend this transformation to ALP systems, and then we will show how to transform OO systems into ALP systems. In each of these cases, the OO system is an idealized system, corresponding to no specific OO language in particular. Thus the claims made about these transformations need to be qualified by this limitation.

OO grouping of sentences into classes can be applied to any language that has an explicit or implicit class structure, including sentences written in formal logic. However, as we have just observed, an arbitrary sentence can be about many different individuals or classes, making it hard to choose a single individual or class to associate with the sentence.

But it is easier to choose a class/sort for clauses in logic programs that define input-output predicates. For such programs, it can be natural to nominate one of the input arguments of the conclusion of a clause (or, more precisely, the sort of that argument) to serve as the "owner" of the clause. The different instances

---

[4] Inheritance can be inhibited by the use of abnormality predicates. For example, the clauses Fly(X) if Bird(X) and not Abnormal(X), Bird(X) if Penguin(X), Walk(X) if Penguin(X), Swim(X) if Penguin(X), Abnormal(X) if Penguin(X) inhibit the inheritance of flying.

[5] The problem of choosing a class or class hierarchy to contain a given sentence is similar to the problem of choosing a folder to store a file. Search engines like Google make it possible to store information in one structure but to access it without reference to the place it is stored. It also makes it possible to store information in an unstructured way, without any penalty in accessing it. Email clients offer similar, but more limited facilities to order emails by such different attributes as sender, date, subject, size, etc.

of the nominated input argument behave as sub-classes and objects, which use the clause as a method to reduce goals to sub-goals.

Whereas arbitrary messages in OO systems may not have a well-defined intuitive interpretation, messages in OO systems that implement input-output logic programs either send requests to solve goals and sub-goals or send back solutions. An object responds to a message requesting the solution of a goal by using a clause to reduce the goal to sub-goals. The object sends messages, in turn, requesting the solution of the sub-goals, to the objects that are the owners of the sub-goals. When an object solves a goal or sub-goal, it sends the solution back to the object that requested the solution.

More formally, let a logic program contain the clause:

$$P_0(o_0, t_{01}, \ldots, t_{0m0}) \text{ if } P_1(o_1, t_{11}, \ldots, t_{1m1}) \text{ and } \ldots \text{ and } P_n(o_n, t_{n1}, \ldots, t_{nmn})$$

where, without loss of generality, the sort of the first argument of each predicate is selected as the owner of the clause. We also call that argument the "owner argument". Assume also that each such owner argument oi is an input argument in the sense that at the time the predicate is invoked, as a goal or sub-goal for solution, the argument oi is instantiated to some object (variable-free term).

Assume for simplicity that the sub-goals in the body of the clause are executed, Prolog-fashion, in the order in which they are written. Then the use of the clause to solve a goal of the form $P_0(o'_0, t'_{01}, \ldots, t'_{0m0})$, where o'_0 is a fully instantiated instance of o_0 , is simulated by some sender object o sending the goal in a message to the receiver object o'_0 and by the receiver object:

0. matching the goal with the head of the clause, obtaining some most general unifying substitution Θ_0

1. sending a message to object $o_1\Theta_0$ to solve the goal $P_1(o_1, t_{11}, \ldots, t_{1m1})\Theta_0$

2. receiving a message back from $o_1\Theta_0$ reporting that the goal $P_1(o_1, t_{11}, \ldots, t_{1m1})\Theta_0$ has been solved with substitution Θ_1

. . .

2n. sending a message to object on $\Theta_0\Theta_1 \ldots \Theta_{n-1}$ to solve the goal $P_n(o_n, t_{n1}, \ldots, t_{nmn})\Theta_0\Theta_1 \ldots \Theta_{n-1}$

2n+1. receiving a message back from on $\Theta_0\Theta_1 \ldots \Theta_{n-1}$ that the goal $P_n(o_n, t_{n1}, \ldots, t_{nmn})\Theta_0\Theta_1 \ldots \Theta_{n-1}$ has been solved with substitution Θ_n

2n+2. sending a message back to the sender object o that the goal $P_0(o_0, t_{01}, \ldots, t_{0m0})\Theta_0$ has been solved with substitution $\Theta_0\Theta_1 \ldots \Theta_{n-1}\Theta_n$.

Notice that objects o_i and o_j need not be distinct. The special case of an object sending a message to itself can be short circuited by the object simply solving the sub-goal locally.

If $n = 0$, then the clause represents a relationship with other objects. The relationship is represented only once, associated with the owner object. All such relationships, like all other clauses, are encapsulated in their owner objects and can be accessed by other objects only by sending and receiving messages.

For example, the atomic clauses:

Father(john, bill)
Father(john, jill)
Mother(mary, bill)
Mother(mary, jill)

would be encapsulated within the john and mary objects. The goal Father(john, X) sent to the object john would receive two messages $X = bill$ and $X = jill$ in return.

It would not be possible with this simple implementation to find out who are the parents of bill or jill. This problem can be solved by redundantly nominating more than one argument to serve as the owner of a clause.

Notice that methods can be public or private. Public methods are ones that are known by other objects, which those objects can invoke by sending messages. Private methods are ones that can only be used internally.

3.5   Polymorphism

The mapping from input-output logic programs to OO systems illustrates polymorphism. In the context of OO systems, polymorphism is the property that the "same" message can be sent to and be dealt with by different classes of objects; i.e. except for the name of the recipient, everything else about the message is the same:

$$P(o, t_1, \ldots, t_m) \text{ and } P(o', t_1, \ldots, t_m).$$

Thus different objects can respond to the same message using their own different methods.

Like classes and class hierarchies, polymorphism is neither a unique nor an original feature of OO systems. In the context of logic, polymorphism corresponds to the fact that the same predicate can apply to different sorts of individuals.

3.6   Aspects as Integrity Constraints in Abductive Logic Programming

The mapping from logic programs to OO systems highlights a number of features of logic programming that are not so easily addressed in OO systems. I have already mentioned the problem of appropriately representing relationships, as well as the problem about representing more general logic programs that do not have input-output form. However, a problem that has attracted much attention in software engineering, and which is addressed in ALP, is how to represent *cross-cutting concerns*, which are behaviours that span many parts of a program, but which can not naturally be encapsulated in a single class.

Integrity constraints in ALP have this same character. For example, the concern:

If a person enters a danger zone,
    then the person is properly equipped to deal with the danger.

cuts across all parts of a program where there is a sub-goal in which a person needs to enter a danger zone (such as a fireman entering a fire). In ALP, this concern can be expressed as a single integrity constraint, but in normal LP it needs to scattered throughout the program, by adding to any clause that contains a condition of the form:

a person enters a danger zone

an additional condition:

the person is properly equipped to deal with the danger.

In software engineering the problem of dealing with such cross-cutting concerns is the focus of aspect-oriented programming (AOP) [11]. AOP seeks to encapsulate such concerns through the introduction of a programming construct called an *aspect*. An aspect alters the behavior of a program by applying additional behavior at a number of similar, but different points in the execution of the program.

Integrity constraints in ALP give a declarative interpretation to aspects. ALP provides the possibility of executing such integrity constraints as part of the process of pre-active thinking [?], to monitor actions before they are chosen for execution. This is like using integrity constraints to monitor updates in a database, except that the updates are candidates to be performed by the program itself.

It is also possible to transform logic programs with integrity constraints into ordinary logic programs without integrity constraints [13,21]. This is similar to the way in which aspects are implemented in AOP. However, whereas in AOP the programmer needs to specify the "join points" where the aspects are to be applied, in logic programming the transformations of [13,21] can be performed automatically by matching the conditions of integrity constraints with conditions of program clauses.

4   The Relationship Between OO Systems and ALP Systems

The mapping from input-output logic programs to OO systems can be extended to more general ALP agent systems, and a converse mapping is also possible. These mappings exploit the correspondence between agents and objects, in which both are viewed semantically as individuals, mutually embedded with other individuals in a common, dynamically changing world. Both agents and objects process their interactions with the world, by manipulating sentences in a formal language, encapsulated, and hidden from other individuals.

The biggest difference between logic and objects, and therefore between ALP agents and objects, is their different views of semantic structure. For logic, the world is a relational structure, consisting of individuals and relationships that change over time. Such changes can be modeled by using the possible world semantics of modal logic or by treating situations or events as individuals, but they can also be modeled by using destructive assignment.

With destructive assignment, the world exists only in its current state. Agents perform actions, which initiate new relationships (by adding them) and terminate old relationships (by deleting them), without the world remembering its past. The agents themselves and their relationships with other individuals are a part of this dynamically and destructively changing world.

ALP agents, as well as undergoing destructive changes, can also represent changes internally among their beliefs. Using such syntactic representations of change, they can represent, not only the current state of the world, but also past states and possible future states. We will return to this use of logic to represent change later in the paper.

Whereas logic distinguishes between changes that take place in the semantic structure of the world and changes that are represented syntactically in an agent's beliefs, objects do not. In OO systems, all changes of state are associated with objects. It makes it easy for objects to deal with changes of values of attributes, but more difficult for them to deal with changes of relationships.

The different ways in which logic and objects view the world are reflected in the different ways in which they interact with the world. In OO systems, because the state of the world is distributed among objects as attribute-value pairs, the only way an object can access the current state is by accessing the attribute-value pairs of objects. The only way an object can change the current state is by changing the attribute-value pairs of objects. In some OO languages these operations are carried out by sending and receiving messages. In other OO languages they are performed directly.

ALP agents, on the other hand, interact by observing and acting on the external world. These interactions typically involve observing and changing relationships among arbitrarily many individuals, not only attributes of individual objects. This way of interacting with the world is similar to the way that processes use the Linda tuple-space as a shared environment and to the way that experts use the blackboard in a blackboard expert system.

4.1   Transformation of ALP Systems into OO Systems

Agents. Our earlier transformation of input-output logic programs into OO systems implicitly treats the owners of clauses as agents. In this transformation, the owner of a clause/belief is selected from among the input arguments of the conclusion of the clause. However, in ALP systems, goals and beliefs are already associated with the agents that are their owners. In transforming ALP systems into OO systems, therefore, it is a simple matter just to treat agents as objects and to treat their goals and beliefs as the objects' methods. In some cases, this transformation of agents into objects coincides with the transformation of input-output logic programs into OO systems. However, in many other cases, it is more general.

The ALP semantic structure. An agent's beliefs include its beliefs about relationships between individuals, expressed as unconditional clauses. The individuals included in these relationships need not explicitly include the agent itself, as in the case of a passenger's belief that there is a fire in a train. These beliefs can be used as methods to respond to requests for information from other agents/objects.

In addition to these syntactic representations of relationships as beliefs, an ALP system as a whole is a semantic structure of relationships between individuals. This semantic structure can also be transformed into objects and their associated attribute-values, similarly to the way in which we earlier associated input-output clauses with owner objects and classes.

However, to obtain the full effect of ALP systems, we need to let each of the individuals o_i in a semantic relationship $P(o_1, \ldots, o_n)$ (expressed as an atomic fact) be an object, and to associate the relationship redundantly with each of the objects o_i as one of its attribute-values. The problem of representing such relationships redundantly has been recognized as one of the problems of OO, and representing relationships as aspects in AOP [19] has been suggested as one way of solving the problem.

Notice that an object corresponding to an agent can contain two records of the same relationship, one as an attribute-value representation of a semantic relationship, and the other as a method representing a belief. When the two records are identical, the belief is true. When they are different, the belief is false.

Actions. Actions and observations need to be transformed into sending and receiving messages. An action performed by an agent A that initiates relationships

$$P_1(o_{11}, \ldots, o_{1l1})$$
$$\ldots$$
$$P_n(o_{n1}, \ldots, o_{nln})$$

and terminates relationships

$$Q_1(p_{11}, \ldots, p_{1k1})$$
$$\ldots$$
$$Q_m(p_{m1}, \ldots, p_{mkn})$$

is transformed into a message sent by object A to each object o_{ij} to add $P_i(o_{i1}, \ldots, o_{il1})$ to its attribute-values together with a message sent by A to each object p_{ij} to delete $Q_i(p_{i1}, \ldots, p_{ik1})$ from its attribute-values.

Observations. The transformation of observations into messages is more difficult. Part of the problem has to do with whether observations are active (intentional), as when an agent looks out the window to check the weather, or whether they are passive (unintentional), as when an agent is startled by a loud noise. The other part of the problem is to transform an observation of a relationship between several objects into a message sent by only one of the objects as the messenger of the relationship. To avoid excessive complications, we shall assume that this problem of selecting a single messenger can be done somehow, if necessary by restricting the types of observations that can be dealt with.

An *active observation* by agent A of a relationship $P(o_1, o_2, \ldots, o_n)$ is transformed into a message sent by object A to one of the objects o_j, say o_1 for simplicity, requesting the solution of a goal $P(o_1, o'_2, \ldots, o'_n)$ and receiving back a message from o1 with a solution Θ, where $P(o_1, o_2, \ldots, o_n) = P(o_1, o'_2, \ldots, o'_n)\Theta$[6].

A *passive observation* of the relationship can be modeled simply by some object o_j sending a message to A of the relationship $P(o_1, o_2, \ldots, o_n)$.

The objects that result from this transformation do not exhibit the benefits of structuring objects into taxonomic hierarchies. This can be remedied by organising ALP agent systems into class hierarchies, similar to sort hierarchies in order-sorted logics. The use of such hierarchies extracts common goals and beliefs of individual agents and associates them with more general classes of agents. Translating such extended ALP agent systems into OO systems is entirely straight-forward.

4.2   Transformation of OO Systems into ALP Systems

The semantic structure. Given the current state of an OO system, the corresponding semantic structure of the ALP system is the set of all current object-attribute-values, represented as binary relationships, attribute(object, value), or alternatively as ternary relationships, say as relationship(object, attribute, value).

Agents. We distinguish between *passive objects* that merely store the current values of their attributes and *active objects* that both store their current values and also use methods to interact with the world. Both kinds of objects are treated as individuals in the ALP semantic structure. But active objects are also treated as agents.

Messages. The treatment of messages is dealt with case by case:

Case 1. A passive object sends a message to another object. By the definition of passive object, this can only be a message informing the recipient of one of the passive object's attribute-values. The only kind of recipient that can make use of such a message is an active object. So, in this case, the message is an *observation* of the attribute-value by the recipient. The observation is *active* (from the recipient's viewpoint) if the message is a response to a previous message from the recipient requesting the sender's attribute-value. Otherwise it is *passive*.

Case 2. An active object sends a message to another object requesting one of the recipient's attribute-values. This is simply the first half of an active observation of that attribute-value. The second half of the observation is a message from the recipient sending a reply. If the recipient does not reply, then the observation fails.

Case 3. An active object sends a message to another object changing one of the recipient's attribute-values. The message is simply an *action* performed on the semantic structure.

---

[6] In most OO languages this can be done more directly, without explicitly sending and receiving messages. This more direct access to an object's attribute-values can be regarded as analogous to an agent's observations and actions.

Case 4. An active object sends any other kind of message to another active object. The message is a combination of an *action by the sender* and an *observation by the recipient.* The action, like all actions, is performed on the semantic structure. In this case the action adds a ternary relationship between the sender, the recipient and the content of the message to the semantic structure. The observation, like all observations, syntactically records the semantic relationship as a belief in the recipient's internal state. The observation then becomes available for internal processing, using forward and backward reasoning with goals and beliefs to achieve the effect of methods. The recipient may optionally perform an action that deletes the ternary relationship from the semantic structure.

Methods. Finally, we need to implement methods by means of goals and beliefs (or equivalently, for ALP agents, by integrity constraints and logic programs). Recall that, in our analysis, only active objects (or agents) employ methods, and these are used only to respond to messages that are transformed into observations. Other messages, which simply request or change the values of an object's attributes, are transformed into operations on the semantic structure.

We need a sufficiently high-level characterization of such methods, so they can be logically reconstructed. For this reason, we assume that methods can be specified in the following input-output form:

If observation and (zero or more) conditions,
then (zero or more) actions.

This is similar to event-condition-action rules in active databases [18] and can be implemented directly as integrity constraints in logical form. However, such specifications can also be implemented at a higher level, by means of more abstract integrity constraints (with more abstract conditions and higher-level conclusions), together with logic programs. At this higher level, an agent implements an active object's method by

– recording the receipt of the message as an observation;
– possibly using the record of the observation to derive additional beliefs;
– possibly using the record of the observation or the derived additional beliefs to trigger an integrity constraint of the form:
    if conditions, then conclusion
– verifying any remaining conditions of the integrity constraint and then,
– reducing the derived conclusion of the constraint, treated as a goal, to sub-goals, including actions.

Beliefs, in the form of logic programs, can be used to reason forwards from the observation and backwards both from any remaining conditions of the integrity constraint and from the conclusion of the integrity constraint. In addition to any actions the agent might need to perform as part of the specification of the method, the agent might also send requests to other agents for help in solving sub-goals, in the manner of the object-oriented logic programs of section 3.

All messages that do not request or change attribute-values are treated by the recipient uniformly as observations. If the message is a request to solve a goal, then the recipient records the request as an observation and then determines whether or not to try to solve the goal, using an integrity constraint such as:

If an agent asks me to solve a goal,
and I am able and willing to solve the goal for the agent,
then I try to solve the goal and I inform the agent of the result.

The recipient might use other integrity constraints to deal with the case that the recipient is unable or unwilling to solve the goal.

Similarly, if the message is a communication of information, then the recipient records the communication as an observation and then determines whether or not to add the information to its beliefs. For example:

If an agent gives me information,
and I trust the agent,
and the information is consistent with my beliefs,
then I add the information to my beliefs.

The input-output specification of OO methods that we have assumed is quite general and hopefully covers most sensible kinds of methods. As we have seen, the specification has a direct implementation in terms of abductive logic programming. However, as we have also noted earlier, other implementations in other computer languages are also possible, as long as they respect the logical specification.

Classes. Methods associated with classes of objects and inherited by their instances can similarly be associated with sorts of ALP agents. This can be done in any one of the various ways mentioned earlier in the paper. As remarked then, this requires an extension of ALP agents, so that goals and beliefs can be associated with sorts of agents and acquired by individual agents. There is an interesting research issue here: whether sentences about several sorts of individuals can be represented only once, or whether they need to be associated, possibly redundantly, with owner classes/sorts.

5   Local Versus Global Change

One of the attractions of object-orientation is that it views change in local, rather than global terms. In OO the state of the world is distributed among objects as the current values of their attributes. Change of state is localized to objects and can take place in different objects both concurrently and independently.

Traditional logic, in contrast, typically views change in global terms, as in the possible-worlds semantics of modal logic and the situation calculus. Modal logic, for example, deals with change by extending the static semantics of classical model theory to include multiple (possible) worlds related by a temporal, next-state accessibility relation. Semantically, a change of state due to one or more concurrent actions or events, is viewed as transforming one global possible world into another global possible world. Syntactically, change is represented by using modal operators, including operators that deal with actions and events as parameters, as in dynamic modal logic.

The situation calculus [17] similarly views change as transforming one global possible world (or situation) into another. Semantically, it does so by reifying situations, turning situations into individuals and turning the next-state accessibility relation into a normal relation between individuals. Syntactically, it represents change by using variable-free terms to name concrete situations and function symbols to transform situations into successor situations.

It was, in part, dissatisfaction with the global nature of the possible-worlds semantics that led Barwise and Perry to develop the situation semantics [4]. Their situations (which are different from situations in the situation calculus) are semantic structures, like possible-world semantic structures, but are partial, rather than global.

It was a similar dissatisfaction with the global nature of the situation calculus that led us to develop the event calculus [15]. Like situations in the situation calculus, events, including actions are reified and represented syntactically. The effect of actions/events on the state of relationships is represented by an axiom of persistence in logic programming form:

A relationship holds at a time T_2
if an event happens at a time T_1 before T_2
and the event initiates the relationship
and there is no other event
    that happens at a time after T_1 and before T_2 and
    that terminates the relationship.

Like change of state in OO, change in the event calculus is also localised, but to relationships rather than to objects. Also as in OO, changes of state can take place concurrently and independently in different and unrelated parts of the world. Each agent can use its own local clock, time-stamping observations as they occur and determining when to perform actions, by comparing the time that actions need to be performed with the current time on its local clock.

The event calculus is a syntactic representation, which an agent can use to reason about change. It can be used to represent, not only current relationships, but also past and future relationships, both explicitly by atomic facts and implicitly as a consequence of the axiom of persistence. However, the event calculus does not force an agent to derive current relationships using the persistence axiom, if the agent can observe those relationships directly, more efficiently and more reliably instead.

The event calculus is not a semantics of change. However, in theory, if events are reified, then the use of the event calculus to reason about change should commit an agent to a semantic structure in which events are individuals. But this is the case only if all symbols in an agent's goals and beliefs need to be interpreted directly in the semantic structure in which the agent is embedded. If some symbols can be regarded as defined symbols, for example, then they need not be so interpreted. Alternatively, in the same way that in physics it is possible to hypothesize and reason with the aid of theoretical particles, which can not be observed directly, it may also be possible in the event calculus to represent and reason about events without their being observable and without their corresponding to individuals in the world of experience.

In any case, the event calculus is compatible with a semantic structure in which changes in relationships are performed destructively, by deleting (terminating) old relationships and adding (initiating) new relationships. These destructive changes in the semantic structure are the ones that actually take place in the world, as opposed to the representation of events and the derivations of their consequences using the axiom of persistence, which might take place only in the mind of the agent.

6 Related Work

There is a vast literature dealing with the problem of reconciling and combining logic programming and object-orientation, most of which was published in the 1980s, when the two paradigms were still contending to occupy the central role in Computing that OO occupies today. Most of this early literature is summarized in McCabe's [16].

Perhaps the most prominent approach among the early attempts to reconcile logic programming and objects was the concurrent object-oriented logic programming approach exemplified by [20]. In this approach, an object is implemented as a process that calls itself recursively and communicates with other objects by instantiating shared variables. Objects can have internal state in the form of unshared arguments that are overwritten in recursive calls. Although this approach was inspired by logic programming it ran into a number of semantic problems, mainly associated with the use of committed choice. The problem of committed choice is avoided in ALP systems by incorporating it in the decision making component of individual agents.

In McCabe's language, $L\&O$ [16], a program consists of a labelled collection of logic programs. Each labelled logic program is like a set of beliefs belonging to the object or agent that is the label. However, in $L\&O$, objects/agents interact by sending messages to other agents, asking for their help in solving sub-goals. This is like the object-oriented logic programs of section 3.4. It is also similar to the way multi-agent systems are simulated in GALATEA [7].

ALP systems differ from $L\&O$, therefore, primarily in their use of a shared environment instead of messages. This use of a shared environment is similar to the use of tuple-spaces in Linda [10]. In this respect, therefore, ALP systems are closest to the various systems [1,8] that use a Linda-like environment to coordinate parallel execution of multiple logic programs. ALP systems can be viewed, therefore, as providing a logical framework in which the shared environment in such systems can be understood as a dynamic semantic structure.

A different solution to the problem of reconciling logic and objects is the language LO [2], which is a declarative logic programming language using linear logic. The language is faithful to the semantics of linear logic, which however is quite different from the model-theoretic semantics of traditional logic. Communication between objects in LO is similar to that in Linda.

7   Conclusions

In this paper, I have explored some of the relationships between OO systems and ALP systems, and have argued that ALP systems can combine the semantic and syntactic features of logic with the syntactic structuring and dynamic, local behaviour of objects. I have investigated a number of transformations, which show how OO systems and ALP systems can be transformed into one another. These transformations are relatively straight-forward, and they suggest ways in which the two kinds of system are related. Among other applications, the transformations can be used to embed one kind of system into the other, for example along the lines of [7], and therefore to gain the benefits of both kinds of systems.

However, the transformations also highlight a number of important differences, including problems with the treatment of relations and multi-owner methods in OO systems in particular. On the other hand, they also identify a number of issues that need further attention in ALP systems, including the need to clarify the distinction between active and passive observations, to organise agents into more general agent hierarchies, and possibly to structure the shared semantic environment, to take account of the fact that different agents can more easily access some parts of the environment than other parts. In addition, it would be useful to make the relationships between OO systems and ALP systems explored in this paper more precise and to prove them more formally.

As a by-product of exploring the relationships between logic and objects, the transformations also suggest a relationship between logic and Linda. On the one hand, they suggest that Linda systems can be understood in logical terms, in which tuple-spaces are viewed as semantic structures and processes are viewed as agents interacting in this shared semantic environment. On the other hand, they also suggest that ALP systems can be generalised into Linda-like systems in which different processes can be implemented in different languages, provided that the external, logical specification of the processes is unaffected by their implementation.

Acknowledgements. I am grateful to Ken Satoh at NII in Tokyo, for valuable discussions and for providing a congenial environment to carry out much of the work on this paper. I would also like to thank Jim Cunningham, Jacinto Davila and Maarten van Emden for valuable comments on an earlier draft of this paper.

It is my pleasure to dedicate this paper to Gigina Aiello, for her inspiration in the early stages of this work, when we first started thinking about logic-based multi-agent systems in the Compulog Project in the late 1980s and early 1990s.

References

1. Andreoli, J.M., Hankin, Le Mtayer, D. (eds.): Coordination Programming: Mechanisms, Models and Semantics - Imperial College Press, London (1996)
2. Andreoli, J.M., Pareschi, R.: Linear Objects: Logical Processes with Built-In Inheritance. New Generation Computing, 9 (1991) 445-473

3. Armstrong, D.J.: The Quarks of Object-Oriented Development. Communications of the ACM 49 (2) (2006) 123-128
4. Barwise, J., Perry, J.: Situations and Attitudes. MIT-Bradford Press, Cambridge (1983)
5. Brogi, A., Ciancarini, P.: The Concurrent Language, Shared Prolog. ACM Transactions on Programming Languages and Systems, 13(1) (1991) 99-123
6. Brown, G. Yule, G.: Discourse Analysis. Cambridge University Press (1983)
7. Davila,J., Gomez, E., Laffaille, K., Tucci, K., Uzcategui, M.: Multi-Agent Distributed Simulations with GALATEA. In Boukerche, A., Turner, T., Roberts, D., Theodoropoulos, G. (eds): IEEE Proceedings of Distributed Simulation and Real-Time Applications. IEEE Computer Society (2005) 165-170
8. De Bosschere, K., Tarau, P.: Blackboard-Based Extensions in Prolog. Software - Practice and Experience 26(1) (1996) 49-69
9. Engelmore, R.S., Morgan, A. (eds.): Blackboard Systems. Addison-Wesley (1988)
10. Gelernter, D.: Generative Communication in Linda, ACM Transactions on Programming Languages, (7)1 (1985) 80-112
11. Kiczales, G., Lamping, J., Mendhekar, A., Maeda, C., Lopes, C., Loingtier, J.-M., Irwin, J.: Aspect-Oriented Programming. Proceedings of the European Conference on Object-Oriented Programming, vol.1241, (1997) 220-242
12. Kowalski, R.: How to be Artificially Intelligent. In Toni, F., Torroni, P. (eds.): Computational Logic in Multi-Agent Systems. LNAI 3900, Springer-Verlag (2006) 1-22
13. Kowalski, R., Sadri, F.: Logic Programming with Exceptions. New Generation Computing, (9)3,4 (1991) 387-400
14. Kowalski, R., Sadri, F.: From Logic Programming towards Multi-agent Systems. Annals of Mathematics and Artificial Intelligence. (25) (1999) 391- 419
15. Kowalski, R., Sergot, M.: A Logic-based Calculus of Events. New Generation Computing. (4)1 (1986) 67-95
16. McCabe, F.G.: Logic and Objects. Prentice-Hall, Inc. Upper Saddle River, NJ (1992)
17. McCarthy, J., Hayes, P.: Some Philosophical Problems from the Standpoint of AI. Machine Intelligence. Edinburgh University Press (1969)
18. Paton, N.W., Diaz, O.: Active Database Systems. ACM Computing Surveys 31(1) (1999)
19. Pearce, D.J., Noble, J.: Relationship Aspects. Proceedings of the ACM conference on Aspect-Oriented Software Development (AOSD'06). (2006) 75-86
20. Shapiro, E., Takeuchi, A.: Object-Oriented Programming in Concurrent Prolog, New Generation Computing 1(2) (1983) 5-48
21. Toni, F., Kowalski, R.: Reduction of Abductive Logic Programs to Normal Logic Programs. Proceedings International Conference on Logic Programming, MIT Press (1995) 367-381
22. Williams, J.: Style: Towards Clarity and Grace. Chicago University Press (1990)

Best-First Rippling

Moa Johansson, Alan Bundy, and Lucas Dixon

School of Informatics, University of Edinburgh,
Appleton Tower, Crichton St, Edinburgh EH8 9LE, UK
{moa.johansson, a.bundy, lucas.dixon}@ed.ac.uk

Abstract. Rippling is a form of rewriting that guides search by only performing steps that reduce the differences between formulae. Termination is normally ensured by a defined measure that is required to decrease with each step. Because of these restrictions, rippling will fail to prove theorems about, for example, mutual recursion where steps that temporarily increase the differences are necessary. Best-first rippling is an extension to rippling where the restrictions have been recast as heuristic scores for use in best-first search. If nothing better is available, previously illegal steps can be considered, making best-first rippling more flexible than ordinary rippling. We have implemented best-first rippling in the IsaPlanner system together with a mechanism for caching proof-states that helps remove symmetries in the search space, and machinery to ensure termination based on term embeddings. Our experiments show that the implementation of best-first rippling is faster on average than Isa-Planner's version of traditional depth-first rippling, and solves a range of problems where ordinary rippling fails.

1 Introduction

Rippling is a heuristic used in automated theorem proving for reducing the differences between formulae [5]. It was originally designed for inductive proofs, where we aim to rewrite the inductive conclusion in such a way that we can apply the inductive hypothesis to advance the proof. Only rewrites that reduce differences and keep similarities are allowed. Rewrite rules can be applied both ways around and termination is guaranteed by defining a *ripple measure* that is required to decrease for each step of rewriting. Rippling has been successfully used for automating proofs in a range of domains, for example, hardware verification [8], summing series [21], equation solving [13] and synthesis of higher-order programs [16].

Rippling is however not guaranteed to succeed. *Proof-planning critics* has been proposed as a solution. Critics analyse failed proof attempts to suggest patches such as a generalisation or conjecturing and proving a missing lemma [14]. Sometimes it may also be necessary to perform a rewrite that does not decrease the ripple measure or temporarily increases the differences between given and goal. This is necessary in, for example, proofs involving mutually recursive functions [5] (§5.9). Ordinary rippling is not flexible enough to deal with this. Best-first rippling is suggested as a possible solution to these problems [5]

O. Stock and M. Schaerf (Eds.): Aiello Festschrift, LNAI 4155, pp. 83–100, 2006.
© Springer-Verlag Berlin Heidelberg 2006

(§5.14). The constraints of rippling are turned into a heuristic measure, allowing previously illegal steps if nothing better is available.

We have implemented best-first rippling in IsaPlanner [11], a proof-planner built on top of the interactive theorem-prover Isabelle [18]. IsaPlanner's current implementation of higher-order rippling [12], has been expanded to allow rewrites that normally would be regarded as illegal and discarded. Heuristic scores are assigned to the steps of rippling, and we use best-first search to pick the most promising new state (§4.2). Allowing previously illegal steps introduces a risk of non-termination, which is dealt with by introducing a check on term embeddings (§4.3). During development, we also discovered that the search space for rippling often contained symmetries and developed methods for pruning such branches accordingly (§4.3). Using best-first search often caused the planner to conjecture and prove the same lemma several times. A new search strategy was developed, which delays steps waiting for the same lemma (§4.4).

Our experiments show that best-first rippling can successfully solve a range of problems where the standard depth-first version of rippling fails. We do not find any problems that are solvable by ordinary rippling but not best-first rippling. Overall, the run-times for best-first rippling are, on average, better than for ordinary depth-first rippling, despite the potentially larger search space.

2   Rippling

Rippling works by identifying differences and similarities between two terms: the given and the goal. It then guides rewriting to reduce the differences, aiming to arrive at a sub-goal which can be justified by the given. Application of the given is called *fertilisation*.

The *skeleton* represents the parts of the goal that are similar to the given while *wave-fronts* represent the differences. A *wave-hole* denotes a sub-term inside a wave-front that belongs to the skeleton. In addition, if the given contains a universally quantified variable the corresponding position in the goal is called a *sink*. An example (from [12]) showing how the parts of a goal (here the inductive conclusion), can be annotated with respect to a given (the inductive hypothesis) is shown below[1]:

$$Given: \quad \forall b : nat.\ a + b = b + a$$

$$Goal: \quad \boxed{suc(\underline{a})}^{\uparrow} + \lfloor b \rfloor = \boxed{suc(\lfloor b \rfloor + a)}^{\downarrow}$$

The wave-front is represented by a box, and the wave-hole by underlining. The skeleton, coming from the given, $a + b = b + a$ corresponds to the parts of the goal that are either without annotation or underlined within the wave front. Note that the universally quantified variable b in the given becomes a sink in the goal, annotated by $\lfloor b \rfloor$. There are two strategies for making fertilisation possible, known

---

[1] Note this is *one* way of annotating this goal; in general a goal may be annotated in several different ways.

as *rippling-in* and *rippling-out*. Rippling-out will try to remove the differences completely or move them out of the way, so that the wave-front surrounds the entire term and the wave-hole contains an instance of the given. Rippling-in tries to move differences into sinks. The universally quantified variable in the given can then be instantiated to the contents of the sink and fertilisation is possible. The arrow of the wave-front indicates if the wave-front is to be rippled out ($\uparrow$) or in ($\downarrow$). In order to make the search space smaller, rippling-in is only allowed if there exists a sink or an outward wave-front inside the inward wave-front that eventually may absorb it. We lift this restriction for best-first rippling.

Rippling proceeds by applying rewrite-rules derived from equations, definitions, theorems and lemmas. To ensure that fertilisation will eventually be possible after rewriting, rippling requires the skeleton to be preserved between each step. Termination is guaranteed by defining a *ripple-measure*, based on the positions of the wave-fronts, which is required to decrease for each rewrite step. This also helps reduce the size of the search space, and make it possible to allow rewrite-rules to be applied in both directions, unlike traditional rewriting where only one direction is allowed. There are different implementations of ripple measures. Here, we will use a measure based on the sum of distances from each outward wave-front to the top of the term tree and from each inward wave-front to the nearest sink. This measure will clearly decrease as outward wave-fronts are moved towards the top of the term-tree, and inward wave-fronts towards a sink further down.

Example. As an example illustrating how rippling moves the wave-front outward to allow fertilisation, consider the step case goal of the inductive proof of the commutativity of addition, where the given is the inductive hypothesis. Note that the sinks have been omitted to reduce clutter, as the proof only uses the rippling-out strategy.

$$Given: \quad \forall b : nat. \; a + b = b + a$$

$$Goal: \quad \boxed{suc(\underline{a})}^{\uparrow} + b = b + \boxed{suc(\underline{a})}^{\uparrow}$$

with the rules [2]:

$$suc(X) + Y \equiv suc(X + Y) \tag{1}$$

$$X + suc(Y) \equiv suc(X + Y) \tag{2}$$

$$suc(X) = suc(Y) \equiv X = Y \tag{3}$$

The proof of the step-case goal will proceed as follows:

$$\boxed{suc(\underline{a})}^{\uparrow} + b = b + \boxed{suc(\underline{a})}^{\uparrow}$$

$$\Big\Downarrow \; by \; rule \; 1$$

---

[2] Following the convention for dynamic rippling (§2.1), the rules have not been annotated as wave-rules in static rippling.

$$\boxed{suc(\underline{a+b})}^{\uparrow} = b + \boxed{suc(\underline{a})}^{\uparrow}$$

$$\Big\Downarrow by \ rule \ 2$$

$$\boxed{suc(\underline{a+b})}^{\uparrow} = \boxed{suc(\underline{b+a})}^{\uparrow}$$

$$\Big\Downarrow by \ rule \ 3$$

$$a + b = b + a$$

$$\Big\Downarrow Fertilise$$

$$True$$

Notice how each ripple-rewrite moves the wave-front outwards until we arrive at a state where the goal contains an instance of the given. We can now simply replace this instance with 'True' and conclude the proof. This is called *Strong fertilisation*.

In the case that rule 3 were missing, there would have been no more rewrites possible after the state: $\boxed{suc(\underline{a+b})}^{\uparrow} = \boxed{suc(\underline{b+a})}^{\uparrow}$. We say that the state is *blocked*. It is however still possible to apply the given using substitution, which rewrites the blocked goal to $suc(b+a) = suc(b+a)$. This is called *weak-fertilisation*. The resulting goal is true by reflexivity. In situations where rippling is blocked but weak fertilisation is not possible, we can attempt to apply a critic [14].

2.1 Static and Dynamic Rippling

There are two main approaches for implementing rippling: static and dynamic rippling. They represent and handle annotations in different ways. Rippling as described by Bundy et al. [5] will be referred to as *static rippling*. In static rippling, the rewrite-rules are annotated before rippling starts in such a way that they will ensure measure decrease and skeleton preservation. The annotated rules are called *wave-rules* and can be applied to any goal with matching annotations. Note that a single theorem or definition may give rise to several wave-rules. Basin and Walsh give a formal calculus for static rippling in first-order logic and provide a proof of termination [1]. They represent annotations as function-symbols at the object level of the goal. The object level annotations require a special notion of substitution as standard substitution may produce illegal annotations. Another problem with static rippling in a higher-order setting, as pointed out by Smaill and Green [20], is that the object level annotations are not stable over β-reduction. This makes it impossible to pre-annotate higher-order rewrite rules as they may turn out to be non-skeleton preserving. To overcome these problems, the use of *dynamic rippling* [9,12], and *term embeddings*, for representing annotations [20,9], have been introduced. In dynamic rippling, annotations are stored separately from the goal and rewrite rules are not annotated in advance. Instead, all ways of rewriting the goal with a particular rule are generated after which the annotations are re-computed and measure decrease and skeleton preservation checked. This means that no specialised version of substitution is needed.

Dynamic rippling is more suitable as a starting point for our best-first rippling implementation because it initially generates all possible rewrites, including new subgoals that are non-skeleton preserving and non-measure decreasing. These would normally be discarded, but we will adapt rippling to instead assign them heuristic scores.

3   Proof-Planning

Rippling has been implemented and used within the context of *proof-planning* [3,6]. Proof planning is a technique for guiding the search for a proof in automated theorem proving by exploiting that 'families' of proofs, for example inductive proofs, share a similar structure. Instead of searching the large space of an underlying theorem-prover, the proof-planner can reason about the applicable methods for a conjecture and construct a *proof-plan* consisting of a tree of *tactics*. A tactic is some sequence of steps, known to be sound, that are used for solving a particular problem in a theorem-prover, such as simplification, induction etc.

The *Clam* proof-planner [7], written in Prolog, and the higher-order λ*Clam* [19], written in λProlog, both implement rippling. The Clam-family of proof planners uses a set of *methods* and *methodicals*. Methods specify what conditions have to be true for the method to be applicable to a goal and what will be true after the method has been applied. They also carry a reference to the corresponding tactic that will be used in the theorem-prover when the proof plan is executed. Methodicals combine several atomic methods into larger compound methods.

3.1   IsaPlanner

Recently, a higher-order version of dynamic rippling has been implemented in IsaPlanner [11,10], a proof planner written in Standard ML for the interactive theorem-prover Isabelle [18]. In IsaPlanner, proof planning is interleaved with execution of the proof in Isabelle giving IsaPlanner access to Isabelle's powerful tactics. The resulting proof-plan is then represented as a proof script in the Isar language [22], executable in Isabelle and argued to be more readable than the output from earlier proof-planners such as λ*Clam*. Rippling in IsaPlanner has also been shown to be considerably faster than in λ*Clam* [12].

As opposed to the Clam-family of proof planners, IsaPlanner plans the proof through a series of *reasoning states*. Each reasoning state contains the partial proof plan constructed so far, the next reasoning technique to be applied and contextual information. The reasoning techniques are defined to be functions from a reasoning state to a sequence of new reasoning states. This sequence represents all the ways the technique can be applied to its input state. The contextual information contains knowledge acquired during proof planning, including information about rippling-annotations and skeletons.

IsaPlanner supports several search strategies, including a generic best-first search. Search strategies can be applied globally or locally over a reasoning technique.

IsaPlanner's implementation of rippling is designed in a modular fashion to be easily extendable and can support different versions of rippling with different notions of annotations and ripple measures simultaneously. Our best-first rippling implementation is a module defined in terms of the module for ordinary rippling, thereby making best-first rippling available for any of IsaPlanner's versions of rippling.

4   Best-First Rippling

Ordinary rippling requires each step in the rippling-process to satisfy the restrictions of measure decrease and skeleton preservation, otherwise the step is regarded as invalid. There are however a number of occasions where these 'invalid' ripple-steps would be useful or necessary for the success of rippling. In proofs involving mutually recursive functions, the skeleton might be temporarily disrupted but restored in a later step (see for example [5], §5.9). Another example is a proof where it is necessary to 'unblock' rippling by performing rewrites inside the wave front [4], which might lead to a temporary increase in the ripple-measure.

In best-first rippling, the measure decrease and skeleton preservation requirements are, instead of being strictly enforced, reflected in a heuristic score. The heuristic prefers smaller ripple measures and skeleton preservation but previously invalid steps can then be considered if nothing better is available.

To realise best-first rippling we need dynamic rippling and best-first search. We must consider all rewrites at any given state, evaluate their heuristics scores and compare them with all other open states in the search. The state with the lowest score is the most promising one from which to continue rippling. IsaPlanner implements dynamic rippling and has a generic version of best-first search, making it a suitable platform for implementing best-first rippling.

Example: Breaking the Skeleton. As an example, consider the following problem with mutually recursive definitions of even and odd (here called $evenM$ and $oddM$).

$$Given: \quad evenM(n) \vee oddM(n)$$

$$Goal: \quad evenM(\boxed{suc(suc(\underline{n}))}^{\uparrow}) \vee oddM(\boxed{suc(suc(\underline{n}))}^{\uparrow})$$

with the rules:

$$evenM(0) \equiv True \tag{4}$$

$$evenM(suc(X)) \equiv oddM(X) \tag{5}$$

$$oddM(0) \equiv False \tag{6}$$

$$oddM(suc(X)) \equiv evenM(X) \tag{7}$$

This gives the following best-first rippling proof using two-step induction:

$$evenM(\boxed{suc(suc(\underline{n}))}^{\uparrow}) \vee oddM(\boxed{suc(suc(\underline{n}))}^{\uparrow})$$

$$\Downarrow \; by \; rule \; 5$$

$$oddM(suc(n)) \vee oddM(suc(suc(n))) \qquad (8)$$

$$\Downarrow \; by \; rule \; 7$$

$$evenM(n) \vee oddM(\boxed{suc(suc(\underline{n}))}^{\uparrow})$$

$$\Downarrow \; by \; rule \; 7$$

$$evenM(n) \vee evenM(suc(n)) \qquad (9)$$

$$\Downarrow \; by \; rule \; 5$$

$$evenM(n) \vee oddM(n)$$

Fertilisation is now possible. Note that the skeleton is disrupted in steps 8 and 9 (the subgoals are therefore not annotated), but restored in the following step. These steps are necessary for the completion of this proof but would not be allowed in ordinary rippling.

Example: Non-measure Decrease Required. The proof of the theorem $evenR(suc(suc(0)) * n)$, taken from [4], requires us to modify the argument to the even-function before we can apply fertilisation. As a result, the rewrites will not move the wave-front, only rearrange the terms inside it, and the measure will stay the same over a number of steps. Note that this is the two-step recursively defined even-function, (referred to as $evenR$) as opposed to the mutually recursive version defined above.

The proof uses the following rules from the definitions of addition, multiplication and for the two-step recursive version of $evenR$:

$$evenR(suc(suc\ X)) \equiv evenR(X) \qquad (10)$$

$$X + 0 \equiv 0 \qquad (11)$$

$$X + suc(Y) \equiv suc(X + Y) \qquad (12)$$

$$X * suc(Y) \equiv (X * Y) + suc(Y) \qquad (13)$$

Our given and goal gives the following rippling sequence (with the sum-of distance ripple measure given for each step):

$$Given: \; evenR(suc(suc(0)) * n)$$

$$Goal: \; evenR(suc(suc(0)) * \boxed{suc(\underline{n})}^{\uparrow}) \quad Measure: \; 2$$

$$\Downarrow by \; rule \; 13$$

$$evenR(\boxed{\underline{suc(suc(0)) * n} + suc(suc(0))}^{\uparrow}) \quad Measure: \; 1$$

$$\Downarrow by \; rule \; 12$$

$$evenR(\boxed{suc(\underline{suc(suc(0)) * n} + suc(0))}^{\uparrow}) \quad Measure: \; 1$$

$$\Downarrow by \; rule \; 12$$

$$evenR(\boxed{suc(suc(\underline{suc(suc(0)) * n} + 0))}^{\uparrow}) \quad Measure: \; 1$$

$$\Downarrow by \; rule \; 10$$

$$evenR(\boxed{suc(\underline{suc(suc(0)) * n} + 0)}^{\uparrow}) \quad Measure: \; 1$$

$$\Downarrow by \; rule \; 11$$

$$evenR(suc(suc(0)) * n) \quad Measure: \; 0$$

Strong fertilisation is now applicable as the wave-front has been fully rippled out leaving the ripple measure 0. All but the first and last step in the rippling proof do not change the ripple-measure as the rewrites are applied to terms inside the wave-front.

4.1   Complications with Best-First Rippling

The price for the greater flexibility of best-first rippling is that the search space is considerably larger. The increased number of possibilities to continue rippling also means that rippling will rarely become blocked, which is when applying fertilisation or critics would normally be considered. Furthermore, allowing non-measure decreasing and non-skeleton preserving steps means that best-first rippling will loose the guarantee for termination, as it is possible to become stuck in a loop by applying the same rewrite-rule in opposite directions.

Mutually recursive functions are another source of potential non-termination as it is possible to apply a non-skeleton preserving rewrite rule in a direction such that the subgoal gets larger and larger. Recall the previous example:

$$evenM(\boxed{suc(suc(\underline{n}))}^{\uparrow}) \lor oddM(\boxed{suc(suc(\underline{n}))}^{\uparrow})$$

Here we can rewrite $evenM(\boxed{suc(suc(\underline{n}))}^{\uparrow})$ in two ways, neither of which preserves the skeleton. We can either apply rewrite-rule 5 from left to right or, as rewrites are allowed in both directions, rule 7 from right to left. The latter would give the result

$$oddM(suc(suc(suc(n)))) \vee oddM(suc(suc(n)))$$

where $evenM$ has been transformed into $oddM$ by adding a successor-function rather than removing one. Consider now applying rule 5 from right to left, which produces a state that does embed the skeleton but adds yet another successor function:

$$evenM(\boxed{suc(suc(suc(suc(\underline{n}))))}^{\uparrow}) \vee oddM(\boxed{suc(suc(\underline{n}))}^{\uparrow})$$

Subsequent bad applications could keep alternating between $evenM$ and $oddM$, each time adding another successor-function and hence never terminating. Our solution to these problems uses caching of the visited states and is discussed in §4.3.

4.2   Best-First Heuristic

Best-first rippling requires a heuristic evaluation function for deciding which state is the most promising to evaluate next in the rippling process. Valid ripples should be considered before non-measure decreasing or non-skeleton preserving steps. The ripple measure gives an indication of how far w are from being able to apply fertilisation and conclude the proof.

We have used IsaPlanner's sum-of-distance ripple measure during development and testing. Rippling with this measure has been shown to perform better than with other kinds of measures [10]. As mentioned earlier, best-first rippling has however been implemented in a modular fashion, allowing use of any type of ripple measure.

IsaPlanner's best-first search function expects to be supplied with a heuristic order function used for keeping the agenda sorted in increasing order. It is therefore not necessary to compute and store explicit numerical scores for the states, just determine their relative ordering. Our heuristic function for the best-first search takes two reasoning states and compares them. A state is regarded as *less* than another state if its heuristic score is better, thus placing it closer to the front of the agenda.

The heuristic function for comparing reasoning states can be summarised as follows:

- States to which strong fertilisation can be applied are always preferred over continued rippling.
- Skeleton preserving states are always given a better score than non-skeleton preserving states.
- When both states preserve the skeleton, the state with the best ripple measure is given the lower score. If the states have the same ripple measure, they are given equal heuristic scores.

– If neither state preserves the skeleton, the reasoning state with the smallest goal-term scores better.

Strong fertilisation should be preferred over everything else as it applies the inductive hypothesis and concludes the proof. Skeleton preservation is always preferred over non-preservation as we only want to apply non-skeleton preserving steps when there are no other options. States that do embed the skeleton are ordered based on the ripple-measures. In comparing ripple-measures, we need to take into account that, as IsaPlanner employs dynamic rippling, each reasoning state might have several ripple measures, one for each way a skeleton embeds. IsaPlanner also supports rippling with multiple skeletons, each of which may embed in different ways. For comparisons, we use the best ripple-measure of each state.

The heuristic also handles non-rippling states, such as setting up a rippling attempt or applying fertilisation. Non-rippling steps are simply preferred before more rippling as a fixed number of non-rippling steps will either result in a solution (if fertilisation is successful) or a new ripple-state to which our standard heuristic is applicable if we have to prove a lemma. Little or no search is needed.

Because best-first rippling does not become blocked as often as ordinary rippling does, we considered introducing some heuristic measure allowing the application of weak fertilisation and critics before we run out of applicable rules. We developed a variant of best-first rippling where weak-fertilisation and IsaPlanner's lemma calculation critics were applied eagerly to states where none of the children were skeleton-preserving, i.e. the state would have been blocked in ordinary rippling. The non-skeleton preserving children are also kept in the agenda, but given a worse heuristic score than to weak fertilise and/or conjecture a lemma.

4.3   Termination and Reduction of Search Space Size

As mentioned before, allowing rippling with non-measure decreasing wave-rules means that best-first rippling is no longer guaranteed to terminate. The same wave-rule now can be applied in opposite directions, causing loops, and it is possible to apply rewrites that just blow up the size of the goal-term as described in §4.1. Another source of inefficiency is the many symmetric branches in the search tree.

To deal with these problems, the best-first implementation caches the visited states of a ripple sequence. We filter out any new subgoals that are identical to subgoals previously seen anywhere in the search tree, thereby pruning symmetric branches. The termination and looping problem is dealt with by introducing an *embedding check*, as used in IsaPlanner's lemma conjecturing ([10] Chapter 9). If a previous goal-term embeds into the new sub-goal it is removed, which filters rewrites that would otherwise cause divergence. Kruskal's Theorem [15], states that there exists no infinite sequence of trees such that an earlier tree does not embed into a later tree. Therefore, the embedding check will restore termination, which was lost as we relaxed the restriction of ripple-measure decrease. We have

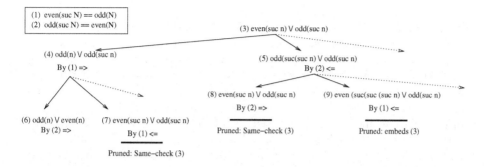

Fig. 1. Partial search tree for best-first rippling showing how branches are pruned to avoid loops and redundant rewrites. Note that the two rules are allowed to be applied in both directions.

chosen to only check embeddings against other states on the same branch, if checked against states on alternative OR-branches we could potentially prune useful states. This approach appears to work well in practice. Figure 1 illustrates how unproductive branches are pruned to reduce the size of the search space.

4.4   Delaying Parts of the Search

We discovered that a common problem arising when using best-first or breadth-first search for rippling is that the same lemma might be conjectured independently at different places in the search space, causing the planner to pursue several simultaneous attempts on the same lemma.

In IsaPlanner's standard depth-first rippling this is not an issue. When a blocked state is encountered, a lemma is conjectured and proved before backtracking to try more rippling in the original proof attempt. Lemmas that have already been proved to be true (or failed) are cached, allowing later blocked states requiring the same lemma to use the previous result, thus saving time by avoiding symmetric parts of the search space. When using best-first search, it may be the case that after a lemma has been conjectured and a proof attempt begun, some state in the original proof attempt has a better heuristic score so rippling is continued from there. If this second ripple also becomes blocked and requires a lemma which we already have started a proof of elsewhere, we want to prevent beginning a second attempt. Instead, the second reasoning state should be suspended until the lemma has been proved. After the lemma is proved, not only the state from which it was originally conjectured, but also any other states waiting for that particular lemma, should be resumed.

Beginning several attempts of the same lemma was one of the major sources for inefficiencies in our initial implementation of best-first rippling. Initially, the problem was tackled by giving rippling in a lemma attempt a better heuristic score than rippling the original conjecture. This is however not always desirable;

if a bad lemma is conjectured, we do want the option to abandon it and explore other possibilities. Experiments also suggested that this approach may miss solutions to some problems and generally lead to longer run-times. We chose to instead create a new generic search strategy in IsaPlanner. This strategy inspects all new states and may temporarily remove them from the agenda if marked as delayed. Similarly, the strategy checks if the current state wishes to resume some delayed states, which are then returned to the agenda. IsaPlanner's lemma conjecturing machinery was augmented with a cache for lemmas-in-progress in addition to the existing caching of completed proof attempts. The lemma conjecturing critic inspects the cache and if an attempt is already in progress, the reasoning state is marked as delayed and not evaluated further until the proof attempt of the relevant lemma is finished.

4.5   Storing Skeletons

Ordinary rippling will discard any skeletons that cannot be embedded in the current goal term as they are not needed any more. Best-first rippling on the other hand, needs to keep all skeletons. After applying a non-skeleton preserving step, the previous skeleton must be kept so we can keep track of whether or not the skeleton is restored in subsequent steps.

The skeletons and their possible embeddings are stored in IsaPlanner's contextual information for rippling. Previously, only the list of possible embeddings of a skeleton was stored. When a skeleton failed to embed, all references to the skeleton were removed, making it impossible to later check if the skeleton could embed into some new state. For best-first rippling, the contextual information for rippling has been modified to store a list of pairs consisting of both the skeleton and a list of embeddings of that skeleton (as opposed to only the embeddings list). A skeleton not embedded in the current subgoal will have an empty list of embeddings, but will still be kept.

5   Evaluation and Results

Best-first rippling has been evaluated by comparing it to IsaPlanner's implementation of ordinary rippling, which uses depth-first search. We measured the number of successfully solved problems as well as run-times on both successful and failed proof attempts. Our test-problems included a set of benchmarks for IsaPlanner, consisting of 55 theorems in Peano arithmetic and about lists, to test the performance of best-first rippling compared to ordinary rippling on standard problems. Best-first rippling has a larger search space and performs some extra work computing heuristic scores, so we expected it to be slower than ordinary depth-first rippling. The benchmarks also included a range of non-theorems, allowing us to test the robustness of best-first rippling. Ideally we would like to exhaust the search space quickly when no solution can be found, rather than see non-termination. In addition to IsaPlanner's benchmarks, we also tested a set of 39 problems where we would expect to see the full benefits of best-first rippling,

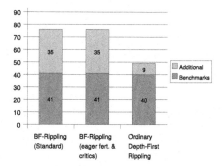

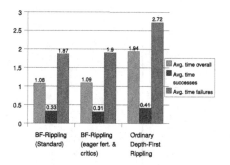

Fig. 2. Number of successes on the 94 theorems in the test set (55 benchmarks and 39 additional)

Fig. 3. Average run-times in seconds

including proofs about mutually recursive functions, proofs involving destructor-style functions (such as the predecessor function in Peano arithmetic) and proofs where measure increasing steps are required. The mutually recursive problems typically require induction schemes reflecting the depth of the nested recursive function definitions. As an example, recall the mutually recursive definition of *even* and *odd* from §4. The two functions are defined in terms of each other so we use two-step induction. In these cases, the induction scheme was supplied manually to IsaPlanner/Isabelle, as inference of induction schemes is currently limited to standard recursively defined data-types.

We also compared a version of best-first rippling that applies critics when it is blocked, with a version that eagerly tries to apply critics or weak-fertilisation when no more skeleton-preserving steps are available. This was expected to indicate whether applying critics is more efficient than searching the larger space arising from allowing non-skeleton preserving steps.

The experiments were conducted on a standard 2 GHz Intel Pentium4 PC with 512 MB of memory running Isabelle2005. Each problem had a timeout limit of 30 seconds.

The number of successful proofs for the three versions of rippling are displayed in figure 2. Both best-first rippling with eager application of critics and the variant without it, managed to find proofs for 76 of the 94 theorems. Ordinary depth-first rippling succeeded to find 49 proofs, 40 from IsaPlanner's benchmarks compared to 41 for best-first rippling. The additional benchmark problem solved by best-first rippling was $rev(l) = qrev(l, [\ ])$, a problem expected to fail as Isa-Planner lacks a generalisation critic for accumulator variables, but here solved as a side effect of our caching mechanism[3]. On the additional set, ordinary rippling proved only 10 theorems compared to 35 for best-first, which was expected as these were chosen from classes of problems known to be difficult for ordinary rippling.

---

[3] The interested reader can find the proof on the project website: `http://dream.inf.ed.ac.uk/projects/bfrippling`.

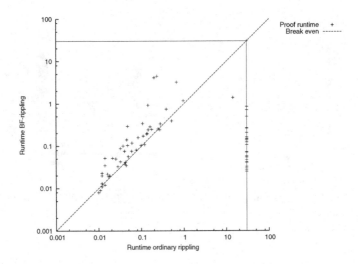

Fig. 4. Each scatter-plot represents a conjecture, with the x-value being the runtime for ordinary rippling and the y-value the runtime of standard best-first rippling. The vertical and horizontal lines marks the timeout limit of 30 seconds. Failed proof-attempts have also been plotted along these lines for clarity. A logarithmic scale is used for better visualisation.

Figure 3 shows the average run-times for proof-attempts while figure 4 shows the time spent on each proof for best-first and ordinary rippling. Ordinary rippling is slightly faster on most problems both techniques can solve but the differences are small. Best-first rippling is however faster on average, due to a few outliers for ordinary rippling. Ordinary rippling fails or times out more often than best-first rippling. As a result, best-first rippling is faster than depth-first rippling overall, and also spends less time on conjectures it cannot prove thanks to the caching and embedding-check.

The differences in runtime appears to be small between the two variants of best-first rippling. Conjecturing lemmas eagerly when no skeleton-preserving steps are available appears to make little difference to the run-times of the mutually recursive problems in our test set. We also notice that best-first rippling spent less time on failed proof attempts, including the non-theorems in the test set, despite the larger number of allowed rewrites.

The full collection of test problems, results and function definitions can be found on-line at `http://dream.inf.ed.ac.uk/projects/bfrippling`. The source code is available from the IsaPlanner website: `http://sourceforge.net/projects/isaplanner/`.

To summarise the results; best-first rippling proves a number of theorems where ordinary rippling is too restricted to succeed, as expected. Despite the larger search-space of best-first rippling the differences in run-times compared to ordinary rippling are small. Best-first rippling appears to be more robust when presented with non-theorems, less time is spent on failed proof-attempts compared to ordinary rippling.

6   Related Work

6.1   Depth-First Rippling

The main difference between our work and IsaPlanner's previous implementations of ordinary depth-first rippling [10], is that best-first rippling relaxes the requirements that each state must preserve the skeleton and decrease the ripple measure. These requirements guarantees the termination of ordinary rippling, something that is lost for best-first rippling. Our implementation instead uses mechanisms for caching of visited states to avoid loops and a check on term embeddings to restore termination (see Kruskal's Theorem [15]). This works well in practise and has the additional advantage of pruning the search space of symmetric branches.

6.2   Best-First Rippling in λClam

James Brotherston implemented a best-first methodical in the λClam proof planner [2]. The best-first methodical use a greedy search strategy, considering only the best option at the current node, not previous nodes higher up in the tree. Higher branches in the search tree are only investigated on backtracking. Applied to rippling[4], Brotherston identifies this as a problem as it does not allow switching focus to the most promising area of the search. Our best-first search strategy is not greedy and we can easily switch focus to different parts of the search tree as IsaPlanner's reasoning states, held in the agenda, contains the necessary local contextual information about the proof-plan and next reasoning technique.

6.3   Best-First Proof-Planning

Manning et al. presents an implementation of best-first proof-planning in *Clam* [17]. A best-first heuristic is employed to make choices between three different proof planning methods; generalisation, simplification and induction, as a fixed ordering sometimes causes unnecessarily complicated proofs or even causes failure. Our work differs from that of Manning as we are applying best-first search *within* the rippling technique. IsaPlanner applies induction and rippling first, then attempts simplification or generalisation if the ripple becomes blocked. Despite this, all proofs in [17] are solvable by best-first rippling, although perhaps not in the most efficient way.

7   Further Work

As a side-effect of the caching mechanism, best-first rippling manages to prove the conjecture $rev(l) = qrev(l, [\,])$ where we would expect rippling to fail without a generalisation critic that can introduce an accumulator variable before induction and rippling is attempted. Best-first rippling does however fail to prove more

---

[4] Personal communication: internal Blue Book Note series, numbers 1405, 1409, 1425.

complicated theorems involving similar tail-recursive functions. Such problems can be solved using a critic to analyse the failed proof attempt in order to suggest a generalisation. Another limitation of the current implementation is that the user is required to specify if an induction scheme other than standard one is required. The *Clam* proof-planner had a number of critics for finding lemmas, forming generalisations, case-splits and revising the induction scheme [14]. IsaPlanner has currently only one critic, for lemma calculation. We plan to implement additional critics in IsaPlanner. This is expected to allow a larger number of problems to be solved automatically, including many of the problems from the test set where both best-first and ordinary depth-first rippling currently fail.

The caching techniques we have discussed could also benefit ordinary rippling. In particular, pruning states already seen from the search space removes symmetric branches which would potentially improve run-times.

Our test-set mainly consisted of relatively easy theorems. Further experiments will evaluate best-first rippling on harder problems. We also plan to undertake a larger comparison between rippling and regular rewriting.

8   Conclusions

We have shown that our implementation of best-first rippling is able to automatically prove a number of theorems where IsaPlanner's previous implementation of depth-first rippling fails, for example, proofs about mutually recursive functions and proofs requiring a temporary increase in the ripple measure. Rippling has been allowed more flexibility by recasting the measure decrease and skeleton preservation requirements into heuristic scores. In allowing these steps we do however lose the guarantee of termination for rippling. Our solution to this problem introduces an embedding check (§4.3), where new subgoals in which we can embed previously seen cached goals on the same branch are pruned. This cuts out branches where subsequent applications of non-skeleton preserving rewrites leads to divergence as described in §4.1 and restores termination. We also found that the search space often would contain symmetries, where the same state occurs in several different places. To improve efficiency, any goal identical to a cached goal is simply pruned.

Using best-first search rather than depth-first search means that it is possible to switch between rippling in a lemma attempt and rippling in the original proof, depending on which seems more promising. This often gave rise to the same lemma being conjectured from different blocked states. Our new search strategy suspends any states requiring a lemma for which a proof is already in progress. When a lemma is proved, all states waiting for it are resumed.

Our test results show that best-first rippling not only is capable of solving a range of problems not solvable by ordinary rippling, but also has faster run-times overall thanks to the combination of efficiency measures described above and the guidance from best-first search. We also compared two versions of best-first rippling to verify if it is beneficial to apply critics before best-first rippling is blocked, as best-first rippling might not become blocked as often as ordinary rippling due to the larger search space. On our test set, we did however find that applying critics eagerly when no more skeleton preserving states were available, made little difference.

References

1. D. Basin and T. Walsh. A calculus for and termination of rippling. *Journal of Automated Reasoning*, 1-2(16):147–180, 1996.
2. J. Brotherston and L. Dennis. *LambdaClam v.4.0.1 User/Developer's manual*. Available online: http://dream.inf.ed.ac.uk/software/lambda-clam/.
3. A. Bundy. The use of explicit plans to guide inductive proofs. In *9th International Conference on Automated Deduction*, pages 111–120, 1988.
4. A. Bundy. The termination of rippling + unblocking. Informatics research paper 880, University of Edinburgh, 1998.
5. A. Bundy, D. Basin, D. Hutter, and A. Ireland. *Rippling: Meta-level Guidance for Mathematical Reasoning*. Cambridge University Press, 2005.
6. A. Bundy, F. van Harmelen, J. Hesketh, and A. Smaill. Experiments with proof plans for induction. *Journal of Automated Reasoning*, 7:303–324, 1992.
7. A. Bundy, F. van Harmelen, C. Horn, and A. Smaill. The Oyster-Clam system. In *10th International Conference on Automated Deduction*, number 449 in LNAI, pages 647–648, 1990.
8. F. Cantu, A. Bundy, A. Smaill, and D. Basin. Experiments in automating hardware verification using inductive proof planning. In *First International Conference on Formal Methods in Computer-Aided Design*, volume 1166 of *LNCS*, pages 94–108. Springer Verlag, 1996.
9. L. A. Dennis, I. Green, and A. Smaill. Embeddings as a higher-order representation of annotations for rippling. Technical Report Computer Science No. NOTTCS-WP-SUB-0503230955-5470, University of Nottingham, 2005.
10. L. Dixon. *A proof-planning framework for Isabelle*. PhD thesis, School of Informatics, University of Edinburgh, 2005.
11. L. Dixon and J. Fleuriot. IsaPlanner: A prototype proof planner in Isabelle. In *Proceedings of CADE'03*, pages 279–283, 2003.
12. L. Dixon and J. Fleuriot. Higher-order rippling in IsaPlanner. In *Proceedings of TPHOLs'04*, pages 83–98, 2004.
13. D. Hutter. Coloring terms to control equational reasoning. *Journal of Automated Reasoning*, 18(3):399–442, 1997.
14. A. Ireland and A. Bundy. Productive use of failure in inductive proof. *Journal of Automated Reasoning*, 16:79–111, 1996.
15. J. B. Kruskal. Well-quasi-ordering, the tree theorem, and Vazsonyi's conjecture. *Transactions of the American Mathematical Society*, 1960.
16. D. Lacey, J. Richardson, and A. Smaill. Logic program synthesis in a higher-order setting. *Computational Logic*, 1861:87–100, 2000.
17. A. Manning, A. Ireland, and A. Bundy. Increasing the versatility of heuristic based theorem provers. In A. Voronkov, editor, *International conference on Logic Programming and Automated Reasoning LPAR'93*, number 698 in Lecture Notes in Artificial Intelligence, pages 194–204. Springer Verlag, 1993.
18. T. Nipkow, L.C. Paulson, and M. Wenzel. *Isabelle/HOL - A proof assistant for higher-order logic*. Number 2283 in Lecture Notes in Computer Science. Springer Verlag, 2002.

19. J. Richardson, A. Smaill, and I. Green. System description: Proof planning in higher-order logic with Lambda-Clam. In *15th International Conference on Automated Deduction*, number 1421 in LNAI, pages 129–133, 1998.

20. A. Smaill and I. Green. Higher-order annotated terms for proof search. In *Theorem Proving in higher-order logics: 9th international conference*, volume 1275 of *Lecture Notes in Computer Science*, pages 399–413. Springer Verlag, 1996.

21. T. Walsh, A. Nunes, and A. Bundy. The use of proof plans to sum series. In *11th Conference on Automated Deduction*, number 607 in LNCS, pages 325–339. Springer Verlag, 1992.

22. M. Wenzel. Isar - a generic interpretative approach to readable formal proof documents. In *Proceedings of TPHOLs'99*, volume 1690 of *Lecture Notes in Computer Science*, pages 167–184. Springer Verlag, 1999.

Partial Solutions with Unique Completion

Marco Cadoli and Marco Schaerf

Dipartimento di Informatica e Sistemistica
Università di Roma "La Sapienza"
Via Salaria 113, I-00198 Roma, Italy
{cadoli, schaerf}@dis.uniroma1.it

Abstract. In this paper we investigate the computational complexity of combinatorial problems with *givens*, i.e., partial solutions, and where a unique solution is required. Examples for this article are taken from the games of Sudoku, N-queens and related games. We will show the computational complexity of many decision and search problems related to Sudoku, a number of similar games and their generalization. Furthermore, we propose a logical description of several such problems that can lead to a formulation in the language of Quantified Boolean Formulae (QBF) and, hence, their mechanization via a QBF solver. Some experiments on finding the minimum number of givens necessary/sufficient to guarantee uniqueness of solution are shown.

1 Introduction

Sometimes a combinatorial problem comes with a *partial solution*, i.e., an assignment to some of the variables is part of the input. A typical case is that of planning, where the initial and goal states are part of the input, and the sequence of states that is searched must, respectively, start and finish with them. Another example is frequency assignment in a telecommunications network, where the specification requires that some of the nodes have their frequency assigned already.

Maybe the best example concerns *Sudoku*, a solitaire game which has recently become very popular in newspapers and magazines: a *Sudoku of order n^2* is an $n^2 \times n^2$ matrix of n^2 symbols in which each symbol occurs exactly once in each row, each column, and each of the n^2 $n \times n$ boxes (typically indicated by the slightly heavier lines) of the matrix. Figure 1(b) shows a 4×4 $(n = 2)$[1] Sudoku board with a partial solution (the numbers are called *givens*). Sudoku completion, and its precursor *Latin square completion* (cf. Section 4) are NP-complete problems [13].

A peculiarity of Sudoku is the *promise*, provided by the newspaper, that the completion is *unique*. This promise typically facilitates people in the search for the solution. The literature on structural complexity offers several examples of *promise problems*, i.e., problems which have the guarantee of some property, like the existence of a solution. Such problems are abstracted in the complexity class TFNP [9] of *total functions in NP*. Other complexity classes are mentioned in [1].

---

[1] Typically newspapers propose 9×9 $(n = 3)$ boards and between 20 and 30 givens. For the sake of readability, in this paper we show only 4×4 boards.

O. Stock and M. Schaerf (Eds.): Aiello Festschrift, LNAI 4155, pp. 101–115, 2006.
© Springer-Verlag Berlin Heidelberg 2006

This paper deals with instances of problems which come with the promise of the existence and uniqueness of a solution, like Sudoku in the newspapers. Figure 1(c-d) show two boards with such a property. Following the current terminology, cf. `en.wikipedia.org/wiki/Sudoku`, we call such instances *proper*. Of course, some boards are not proper (cf., e.g., Figure 1(b)), and some are not even solvable, (cf., e.g., Figure 1(a)).

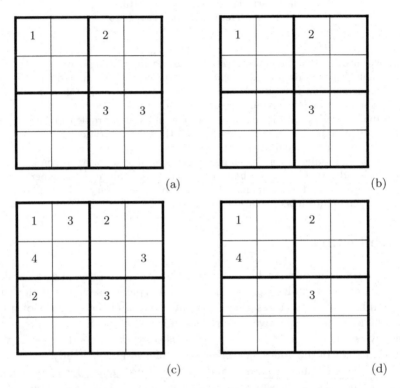

Fig. 1. 4 × 4 Sudoku boards with partial assignments (givens): (a) with no solution, i.e., a CSP-PA, not PS (cf. Def. 3, 4), (b) with multiple solutions, i.e., a CSP-PS, not PPS (cf. Def. 4, 5), (c) non minimal, with unique solution, i.e., a CSP-PPS, neither McPPS nor MsPPS (cf. Def. 5, 6), (d) cardinality-minimal, i.e., a CSP-McPPS (which implies MsPPS) (cf. Def. 6)

A number of computational problems arise, which, to the best of our knowledge, have never previously been studied. As an example, given an instance with a partial solution, decide whether it is proper, and if not, find a set of further assignments (possibly minimal) such that it becomes proper. The formal definitions and the statements of the computational problems are listed in Section 2. Some results concerning computational complexity of the problems are presented in Section 3. Section 4 reports an analysis of the minimum number of assignments that must be given in order to have the possibility, or the necessity, of proper instances. This problem has received some attention: as an example, at the time of writing this paper, it is known (cf. en.wikipedia.org/wiki/Mathematics_of_Sudoku) that there are sets of 17 givens that are proper for the 9×9 Sudoku, and it is currently unknown whether there are such sets of 16 givens. We formally define problems about the minimum number of givens necessary/sufficient to guarantee uniqueness of the solution and formulate them in terms of formulae in second-order logic; then we show an experimental analysis. Section 5 concludes the paper and reports current research.

2   Preliminaries

In this section we briefly recall the definition of constraint satisfaction problems and then define the decision and search problems whose complexity we are going to analyze in the sequel.

First of all, we define the basic form of constraint satisfaction problems and their solutions. Similar definitions can be found in, e.g., [5]. In the sequel, given two sets X and Y, we denote with $X \Delta Y$ their difference, that is the elements that belong to one of the seta, but not to the other. More precisely, $X \Delta Y = \{a \in (X \cup Y) | ((a \in X) \wedge (a \notin Y)) \vee ((a \in Y) \wedge (a \notin X))\}$.

Definition 1 (CSP). *Let $\mathbb{D}$ be a finite set of size at least 2. A V-tuple t, where V represents a finite set of variables, is a mapping which associates a value $t_x \in \mathbb{D}$ to every $x \in V$. A V-relation is a set of V-tuples. A* Constraint Satisfaction Problem *(CSP) is a triple $\langle X, D, C \rangle$ where:*

- *X is a finite set of variables,*
- *D associates to every variable $x \in X$ a domain $D_x \subseteq \mathbb{D}$ and*
- *C is a finite set of constraints, each of which is a V-relation for some $V \subseteq X$.*

Definition 2 (Solution of a CSP). *Given a V-tuple t and a subset $U \subseteq V$ of its variables, we denote by $t|_U$ the* restriction *of t to U, which has the same value as t on the variables of U and is undefined elsewhere. The explicit assignment of the value of a V-tuple t on a variable $x \in V$ to value a is written $t[x := a]$.*

An X-tuple t satisfies *a V-relation $c \in C$ if $t|_V \in c$. We denote by $Sol(c)$ the set of X-tuples which satisfy c. The set $\bigcap_{c \in C} Sol(c)$ of X-tuples which satisfy all the constraints is called the* solution space, *and denoted $Sol(C)$. The set of X-tuples t such that $t_x \in D_x$ for all variables x is called the* search space *and noted S_D, or simply S if the domain is implicit from the context. Given a tuple t we denote with $V(t)$ the set of variables $x \in X$ to which t assigns a value, while t is undefined on the variables in $X \Delta V(t)$. For the sake of simplicity, the sets X and C will be considered as globally defined and shall therefore be omitted from the parameters of most definitions; only the search space will be explicitly mentioned.*

Using the above definitions we can formally define a number of properties of CSPs that formalize the decision and search problems outlined in Section 1. The first two definitions distinguish the notion of partial assignment and partial solution.

Definition 3 (CSP with partial assignment). *Given a CSP $\Sigma = \langle X, D, C \rangle$, we define a CSP with Partial Assignment (CSP-PA) as a pair $\Pi = \langle \Sigma, \tau \rangle$, where τ is a U-tuple for some set $U \subseteq X$.*

From now on we call τ the *partial assignment*.

Definition 4 (CSP with partial solution). *Given a CSP $\Sigma = \langle X, D, C \rangle$, we define a CSP with Partial Solution (CSP-PS) as a pair $\Pi = \langle \Sigma, \tau \rangle$, where τ is a U-tuple for some set $U \subseteq X$ and the CSP admits a solution that is compatible with τ.*

A more interesting notion in our context is the definition of *properness*, already outlined in the Introduction.

Definition 5 (CSP with proper partial solution). *Given a CSP $\Sigma = \langle X, D, C \rangle$, we define a CSP with Proper Partial Solution (CSP-PPS) as a pair $\Pi = \langle \Sigma, \tau \rangle$, where τ is a U-tuple for some set $U \subseteq X$ and the CSP admits exactly one solution that is compatible with τ.*

In order to clarify the above definitions we introduce an example:

Example 1. Consider a CSP $\Sigma = \langle X, D, C \rangle$ over Boolean variables, where $X = \{a, b, c, d, e\}$, $D = \{true, false\}$ and $C =$

$$
\begin{array}{ll}
a & \wedge \\
a \rightarrow b & \wedge \\
(c \vee d) \leftrightarrow e &
\end{array}
$$

In this setting, a U-tuple τ is a partial assignment to the Boolean variables $\{a, b, c, d, e\}$. For example, let $\tau_1 = \{a, \neg b\}$, $\tau_2 = \{a, b\}$, and $\tau_3 = \{a, \neg e\}$. It is easy to verify that the CSP with τ_1 is a partial assignment but not a partial solution, τ_2 is both a partial assignment and a partial solution, but not a proper one, while τ_3 satisfies all three definitions.   □

As mentioned in the Introduction, in the context of 3×3 Sudoku, it is an open problem to decide what is the minimal number of givens that are needed to guarantee the existence of a proper Sudoku instance. We define two forms of minimality, one with respect to cardinality, and the other with respect to set containment. The first form is more intuitive and, in the context of Sudoku, defines when a set of givens contains the least number of elements necessary. The set-containment minimality defines when a set (of givens) is minimal in the sense that no subset of it is a proper instance. Since the definition of the CSP includes the variables that we want to minimize as well as other variables, the minimization will only apply to a given subset (denoted as M in the sequel) of all the variables (X) of the CSP.

Definition 6 (Minimal proper partial solution). *Given a CSP $\Sigma = \langle X, D, C \rangle$, a set of variables $M \subseteq X$ and a U-tuple τ (for some set $U \subseteq M$) such that the CSP-PA $\Pi = \langle \Sigma, \tau \rangle$ is proper, we define two forms of Minimal Proper Partial Solution:*

  – Minimality w.r.t. cardinality (CSP-McPPS): for all sets $V \subseteq M$ such that $|V| < |U|$ there is no V-tuple τ_1 such that the CSP-PA $\Pi = \langle \Sigma, \tau_1 \rangle$ is proper.

- Minimality *w.r.t.* set containment (CSP-MsPPS): *for all sets $V \subset U$ there is no V-tuple τ_1 such that the CSP-PA $\Pi = \langle \Sigma, \tau_1 \rangle$ is proper.*

Example 2 (Example 1, continued). τ_3 is not minimal *w.r.t.* set containment, since $\tau_4 = \{\neg e\}$ is proper. Actually, τ_4 is minimal *w.r.t.* cardinality, hence it is minimal *w.r.t.* set containment. $\tau_5 = \{c, d\}$ is proper and minimal *w.r.t.* set containment, but not *w.r.t.* cardinality. □

The above definitions naturally lead to a number of decision and search problems that we start investigating in this paper. More precisely, we investigate the following decision problems:

Definition 7 (Decision problems)

Satisfiability: *Given a CSP-PA, is it a CSP-PS?*
Properness: *Given a CSP-PS, is it a CSP-PPS?*
C-minimality: *Given a CSP-PPS, is it a CSP-McPPS?*
S-minimality: *Given a CSP-PPS, is it a CSP-MsPPS?*

Moreover, there are a number of interesting search problems in this framework, here we only mention the most relevant ones.

Definition 8 (Search problems)

Minimal repair: *Given an unsatisfiable CSP-PA, find a "minimal delete" CSP-PS. That is, this problem amounts to find a minimal set of assignments that must be withdrawn in order to make the CSP satisfiable.*
Minimal add: *Given a CSP-PS, find a "minimal add" CSP-McPPS (CSP-MsPPS). That is, this problem amounts to find a minimal set of assignments that must be included in order to make the CSP minimally proper.*
Minimal delete: *Given a CSP-PPS, find a "minimal delete" CSP-McPPS (CSP-MsPPS). That is, this problem amounts to find a minimal set of assignments that must be withdrawn in order to make the CSP minimally proper.*

In the next section we will analyze the computational complexity of the above defined decision and search problems and hint at how they can be solved using standard solvers for SAT and QBF.

3   Results on Computational Complexity

In this section we investigate the computational complexity of the decisional (subsection 3.1) and search problems (subsection 3.2) listed in Section 2. From now on, we assume that the input is given as a set of constraints C over a set of variables X. We also assume that the problem of checking whether $t \in Sol(C)$ is polynomial in the size of the representation of the input. Additionally, we assume that the size of D is fixed. Such properties hold for propositional logic and for CSPs, in the sense of [5].

3.1   Decision Problems

We first analyze the computational complexity of the simplest decision problem, that is deciding whether a constraint satisfaction problem with a given partial assignment has a solution. This problem is well-known in the literature and this result is not new.

Theorem 1. *Given a CSP $\Sigma = \langle X, D, C, \rangle$ and a partial assignment τ, deciding whether the CSP-PA $\Pi = \langle \Sigma, \tau \rangle$ is a CSP-PS (i.e., it has a solution) is NP-complete.*

Proof. Trivial, in fact let the partial assignment be empty. This problem reduces to the satisfiability of a CSP, that in our setting is well-known to be NP-complete. Moreover, the problem obviously belongs to NP, since we only need to guess a solution and check whether it satisfies the constraints. □

We now investigate the computational complexity of checking properness of a constraint satisfaction problem with partial solution.

Theorem 2. *Given a CSP $\Sigma = \langle X, D, C, \rangle$ and a partial assignment τ such that $\Pi = \langle \Sigma, \tau \rangle$ is a CSP-PS, deciding whether it is a CSP-PPS (i.e., it has a unique solution) is coNP-hard and in US.*

Proof. coNP-hardness: we reduce the problem of checking whether a propositional formula ϕ is unsatisfiable to our problem. Given a propositional formula ϕ defined on a set of variables X, and a fresh propositional variable $r \notin X$, we define the formula ψ as $(\phi \wedge \neg r) \vee (\bigwedge_{x \in X} x \wedge r)$. Note that ψ is satisfiable, because the second disjunct is satisfied by the assignment "alltrue" A such that $A(y) = \mathsf{true}$ for each $y \in X \cup \{r\}$. Therefore ψ can be readily defined as a CSP-PS Π_ψ (with empty τ). Moreover, ϕ is unsatisfiable iff A is the unique model of ψ, because if ϕ is unsatisfiable then ψ has no other models, and if there exists a model M (over X) of ϕ, then both A and $M \cup \{\neg r\}$ are distinct models of ψ, hence ψ is not uniquely satisfiable. Summing up, ϕ is unsatisfiable iff Π_ψ is a CSP-PPS.

Membership to US [2]: US is defined as the class of decision problems solvable by an NP machine such that the answer is "yes" iff exactly one computation path accepts. Therefore our problem belongs to US by definition. Note that US $\subseteq D^p$ [8]. □

The above results characterize the basic problems, we now analyze the complexity of the more difficult problems involving minimality, that is cardinality-minimal and subset-minimal solutions. It is somehow surprising that these two problems belong to different complexity classes, as we show in the following theorems:

Theorem 3. *Given a CSP $\Sigma = \langle X, D, C \rangle$, a set of variables $M \subseteq X$ and a U-tuple τ (for some set $U \subseteq M$) such that $\Pi = \langle \Sigma, \tau \rangle$ is a CSP-PPS, deciding whether it is a CSP-MsPPS (i.e., it is a cardinality minimal CSP with a proper partial solution) is in Π_2^p.*

Proof. This problem can be reformulated as follows: for all sets of variables L (such that $|L| < |U|$) and corresponding assignment τ', there exists more than one solution (they are not proper). More precisely, $\forall L$ such that $|L| < |U|$ $\exists \tau_1, \tau_2$ such that $U \subseteq V(\tau_1), U \subseteq V(\tau_2), \tau_1 \neq \tau_2, \tau_1 \models \Pi$ and $\tau_2 \models \Pi$. Since all these checks can be accomplished in polynomial time, the problem belongs to Π_2^p, cf. [11,8].   □

Theorem 4. *Given a CSP $\Sigma = \langle X, D, C \rangle$, a set of variables $M \subseteq X$ and a U-tuple τ (for some set $U \subseteq M$) such that $\Pi = \langle \Sigma, \tau \rangle$ is a CSP-PPS, deciding whether it is a CSP-MsPPS (i.e., it is a set containment minimal CSP with a proper partial solution) is in Δ_2^p.*

Proof. This problem can be reformulated as follows: none of the variables in U can become undefined in τ without losing properness. More precisely, $\Pi = \langle \Sigma, \tau \rangle$ is a CSP-MsPPS if and only if for all variables $v \in U$ $\exists \tau_1, \tau_2$ such that $V(\tau) \subseteq V(\tau_1), V(\tau) \subseteq V(\tau_2), \tau_1 \neq \tau_2, \tau_1 \models \Pi$ and $\tau_2 \models \Pi$. Since the first quantification ($\forall v \in U$) can be replaced by $|U|$ calls to an NP-oracle and all the other checks can be accomplished in polynomial time, the problem belongs to Δ_2^p.   □

Notice that problems in Δ_2^p can be solved by calling a polynomial number of times a solver for NP-complete problems, such as a SAT-solver, while a problem in Π_2^p requires a solver for Quantified Boolean Formulae (QBF-solver) to be addressed effectively.

3.2   Search Problems

In this section we investigate the computational complexity of the search problems defined in Section 2. More precisely, we consider five search problems, since the definitions of *minimal add* and *minimal delete* lead to two different search problems, one for each minimality criterion.

The first problem we analyze is the complexity of finding a cardinality-minimal set of assignments (givens) that we need to retract from an inconsistent CSP in order to make it proper.

Theorem 5. *Let $\Sigma = \langle X, D, C \rangle$ be a CSP, $M \subseteq X$ be a set of variables and τ be a U-tuple (for some set $U \subseteq M$) such that $\Pi = \langle \Sigma, \tau \rangle$ is a CSP-PA.*

With a polynomial number of calls to a Σ_2^p-oracle we can compute the cardinality of the smallest set of variables $L \subseteq M$, such that $\Pi_1 = \langle \Sigma, \tau_1 \rangle$, where $\tau_1 = \tau$ for all variables in $U \Delta L$ and is undefined otherwise, is a CSP-PS (i.e., it has a solution).

Proof. We guess an X tuple v, we denote with τ_1 the restriction of v to the variables in $U \Delta L$ (that is, $\tau_1 = v|_{U \Delta L}$) and check:

1. $\Pi_1 = \langle \Sigma, \tau_1 \rangle$ is a CSP-PS
2. for all sets τ_2 such that $V(\tau_1) \subset V(\tau_2) \subseteq V(\tau)$, $\Pi_2 = \langle \Sigma, \tau_2 \rangle$ is unsatisfiable (not a CSP-PS).

The first check can be computed with a single call to an NP-oracle, while the second one requires a call to an NP-oracle that guesses a new assignment V_2 and checks condition 1. Hence, the problem can be solved by using an NP^{NP}-oracle, that is a Σ_2^p-oracle. $\qquad\square$

We now investigate the complexity of the minimal add problems. In this problem we are looking for a (cardinality or set-containment) minimal set of variables, and their assignments, that will make the CSP proper. The variables we add must belong to the set M and not be already contained in the current set of variables U, therefore, we only look for a set of variables L such that $L \subseteq (M \Delta U)$.

Theorem 6. *Let $\Sigma = \langle X, D, C \rangle$ be a CSP, $M \subseteq X$ be a set of variables and τ be a U-tuple (for some set $U \subseteq M$) such that $\Pi = \langle \Sigma, \tau \rangle$ is a CSP-PS.*

   With a polynomial number of calls to a Σ_2^p-oracle we can compute an assignment τ_1 to a cardinality-minimal set of variables $L \subseteq (M \Delta U)$ such that $\Pi_1 = \langle \Sigma, \tau_2 \rangle$, where $\tau_2 = \tau$ for all variables in U, $\tau_2 = \tau_1$ for all the variables in L and is undefined otherwise, is a CSP-McPPS.

Proof. We guess an X-tuple v and a set L, we denote with $\tau_1 = (v|_L)$ and with $\tau_2 = (v|_{U \cup L})$ and check:

1. $\Pi_1 = \langle \Sigma, \tau_1 \rangle$ is a CSP-PS
2. for all sets $L_1 \subset L$ and L_1-tuples τ_3 we have that $\Pi_2 = \langle \Sigma, \tau_3 \rangle$ is unsatisfiable (not a CSP-PS).

The first check can be computed with a single call to an NP-oracle, while the second one requires a call to an NP-oracle that guesses a new assignment V_2 and checks conditions 1 and 2. $\qquad\square$

Theorem 7. *Let $\Sigma = \langle X, D, C \rangle$ be a CSP, $M \subseteq X$ be a set of variables and τ be a U-tuple (for some set $U \subseteq M$) such that $\Pi = \langle \Sigma, \tau \rangle$ is a CSP-PS.*

   With a polynomial number of calls to a Σ_2^p-oracle we can compute an assignment τ_1 to a subset-minimal set of variables $L \subseteq (M \Delta U)$ such that $\Pi_1 = \langle \Sigma, \tau_2 \rangle$, where $\tau_2 = \tau$ for all variables in U, $\tau_2 = \tau_1$ for all the variables in L and is undefined otherwise, is a CSP-MsPPS.

Proof. This proof is similar to the previous one, we guess an X-tuple v and a set L, we denote with $\tau_1 = (V_1|_L)$ and with $\tau_2 = (V_1|_{U \cup L})$ and check:

1. $\Pi_1 = \langle \Sigma, \tau_1 \rangle$ is a CSP-PS
2. for all sets $L_1 \subset L$ and L_1-tuples τ_3, $\Pi_2 = \langle \Sigma, \tau - \tau_2 \rangle$ is unsatisfiable (not a CSP-PS).

The first check can be computed with a single call to an NP-oracle, while the second one requires a call to an NP-oracle that guesses a new set and assignment and checks condition 1. $\qquad\square$

The same complexity upper bounds can be shown for the minimal delete problems. In this problem we are looking for a (cardinality or set-containment) minimal set of variables, and their assignments, to be deleted in order to make the CSP proper. The variables we delete must belong to the set M and to the current set of variables U. Since $U \subseteq M$, we look for a set of variables L such that $L \subseteq U$.

Theorem 8. *Let $\Sigma = \langle X, D, C \rangle$ be a CSP, $M \subseteq X$ be a set of variables and τ be a U-tuple (for some set $U \subseteq M$) such that $\Pi = \langle \Sigma, \tau \rangle$ is not satisfiable (i.e., not a CSP-PS).*

With a polynomial number of calls to a Σ_2^p-oracle we can compute a cardinality-minimal set of variables $L \subseteq U$ such that $\Pi_1 = \langle \Sigma, \tau_2 \rangle$, where $\tau_2 = \tau$ for all variables in $U \Delta L$ and is undefined otherwise, is a CSP-McPPS.

Proof. We guess the set of variables L and check:

1. $\Pi_1 = \langle \Sigma, \tau_2 \rangle$ is a CSP-PS
2. for all sets $L_1 \subseteq U$ such that $|L_1| < |L|$, $\Pi_2 = \langle \Sigma, \tau_3 \rangle$, where $\tau_3 = \tau$ for all variables in $U \Delta L_1$ and is undefined otherwise is unsatisfiable (not a CSP-PS).

The first check can be computed with a single call to an NP-oracle, while the second one requires a call to an NP-oracle that guesses a new set of variables and checks condition 1. □

Theorem 9. *Let $\Sigma = \langle X, D, C \rangle$ be a CSP, $M \subseteq X$ be a set of variables and τ be a U-tuple (for some set $U \subseteq M$) such that $\Pi = \langle \Sigma, \tau \rangle$ is not satisfiable (i.e., not a CSP-PS).*

With a polynomial number of calls to a Σ_2^p-oracle we can compute a subset-minimal set of variables $L \subseteq U$ such that $\Pi_1 = \langle \Sigma, \tau_2 \rangle$, where $\tau_2 = \tau$ for all variables in $U \Delta L$ and is undefined otherwise, is a CSP-MsPPS.

Proof. We guess the set of variables L and check:

1. $\Pi_1 = \langle \Sigma, \tau_2 \rangle$ is a CSP-PS
2. for all sets $L_1 \subseteq U$ such that $L_1 \subseteq L$, $\Pi_2 = \langle \Sigma, \tau_3 \rangle$, where $\tau_3 = \tau$ for all variables in $U \Delta L_1$ and is undefined otherwise is unsatisfiable (not a CSP-PS).

The first check can be computed with a single call to an NP-oracle, while the second one requires a call to an NP-oracle that guesses a new set of variables and checks condition 1. □

Summing up, we have shown that all the above problems belong to PSPACE and, more precisely, to various levels of the Polynomial Hierarchy [11]. Therefore, all of the above problems can be reduced to the problem of deciding the truth of an appropriate Quantified Boolean Formula (QBF). Since the technology of QBF solvers has dramatically improved in recent years, our results indirectly show a promising way to tackle all these combinatorial problems by reducing them to QBFs and feeding them to a QBF solver.

4 Single-Parameter Problems

In this section we focus on "single-parameter problems", i.e., on CSPs whose instances are conveniently described by a single integer parameter denoting the number of variables to be assigned.

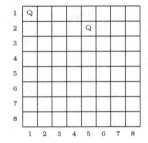

Fig. 2. A placement of two queens on an 8×8 board which can be completed in a unique way

The well-known *queens problem* gives a simple example: Is it possible to place n non-attacking queens on a $n \times n$ chessboard? Since each queen must be assigned a different column, there are n different variables to be assigned, and the single parameter is n. The problem has solutions for each $n \geq 4$.

Apart from this "basic" formulation, we also consider some variations of the queens problem, with the goal of adding more constraints and thus decreasing the number of solutions:

- queensTopLeftAngle: basic + "top left angle of the board must be occupied",
- queensCenter: basic + "center of the board must be occupied", where "center" is defined as the central square, if n is odd, and one out of the four central squares, otherwise.

We take n as the single parameter also for the two variations.

Given a single-parameter problem, we may be interested in knowing how many variables must be fixed before having the possibility of incurring into a CSP-PPS (cf. Definition 5), or, in other words, in finding the cardinality of a CSP-McPPS (cf. Definition 6). Of course, this depends on the number of queens/variables. As an example, how many queens must be placed in the 8-queens problem (a CSP with partial solution) so that there is at least one configuration which has a unique completion? The next definition captures this notion.

Definition 9 ($\sigma_\pi(n)$). *Let $\pi(n)$ be a problem with single parameter n. $\sigma_\pi(n) = m$ if m is the smallest integer such that there is a set of givens of $\pi(n)$ of cardinality m which is proper.*

Since no placement of a single queen on an 8×8 board can be completed in a unique way, $\sigma_{queens}(8) > 1$. Moreover, since there is a placement of just two queens which can be completed in a unique way (cf. Figure 2) $\sigma_{queens}(8) \leq 2$, hence $\sigma_{queens}(8) = 2$.

Some general properties of the $\sigma_\pi(n)$ function is that it is undefined when π has no solution, that equals 1 when π has one solution, and that it decreases if more constraints are added to π. Monotonicity with respect to n is not guaranteed.

The function $\sigma_\pi(n)$ characterizes the notion of the *possibility* of a proper partial solution. For capturing the notion of the *necessity* of a proper partial solution we need the next definition.

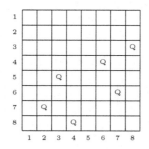

Fig. 3. A placement of six queens on an $8{\times}8$ board which can be completed in two different ways

Definition 10 ($\Sigma_\pi(n)$). *Let $\pi(n)$ be a problem with single parameter n. $\Sigma_\pi(n) = M$ if M is the smallest integer such that all sets of givens of $\pi(n)$ of cardinality M (or higher) are either unfeasible or proper.*

Since there are legal placements of six distinct queens on an $8{\times}8$ board which can be completed in two different ways (cf. Figure 3), $\Sigma_{queens}(8) > 6$.

Moreover, for each n it holds that $\Sigma_{queens}(n) \leq n - 1$ since placing legally n queens uniquely determines the last one. As a consequence, $\Sigma_{queens}(8) = 7$.

Some general properties of the $\Sigma_\pi(n)$ function is that it equals 0 when π has no solution, that it is greater than or equal to $\sigma_\pi(n)$, and that it decreases if more constraints are added to π. Monotonicity with respect to n is not guaranteed.

It is interesting to characterize the functions $\sigma_\pi(n)$ and $\Sigma_\pi(n)$ by means of logical formulae. Let's start with the latter, which is simpler. Since we deal with CSP's in the complexity class NP, by Fagin's theorem [6] we assume that problem π can be characterized by means of a formula in Existential Second Order logic (ESO) such as the following:

$$\exists\, S_n \ \pi(n, S_n), \tag{1}$$

where S_n is a *guessed* predicate representing the search space for the instance with n variables, and $\pi(n, S_n)$ is a function-free first-order formula expressing the constraints of the problem with n variables. As an example, the queens problem can be represented by means of the following ESO formula:

$$\exists\ queens/2 \in [1..n, 1..n] \tag{2}$$
$$\forall\, r \in 1..n\ \exists\, c \in 1..n\ queens(r,c) \tag{3}$$
$$\forall\, c \in 1..n\ \exists\, r \in 1..n\ queens(r,c) \tag{4}$$
$$\forall\, r, c_1, c_2 \in 1..n\ queens(r, c_1) \wedge queens(r, c_2) \rightarrow c_1 = c_2 \tag{5}$$
$$\forall\, r_1, r_2, c_1, c_2 \in 1..n\ r_1 < r_2 \wedge c_1 < c_2 \wedge \tag{6}$$
$$(queens(r_1, c_1) \wedge queens(r_2, c_2)) \vee (queens(r_1, c_2) \wedge queens(r_2, c_1))) \rightarrow$$
$$(abs(r_1 - r_2) \neq abs(c_1 - c_2))$$

The guessed predicate in (2) is forced to be a total, surjective and monodrome relation, i.e., a permutation, by means of constraints (3), (4), and (5), respectively. Diagonal attacks are forbidden by constraint (6).

Table 1. Values of $\sigma(n)$ and $\Sigma(n)$ for the queens problem and variations ('-' means that the problem has no solutions)

| | queens | | queensTopLeftAngle | | queensCenter | |
n	$\sigma(n)$	$\Sigma(n)$	$\sigma(n)$	$\Sigma(n)$	$\sigma(n)$	$\Sigma(n)$
4	1	1	-	0	-	0
5	2	2	1	2	1	2
6	1	1	-	0	-	0
7	2	6	1	2	2	6
8	2	7	1	4	2	6
9	2	8	2	7	2	8
10	2	9	2	9	2	9
11	3	10	2	10	2	10
12	3	11	2	11	2	11

The relation $\Sigma_\pi(n) > k$, where k is a non-negative integer less than or equal to n, can also be easily expressed by means of an ESO formula as follows:

$$\exists\, S_n^1\, S_n^2\, givens_n \tag{7}$$
$$|givens_n| = k \,\wedge \tag{8}$$
$$givens_n \subseteq S_n^1 \wedge givens_n \subseteq S_n^2 \,\wedge \tag{9}$$
$$\pi(n, S_n^1) \wedge \pi(n, S_n^2) \,\wedge \tag{10}$$
$$S_n^1 \neq S_n^2. \tag{11}$$

In fact $\Sigma_\pi(n) > k$ holds if and only if there are *witness* predicates (cf. (7)), i.e., two copies (S_n^1 and S_n^2) of the guessed predicates for (1) and a set of givens ($givens_n$) such that: $givens_n$ has the appropriate size (8), S_n^1 and S_n^2 agree with $givens_n$ (9), S_n^1 and S_n^2 are solutions (10), and S_n^1 and S_n^2 are different (11).

Since $\Sigma_\pi(n) > k$ can be expressed by means of an ESO formula, its valid-ity can be computed by means of an NP computation. As a consequence, the effective value of $\Sigma_\pi(n)$ can be computed by less than or equal to n NP com-putations. Table 1 reports the value of $\Sigma_\pi(n)$ for the queens problem and the above mentioned variations. The values have been obtained experimentally by 1) encoding the problem (7-11), customized by constraints similar to (2-6), in the NP-SPEC language, 2) a translation into SAT by means of the SPEC2SAT system [3], and 3) a solution of the generated SAT instance by means of the ZCHAFF SAT solver [10].

The results show that $\Sigma(n)$ seem to "converge" to $n - 1$, and that this hap-pens faster when the problem has fewer constraints, i.e., for the basic version. Unfortunately we have no proof that this happens for each n. We note that $\Sigma_{queens}(n) \leq n - 1$ trivially holds, since placing $n - 1$ queens forces the posi-tion for the last one. The fact that $\Sigma_{queens}(n)$ actually equals $n - 1$ could be interpreted by saying that there are so many solutions that it is always nec-essary to specify a partial solution almost completely to be sure that there is only one completion. This property seems to hold even for more constrained and less symmetric versions of the problem, i.e., "queensTopLeftAngle". Note that $\Sigma_{queens}(n) < n - 1$ is proven by finding two boards with $n - 2$ queens placed and two distinct completions. More knowledge on this problem could be obtained by analyzing the structure of such boards, and also by studying other variants of the problem, e.g., "queensTopAngle" (one of the top angles of the board must be occupied), or "queensAngle" (one of the angles of the board must be occupied).

For expressing the relation $\sigma_\pi(n) > k$ we need a more complicated formula, which is actually not ESO. The formula is as follows:

$$\neg\forall\ givens_n\ \exists\ S_n^1\ S_n^2 \tag{12}$$

$$|givens_n| = k \land \tag{13}$$

$$givens_n \subseteq S_n^1 \land givens_n \subseteq S_n^2 \land \tag{14}$$

$$\pi(n, S_n^1) \land \pi(n, S_n^2) \land \tag{15}$$

$$S_n^1 \neq S_n^2. \tag{16}$$

In fact $\sigma_\pi(n) > k$ holds if and only if it is not true that for each set of givens ($givens_n$) of the appropriate size (13) there are witness predicates (cf. (12)), i.e., two copies (S_n^1 and S_n^2) of the guessed predicates for (1) such that: S_n^1 and S_n^2 agree with $givens_n$ (14), S_n^1 and S_n^2 are solutions (15), and S_n^1 and S_n^2 are different (16).

Note that formula (12-16) differs from (7-11) on the type of the quantifiers for the predicates. Therefore it is possible to express the problem $\sigma_\pi(n) \leq k$ with an EUSO (Existential Universal Second Order logic) formula. EUSO is a subclass of second-order logic which characterizes the complexity class Σ_2^p [11], i.e., a class at the second level of the Polynomial Hierarchy which is widely conjectured to contain NP properly (cf. also Theorem 3).

Since the problem of determining whether $\sigma_\pi(n) \leq k$ or not is in Σ_2^p, it is impossible to solve it by means of a single call to a SAT solver. We computed the value of $\sigma_\pi(n)$ for the queens problem and the previously mentioned variants by means of a program written in OPL [12] (a constraint modelling and programming system by Ilog, www.ilog.com) and OPLSCRIPT (a script language with a C++-like syntax which can invoke OPL). In particular, a procedure in OPLSCRIPT asks OPL to solve the problem guessing k givens, and then to find a different solution with the same givens.

Table 1 reports results of experiments also on $\sigma_\pi(n)$ for the queens problem and variations. $\sigma_\pi(n)$ is much smaller than $\Sigma_\pi(n)$ and, as expected, gets smaller when more constraints are added.

As mentioned in the Introduction, it is known that $\sigma_{Sudoku}(81) \leq 17$, and it is currently unknown whether $\sigma_{Sudoku}(81) \leq 16$ or not. With techniques similar to those described before for the queens problem, we have proven that $\sigma_{Sudoku}(16) = 4$, i.e., that in the "4 × 4" Sudoku no set of 3 givens with unique completion exists, and that there are sets of 4 givens with unique completion (cf., e.g., Figure 1(d)).

We plan to enhance the experimental analysis by studying other single-para-meter problems. Some popular such problems are listed as follows, and some of them are taken from the CSPLib library of CSP problems [7], www.csplib.org.

- Golomb ruler [CSPLib prob006]: A *Golomb ruler* may be defined as a set of m integers $0 = a_1 < a_2 < ... < a_m$ such that the $m(m-1)/2$ differences $a_j - a_i, 1 \leq i < j \leq m$ are distinct. Such a ruler is said to contain m marks and is of length a_m. The objective is to find optimal (minimum length, which are known for $m \leq 20$) rulers. The single parameter is m.

Apart from this "basic" formulation, we may also consider some variations, with the goal of relaxing constraints and increasing the number of solutions:

- ruler's size = minimum + number of marks;
- ruler's size = 2 × minimum.

- Latin square [L. Euler, 1783]: a *Latin square of order* n is an $n \times n$ matrix of n symbols in which each symbol occurs exactly once in each row and each column of the matrix. The single parameter is n^2, which equals the number of variables.
- Sudoku [en.wikipedia.org/wiki/Sudoku]: cf. Section 1. The single parameter is n^4, which equals the number of variables.
- Ramsey problem [CSPLib prob017]: colour the edges of a complete graph with n nodes using at most k colours, in such a way that there is no monochromatic triangle in the graph, i.e., in any triangle at most two edges have the same colour. Since with 3 colours the problem has a solution if $n < 17$, we fix k to 3, and the single parameter is $n \cdot (n-1)/2$, which equals the number of edges, i.e., the number of variables.
- Maximum density still life [CSPLib prob032]: this problem arises from the Game of Life, which is played on a squared board, considered to extend to infinity in all directions; the configuration of live and dead cells at time t leads to a new configuration at time $t+1$ according to the rules of the game; a stable pattern, or *still-life*, is not changed by these rules. What is the densest possible still-life pattern, i.e., the pattern with the largest number of live cells, that can be fitted into an $n \times n$ section of the board, with all the rest of the board dead? The single parameter is n^2.

5   Conclusions, Current and Future Work

In this paper we have presented a preliminary analysis of many combinatorial problems that are required to have a unique solution. We have defined a number of computational problems, discussed their complexity, presented a logical formulation and hinted at a reduction in the language of QBF's.

In the future we plan to continue both the complexity and the experimental analysis. As for the complexity analysis, it would be nice to obtain tighter results, by, e.g., finding lower bounds. As for the experiments, we mentioned a number of single-parameter problems that could be considered. Also, the computation of $\sigma_\pi(n)$ could be done by means of more efficient methods, e.g., by translating the problem into the evaluation of a QBF, for which several efficient solvers, cf., e.g., [4], exist.

Finally, the experiments showed that some patterns for the $\sigma_\pi(n)$ and $\Sigma_\pi(n)$ seem to emerge for various problems, and it would be worthwhile to confirm those patterns in general.

Acknowledgements. The authors are grateful to the anonymous reviewers for their useful comments.

References

1. Paul Beame, Stephen Cook, Jeff Edmonds, Russell Impagliazzo, and Toniann Pitassi. The relative complexity of NP search problems. *Journal of Computer and System Sciences*, 57(1):3–19, 1998.
2. A. Blass and Y. Gurevich. On the unique satisfiability problem. *Information and Computation*, 55(1–3):80–88, 1982.
3. Marco Cadoli and Andrea Schaerf. Compiling problem specifications into SAT. *Artificial Intelligence*, 162:89–120, 2005.
4. Marco Cadoli, Marco Schaerf, Andrea Giovanardi, and Massimo Giovanardi. An algorithm to evaluate quantified boolean formulae and its experimental evaluation. *Journal of Automated Reasoning*, 28:101–142, 2002.
5. R. Dechter. *Constraint Networks (Survey)*, pages 276–285. John Wiley & Sons, Inc., 1992.
6. R. Fagin. Generalized First-Order Spectra and Polynomial-Time Recognizable Sets. In R. M. Karp, editor, *Complexity of Computation*, pages 43–74. AMS, 1974.
7. I.P. Gent and T. Walsh. Csplib: a benchmark library for constraints. Technical report, Technical report APES-09-1999, 1999. Available from http://csplib.cs.strath.ac.uk/. A shorter version appears in the Proceedings of the 5th International Conference on Principles and Practices of Constraint Programming (CP-99).
8. D. S. Johnson. A catalog of complexity classes. In J. van Leeuwen, editor, *Handbook of Theoretical Computer Science*, volume A, chapter 2, pages 67–161. Elsevier Science Publishers (North-Holland), Amsterdam, 1990.
9. N. Megiddo and C. H. Papadimitriou. On total functions, existence theorems and computational complexity. *Theoretical Computer Science*, 81:317–324, 1991.
10. M. W. Moskewicz, C. F. Madigan, Y. Zhao, L. Zhang, and S. Malik. Chaff: Engineering an efficient SAT solver. In *Proceedings of the Thirtyeighth Conference on Design Automation (DAC 2001)*, pages 530–535. ACM Press, 2001.
11. L. J. Stockmeyer. The polynomial-time hierarchy. *Theoretical Computer Science*, 3:1–22, 1976.
12. Pascal Van Hentenryck. *The OPL Optimization Programming Language*. The MIT Press, 1999.
13. Takayuki Yato and Takahiro Seta. Complexity and completeness of finding another solution and its application to puzzles, 2002. Available at http://www.phil.uu.nl/ oostrom/cki20/02-03/japansepuzzles/ASP.pdf.

A Computerized Referee*

Eugenio G. Omodeo[1], Domenico Cantone[2], Alberto Policriti[3],
and Jacob T. Schwartz[4]

[1] University of Catania, Dipartimento di Matematica e Informatica
cantone@dmi.unict.it
[2] University of Trieste, Dipartimento di Matematica e Informatica
eomodeo@units.it
[3] University of Udine, Dipartimento di Matematica e Informatica
policriti@dimi.uniud.it
[4] New York University, Department of Computer Science, Courant Institute of
Mathematical Sciences
schwartz@cs.nyu.edu

Abstract. The Referee system (aka AetnaNova), accessible on the Web,
ingests bodies of text which it either certifies as constituting a valid
sequence of definitions and theorems, or rejects as defective.

The functionality of this proof verifier and the key issues for its effective use are illustrated, in particular by a case-study referring to bisimulations, and through excerpts from a large-scale script which leads from the
built-in rudiments of set theory to the formal foundations of mathematical analysis. (The latter scenario, although incomplete as yet, already
comprises over 1000 verified proofs, definitions, and 'theories'.)

The paper also discusses enhancements to Referee which are in progress: a new inference mechanism, named proof-by-structure, whose addition should make proofs lighter and more readable; an interface to external provers; and an automatic proof optimizer (currently being tested),
aimed at speeding up proof verification.

Keywords: Automated proof verification, Set theory.

$$\therefore \ \textit{To Gigina} \ \because$$

Introduction

The computer scientist's primary interest in automated tools for proof checking
arises from evidence that the logical armory necessary for checking mathematical
proofs can also be used to check the compliance of software systems with their
specifications.

This paper outlines a computerized system for verifying script-files composed
of formalized mathematical proofs. This system, named Referee or Ref for short,
entered its test and experimentation phase several years ago. The design of this

---

* Research partially funded by INTAS project *Algebraic and deduction methods in
non-classical logic and their applications to Computer Science.*

O. Stock and M. Schaerf (Eds.): Aiello Festschrift, LNAI 4155, pp. 117–139, 2006.

system draws some of its ideas from a decades-long tradition (cf. [1]) of systems of this kind. Ref uses, e.g., a natural deduction style, a kind of semantic attachment which we call *proof by computation*, and modularization constructs aimed at proof reuse (cf. [18]). On the other hand, some of its features, in particular its use of a version of the Zermelo-Fraenkel set theory rather than of a fragment of second-order predicate logic, were inspired by the work of logicians on the foundations of mathematics. For this reason, the primary benchmark which we are using for Ref is the formal reconstruction of a result of mathematical analysis— the Cauchy Integral Theorem on analytic functions—whose scenario has already reached a considerable size. We have developed the material necessary to reach a proof of such a theorem from the bare rudiments of set theory. Up to now we have checked about one half of its overall bulk and it is clear that Ref is an attractive basis for checking broad areas of mathematical analysis. We have also begun testing its use for discrete mathematics, which is an essential preamble to algorithmics.[1]

We begin by describing the key ingredients of our system in broad-brush terms, addressed more to potential users than to scholars in the automated deduction field. Then we outline a Ref scenario (much less extensively developed than the one on the Cauchy Integral Theorem), aimed at verifying the theorems on simulations and bisimulations which lie behind any efficient stable partitioning algorithm (cf. [15,14]). An appendix provides a tableau-fashioned account of a decision algorithm, **MLSS**, which plays a key role in Ref's architecture.

1   The **Referee** Proof Verifier

The Ref verifier is fed script files, called *scenarios*, consisting of successive definitions, theorems, and auxiliary commands, which Ref either certifies as constituting a valid sequence or rejects as defective. In the case of rejection, the verifier attempts to pinpoint the troublesome locations within a scenario, so that errors can be located and repaired. Step timings are produced for all correct proofs, to help the user in spotting places where appropriate modifications could speed up proof processing.

The bulk of the text normally submitted to the verifier consists of theorems and proofs. Some theorems (and their proofs) are enclosed within so-called *theories*, whose external conclusions these internal theorems serve to justify. This lets scenarios be subdivided into modules, which increases readability and supports proof reuse (cf. [18]). The following example, which appears early in Ref's main proof scenario, illustrates the syntactic form of Ref proofs:

-- Next we prove a first basic property of ordinals.

---

[1] The URL http://www.settheory.com/Setl2/Ref_user_manual.html gives access to the Ref user's manual. Among others, this document explains how to register as a user of the Ref system; and it also provides a link to the scenario of the Cauchy Integral Theorem presented in keyboard form.

THEOREM 11: [Members of ordinals are ordinals] $\mathcal{O}(S)$ & $T \in S \to \mathcal{O}(T)$.

PROOF: Suppose_not(s, t) $\implies$ AUTO

> -- We proceed by contradiction. If our theorem is false, there is an ordinal s having a member t which is not an ordinal.

Use_def($\mathcal{O}$) $\implies$ $Stat1:$ $\neg(\langle\forall x \in t \,|\, x \subseteq t\rangle$ &
$\qquad\qquad\qquad\qquad\langle\forall x \in t, y \in t \,|\, x \in y \lor y \in x \lor x = y\rangle)$

> -- Hence, by definition of ordinal, t must either have a member a not included in t, or a pair b, c of distinct members not related by membership.

$\langle a, b, c\rangle \hookrightarrow Stat1 \implies$ AUTO

> -- But since s is an ordinal, it must include its member t, so that the second case is impossible.

Use_def($\mathcal{O}$) $\implies$ $Stat2:$ $\langle\forall x \in s \,|\, x \subseteq s\rangle$ &
$\qquad\qquad\qquad Stat3:$ $\langle\forall x \in s, y \in s \,|\, x \in y \lor y \in x \lor x = y\rangle$
$\langle t\rangle \hookrightarrow Stat2 \implies$ AUTO
Suppose $\implies$ $b, c \in t$ & $\neg(b \in c \lor c \in b \lor b = c)$
$\langle b, c\rangle \hookrightarrow Stat3 \implies$ AUTO
Discharge $\implies$ $Stat4:$ $a \not\subseteq t$ & $a \in t$

> -- Thus we need only consider the first case, in which a is a member but not a subset of t. In this case there plainly exists a d in a but not in t. Plainly a is a member of s, and thus a subset of s; so d is also a member of s.

$\langle d\rangle \hookrightarrow Stat4 \implies$ $d \in a$ & $d \notin t$
$\langle a\rangle \hookrightarrow Stat2 \implies$ $a \subseteq s$
ELEM $\implies$ $d \in s$

> -- By definition of ordinal, it follows that d either equals t, is a member of t, or that t is a member of d. But all three of these cases are impossible, since any would imply the existence of a membership cycle. This contradiction proves our theorem.

$\langle d, t\rangle \hookrightarrow Stat3 \implies$ $d \in t \lor t \in d \lor t = d$
$\langle Stat4\rangle$ Discharge $\implies$ QED

As this example illustrates, a theorem's proof consists of a sequence of *statements* (also called *inference steps*), each of which consists of a *hint* portion (e.g.: Use_def($\mathcal{O}$), $\langle a, b, c\rangle \hookrightarrow$Stat1, Discharge, ELEM) separated by the sign $\implies$ from the *assertion* of the statement. Each assertion must be a syntactically well-formed formula in Ref's set-theoretic language; each hint must reference one of the basic inference mechanisms that Ref provides, and may also supply this inference mechanism with auxiliary parameters (e.g.: Use_def($\mathcal{O}$), Suppose_not(s, t)), including the context of preceding statements in which it should operate (e.g., $\langle Stat4\rangle$ Discharge draws a contradiction from the conjunction of all assertions following the label $Stat4$). When no ambiguity or obscurity ensues from this, an assertion can be represented laconically by the keyword AUTO. Thus, in the above proof: when AUTO occurs in the initial Suppose_not–statement, it obviously stands for the assertion $\mathcal{O}(s)$ & $t \in s$ & $\neg\mathcal{O}(t)$, contrary to the sought conclusion; when it occurs in the $\langle a, b, c \rangle \hookrightarrow$Stat1–statement, it stands for the formula

$$(a \in t \,\&\, a \not\subseteq t) \vee (b, c \in t \,\&\, \neg(b \in c \vee c \in b \vee b = c)),$$

because this is what results from the assertion bearing the label *Stat1* when its bound occurrences of variables get replaced by the new constants a, b, c; dually, in the $\langle t \rangle \hookrightarrow$Stat2– and in the $\langle b, c \rangle \hookrightarrow$Stat3–statement AUTO stands for $t \in s \rightarrow t \subseteq s$ and for $b, c \in s \rightarrow (b \in c \vee c \in b \vee b = c)$, respectively.

Fifteen inference mechanisms currently constitute the inferential armory of Ref. In the five sections which follow, after outlining the inference mechanisms give our verifier most of its special flavor, we explain its notion of *inference step context* and describe how contexts can be restricted by means of statement labels.

2   The ELEM Primitive and 'Blobbing'

Among Ref's inference primitives, ELEM is the most central (its use being, often, tacitly combined with other forms of inference). ELEM implements *multilevel syllogistic* [11,12], a decision algorithm which determines whether a given unquantified set-theoretic formula involving individual variables (which designate sets) and a restricted collection of set operators is satisfiable. (A tableau-fashioned account of a decision procedure for this fragment of set theory is given in Sec. 8). Using the ELEM algorithm, the Ref verifier can identify many cases in which a conjunction constructed by negating one statement of a proof and conjoining a selection of earlier statements is unsatisfiable, so that the statement follows from the preceding context. When not all the constructs appearing in this context (e.g. quantifiers and setformers) are part of Ref's built-in syllogistic, a preprocessing step, called *blobbing*, replaces all parts of the current context whose principal operators are not recognized by the decision algorithm by 'blobs', i.e. by new variables designating either sets (when they occur as terms) or propositions (when they occur as subformulae). This blobbing operation replaces syntactically identical (or recognizably equal) parts of a conjunction by the same variable. It is also able to treat as equal well-formed parts which only differ by the renaming of bound variables in quantifiers or setformers, and also treats existential quantifiers as negated universal quantifiers.

The primary function of blobbing is to reduce all the constructs that appear in proof statements submitted to ELEM to the ones which multilevel syllogistic can handle. Blobbing is also used to introduce other simplifications which extend the power of ELEM beyond that of simple multilevel syllogistic and improve system performance.

Blobbing consists of three subphases: (1) *pre-blobbing*, which makes reductions such as the reduction of any part of the form $\mathcal{F}in(\#X)$ to $\mathcal{F}in(X)$ (justified by

the remark that the cardinality of a set X is finite if and only if X is finite);
(2) *blobbing proper*, during which subterms whose lead constructs are not known
to the multilevel syllogistic algorithm are replaced by set names and quantified
subformulae are replaced by propositional variables; (3) *post-blobbing*, which
drops parts of a purported contradiction when it is clear that they can play no
role in establishing its contradictory nature.

In some cases the verifier provides a few efficiency-oriented variants of the
ELEM deduction primitive. These are invoked by prefixing the keyword ELEM
with a parenthesized label (as we will see again in Section 6) which may include
various special characters. Including the character "*" just before the closing
parenthesis of the prefix suppresses the normal internal examination of special
functions like cons, car, and cdr (the ordered pair constructor $x, y \mapsto [x, y]$ and
its associated projections $p \mapsto p^{[1]}$, $p \mapsto p^{[2]}$, normally treated by the methods
discussed in [8]), i.e. it treats these as unknown functions whose occurrences
must be 'blobbed'. This treats statements like

$$[x, [y, z]] = [x_2, [y_2, z_2]] \,\&\, [x, [y, z]] = [x_3, [y_3, z_3]] \,\&\, [x, [y, z]] = [x_4, [y_4, z_4]]$$

as if they read

$$xyz = xyz_2 \,\&\, xyz = xyz_3 \,\&\, xyz = xyz_4 \,,$$

and so makes deduction of

$$[x_2, [y_2, z_2]] = [x_3, [y_3, z_3]]$$

from the conjunction shown above easy. Without modification of the ELEM primitive's operation this same deduction would require many seconds. This coarse
treatment is of course incapable of deducing the implication

$$[x, [y, z]] = [x_2, [y_2, z_2]] \rightarrow (x = x_2 \,\&\, y = y_2 \,\&\, z = z_2)$$

which it sees as

$$xyz = xyz_2 \rightarrow (x = x_2 \,\&\, y = y_2 \,\&\, z = z_2)\,.$$

3   The Suppose_not, QED, Suppose, Discharge Primitives

Suppose_not statements occur, exclusively and always, as the first inference step
in Ref proofs. They have the form

$$\mathsf{Suppose\_not}(c_1, \ldots, c_n) \Longrightarrow \cdots \,,$$

where $c_1, \ldots, c_n$ are distinct constants local to a proof, which correspond in
number and in positions to the distinct unquantified variables appearing in the
statement T of the corresponding theorem. Such theorem variables are in fact
understood to be universally quantified; and in a proof-by-contradiction the
constants c_i replace them during deduction of a contradiction. Accordingly, the

statement which follows $\Longrightarrow$ in the Suppose_not step must be logically equivalent to the negation of an instantiated version of T. At the end of the proof there must appear a statement of the form

$$\text{Discharge} \Longrightarrow \text{QED}$$

which matches the Suppose_not and indicates that a contradiction was derived by assuming the existence of a counterexample to T.

A Suppose statement has the form

$$\text{Suppose} \Longrightarrow B \cdots ,$$

where the formula B that follows $\Longrightarrow$ can involve no constants save those already available in the part of the proof preceding it (including globally defined constants, constants of the form c_Θ local to the current theory, constants generated within the proof by substitution of an existentially bound variable, and constants generated by application of a THEORY—cf. Sec. 5).

Every step C coming after a Suppose $\Longrightarrow B$ can exploit the temporary assumption B as part of its context, until the following Discharge statement which matches this Suppose statement and so eliminates this assumption, along with all the intermediate steps C which were derived from it.

As already said, Discharge statements always match Suppose and Suppose_not statements within a proof, in the same balanced way in which closed parentheses match open parentheses within an arithmetic expression.

To see how this inference primitive works, let us consider the following proof fragment:

$$C$$
$$\text{Suppose} \Longrightarrow B$$
$$D$$
$$\text{Discharge} \Longrightarrow A .$$

Here the Suppose and Discharge are taken to match each other, so that C represents the overall context available before the Suppose, D represents the context portion derived from the temporary assumption B, and A is the assertion which the Discharge yields. Ref will only regard this derivation as legitimate if it can find an intermediate formula D' implied by $C\&B\&D$ and such that A 'trivially' follows from the formula $C\&(B \to D')$.

4   Proof by Structure

Proof by structure uses a simple auxiliary language of *structure descriptors* to keep track of the top structural levels of sets appearing in scenario proofs. Any special set defined in a scenario, for example $\mathbb{N}$, the set of all integers, or $\mathbb{R}$, the set of all reals, can be used as a primary structure symbol in this language. This descriptor attaches to all members of the set, for instance any integer has the descriptor $\mathbb{N}$. A significant but less basic example is $\mathbb{Z}^+$, the set of all non-negative signed integers, which does not occur in our present scenarios but could easily be defined. Structure descriptors need not be confined to sets, but can also designate classes, like the class $\mathcal{O}$ of all ordinals, the class $\mathcal{F}in$ of all finite sets, the class $\mathcal{I}nf$ of all infinite sets, and the class $\mathcal{V}$ of all sets.

Given any symbols $S, S_1, S_2, \ldots$ representing structures, we can then form:

(i) $\{S\}$: describes a set all of whose elements have the descriptor S. For example, the set $\mathbb{N}$ has the descriptor $\{\mathbb{N}\}$; the set $\mathcal{P}(\mathbb{N})$ of all sets of integers has the descriptor $\{\{\mathbb{N}\}\}$.

(ii) $[S_1, S_2]$: describes a pair whose components have respectively the descriptors S_1 and S_2.

These constructions can be compounded. For example

(e.1) $\{[\mathbb{N}, \mathbb{N}]\}$ describes a set of integer pairs (and so applies to $\mathbb{Z}$, the set $\{[i,j] : i \in \mathbb{N}, j \in \mathbb{N} \mid i = 0 \vee j = 0\}$ of *signed* integers);

(e.2) $\{[\mathbb{N}, \mathcal{V}]\}$ describes a map from integers to elements of any kind, e.g. it describes any finite or infinite(ly denumerable) sequence.

A given set can have several descriptors. For example, a finite sequence of signed integers has the descriptors $\{[\mathbb{N}, \mathbb{Z}]\}$ and $\mathcal{F}in$. Since $\mathbb{Z}$ itself has the descriptor $\{[\mathbb{N}, \mathbb{N}]\}$, a sequence of signed integers also has the descriptor $\{[\mathbb{N}, [\mathbb{N}, \mathbb{N}]]\}$, which in any given situation we may wish either to use or ignore. Infinite sequences of signed integers have the descriptors $\{[\mathbb{N}, \mathbb{Z}]\}$ and $\mathcal{I}nf$. Real numbers in Cantor's representation are equivalence classes of such sequences (cf. Section 5), and accordingly have the descriptors $\{\{[\mathbb{N}, \mathbb{Z}]\}\}$ and $\{\mathcal{I}nf\}$.

A mechanism within the verifier should track the descriptors of variables and expressions appearing in proofs whenever possible. For example, a variable x known to satisfy a clause $x \in \mathbb{N}$ has the descriptor $\mathbb{N}$, a variable known to satisfy $x \in \mathbb{Z}$ has the descriptors $\mathbb{Z}$ and $[\mathbb{N}, \mathbb{N}]$.

Setformers and other basic constructors operate in a known way on the structure descriptors introduced above. Suppose, for example, that s is a set known to have some descriptor $\{D\}$, and that $e(x)$ is an expression having the free variable x which is known to map elements having the descriptor D into elements having the descriptor D'. Then

$$\{e(x) : x \in s \mid P\}$$

has the descriptor $\{D'\}$, while

$$\{[x, e(y)] : x \in s, \, y \in s \mid P\}$$

has the descriptor $\{[D, D']\}$.

When a set s is known to have a descriptor $\{D\}$, any element x for which $x \in s$ has been proved is known to have the descriptor D. If D is a primitive descriptor representing a known set, this will give us the assertion $x \in D$, for example $x \in \mathbb{N}$, which may be needed as an auxiliary hypothesis for the application of some theorem. Similarly any set s having the descriptor $\{[\mathbb{N}, \mathbb{N}]\}$ is known to satisfy $Is\_map(s)$, and also

$$\langle \forall x \in s \mid x^{[1]} \in \mathbb{N} \,\&\, x^{[2]} \in \mathbb{N} \rangle.$$

Conclusions of this kind can often result automatically. This is a principal use for our system of structure descriptors.

Many other basic set-theoretic operations have known effects on descriptors. These often follow from the definitions of the operators in question. For example:

(1) If s has the descriptor $\{D\}$, then so does every one of its subsets, and $\mathcal{P}(s)$ has the descriptor $\{\{D\}\}$.

(2) If s has the descriptor $\{\{D\}\}$, then $\bigcup s$ has the descriptor $\{D\}$. Note hat this follows automatically from the definition $\{x : y \in s, x \in y\}$ of $\bigcup s$, since the bound variable y in the iterator has the descriptor $\{D\}$, so each of the x has the descriptor D, and the set as a whole has the descriptor $\{D\}$.

(3) If s_1 and s_2 both have a descriptor $\{D\}$, then so does $s_1 \cup s_2$.

(4) If s_1 and s_2 both have the descriptor $\mathcal{F}in$, then so does $s_1 \cup s_2$.

(5) If s_1 and s_2 have descriptors $\{D_1\}$ and $\{D_2\}$ respectively, then $s_1 \cap s_2$ has both descriptors $\{D_1\}$ and $\{D_2\}$. Even if s_2 has no descriptor, $s_1 \cap s_2$ and $s_1 \setminus s_2$ have the descriptor $\{D_1\}$, as does any set s for which an assertion $s \subseteq s_1$ has been proved.

(6) If s_1 has the descriptor $\mathcal{F}in$, so do $s_1 \cap s_2$ and $s_1 \setminus s_2$, as does any set s for which an assertion $s \subseteq s_1$ has been proved.

(7) If s_1 and s_2 have the descriptor $\mathcal{F}in$, so does any setformer $\{e : x \in s_1, y \in s_2 \mid P\}$, or any setformer $\{e : x \in s_1 \mid P\}$.

(8) $\#s$ always has the descriptor Card. Since the class of cardinals has the descriptor $\{\mathcal{O}\}$, $\#s$ also has the descriptor $\mathcal{O}$, as does any x known to be a cardinal. If s has the descriptor $\mathcal{F}in$, then $\#s$ has the descriptor $\mathbb{N}$. Since $\mathbb{N}$ itself has the descriptor $\{\mathcal{F}in\}$, $\#s$ also has the descriptor $\mathcal{F}in$.

(9) If s has the descriptor $\{D\}$, then any setformer like $\{x : x \subseteq s \mid P\}$ is known to have the descriptor $\{\{D\}\}$; this result obviously generalizes.

(10) If sets s and t have the descriptors $\{D\}$ and $\{D'\}$ respectively, then their Cartesian product $s \times t$ has the descriptor $\{[D, D']\}$. If s and t both have the descriptor $\mathcal{F}in$, so does $s \times t$.

(11) If s and t have descriptors $\{[D, D']\}$ and $\{[D', D'']\}$ respectively, then $t \circ s$ has the descriptor $\{[D, D'']\}$. If s and t both have the descriptor $\mathcal{F}in$, so does $t \circ s$.

(12) If s and t have descriptors D, D' respectively, then $[s, t]$ has the descriptor $[D, D']$. If u has the descriptor $[D, D']$, then $u^{[1]}$ has the descriptor D and $u^{[2]}$ has the descriptor D'.

(13) If F has the descriptor $\{[D, D']\}$, then its inverse $F^{\leftarrow}$ has the descriptor $\{[D', D]\}$, and any of its domain restrictions $F|_s$ has the descriptor $\{[D, D']\}$. If F has the descriptor $\mathcal{F}in$, then $F^{\leftarrow}$ and $F|_s$ both have the descriptor $\mathcal{F}in$ also.

There may be useful extensions of these ideas to single-valued and one-one maps; also to topological situations, spaces of continuous functions, etc. Note that some of the conclusions derived manually in the present scenarios can result automatically by use of structure descriptors. For example, the cardinal sum of s_1 with s_2 is defined as

$$\#(\{[x, 0] : x \in s_1\} \cup \{[x, 1] : x \in s_2\}),$$

making it obvious that the sum of two integers is an integer. Similarly, the definition of cardinal product, namely

$$\#\{[x, y] : x \in s_1, y \in s_2\}$$

makes it apparent that the product of two integers is an integer. Since the difference of integers is defined by $\#(n \setminus m)$, it also follows immediately that the difference of integers is an integer.

Ordinals also have the descriptor $\{\mathcal{O}\}$, since any element of an ordinal is an ordinal. Any $\bigcup s$ of a set having the descriptor $\{\mathcal{O}\}$ has the descriptor $\mathcal{O}$. It may be worth carrying the set $\text{next}(\mathbb{N})$ as an additional descriptor. If this is done, $\bigcup s$ will be known to have the descriptor $\text{next}(\mathbb{N})$ if s has the descriptor $\{\mathbb{N}\}$, and so to have the descriptor $\mathbb{N}$ (i.e. to be an integer) if there is another s' having the descriptor $\text{next}(\mathbb{N})$ for which a statement $s \in s'$ is available.

In many cases a definition or theorem appearing in a scenario will characterize the action on structure descriptors of one or more of the function symbols appearing in it. The examples given just above illustrate this. Such facts combined with the other rules given above extend the verifier's ability to tack the structures of objects appearing in proofs. For example, if s, t, and u are sets known to have the descriptor $\{\mathbb{N}\}$, then

$$\{(x * y) + z : x \in s, y \in t, z \in u \mid P\}$$

is also known to have the descriptor $\{\mathbb{N}\}$.

The theory of summation yields the fact that $\sum f$ has the descriptor D if f has the descriptors $\{[d, D]\}$ and $\mathcal{F}in$, and if the $\oplus$ operator appearing in the summation can be shown to map pairs of objects having the descriptor D into objects having this same descriptor. Thus, for example, the sum or product of any setformer like

$$\{[[x, y, z], (x * y) + z] : x \in s, y \in t, z \in u \mid P\}$$

is also known to be an integer if s, t, and u are sets known to have the descriptors $\{\mathbb{N}\}$ and $\mathcal{F}in$.

The structure definition mechanism explained above carries over in a useful way to recursively defined functions (in our set-theoretic context, these can be functions defined by transfinite induction). To show why such extension is possible, we first need to note that the system of descriptors extends readily to function symbols, since these are very close semantically to sets of pairs. For example, the descriptor $\{[D_1, D_2]\}$ can be ascribed to any one-parameter function symbol which maps each object having the descriptor D_1 into an object having the descriptor D_2. Similarly, the descriptor $\{[[D_1, D_2], D_3]\}$ can be ascribed to any two-parameter function symbol which yields an object having the descriptor D_3 whenever its two parameters have the respective descriptors D_1 and D_2. (For example, the integer addition operator $+$ has the descriptor $\{[[\mathbb{N}, \mathbb{N}], \mathbb{N}]\}$, but also the descriptors $\{[[\mathcal{V}, \mathcal{V}], \mathcal{O}]\}$ since it always produces an ordinal, and the descriptor $\{[[\mathcal{F}in, \mathcal{F}in], \mathbb{N}]\}$, since it produces an integer for any two finite inputs.) In the three-parameter case, $\left\{\left[[[D_1, D_2], D_3], D_4\right]\right\}$ can be ascribed to any three-parameter function symbol which yields an object having the descriptor D_4 whenever its three parameters have the respective descriptors D_1, D_2, and D_3.

Using these descriptors, we can state the rule for function application as follows: If a one-parameter function symbol f has the descriptor $\{[D_1, D_2]\}$, and x has the descriptor D_1, then $f(x)$ has the descriptor D_2. Similarly, if a two-parameter function symbol f has the descriptor $\{[[D_1, D_2], D_3]\}$, and its two arguments x_1, x_2 have the descriptors D_1, D_2, then $f(x_1, x_2)$ has the descriptor D_3. We leave it to the reader to formulate the rules for more than two arguments.

Function compounding acts in an obvious way on descriptors, for example if f has the descriptor $\{[D_1, D_2]\}$ and g has the descriptor $\{[D_2, D_3]\}$, then $g(f(\cdot))$ has the descriptor $\{[D_1, D_3]\}$. Rules like this make it obvious why

$$\#(\{[x, 0] : x \in s_1\} \cup \{[x, 1] : x \in s_2\}),$$

yields an integer for every pair of integer arguments: the functional expression $\{[x, 0] : x \in s_1\}$ has the descriptor $\{[\mathcal{F}in, \mathcal{F}in]\}$ simply because it is a setformer with s_1 as its only free variable, and likewise for $\{[x, 1] : x \in s_2\}$. Since the union operator $\cup$ has the descriptor $\{[[\mathcal{F}in, \mathcal{F}in], \mathcal{F}in]\}$, it follows immediately that $\{[x, 0] : x \in s_1\} \cup \{[x, 1] : x \in s_2\}$ has the descriptor $\{[[\mathcal{F}in, \mathcal{F}in], \mathcal{F}in]\}$ also. Since $\#$ has the descriptors $\{[\mathcal{F}in, \mathcal{F}in]\}$ and $\{[\mathcal{V}, \mathcal{O}]\}$, $\#(\{[x, 0] : x \in s_1\} \cup \{[x, 1] : x \in s_2\})$ has the descriptors $\{[[\mathbb{N}, \mathbb{N}], \mathcal{F}in]\}$ and $\{[[\mathbb{N}, \mathbb{N}], \mathcal{O}]\}$, and therefore $\{[[\mathbb{N}, \mathbb{N}], \mathbb{N}]\}$. Much the same argument applies to the integer product.

Next consider a transfinite recursive definition of one of the general types we allow, namely

$$f(s, t) =_{\text{Def}} d\left(\left\{g\Big(f(x, h(s, t)), s, t\Big) : x \in s \mid P\Big(x, f(x, h(s, t)), s, t\Big)\right\}, s, t\right),$$

where we assume that the functions d, g, and h have been defined prior to the occurrence of the recursive definition shown. In working with this definition we will want to establish that f has some descriptor $\{[[D_1, D_2], D_3]\}$, i.e. that it yields an element having descriptor D_3 for any input arguments with descriptors D_1, D_2 respectively.

This conclusion will be valid under the following circumstances: we need to know that the null set has descriptor D_1, that one can ascribe the descriptor $\{D_1\}$ to any set which has the descriptor D_1, and that there exists a descriptor D' such that

(a) h has the descriptor $\left\{\left[[D_1, D_2], D_2\right]\right\}$;

(b) g has the descriptor $\left\{\left[[D_3, D_1], D_2\right], D'\right]\right\}$;

(c) d has the descriptor $\left\{\left[[[\{D'\}, D_1], D_2], D_3\right]\right\}$.

Then in the ground case of the transfinite recursive definition $f(\emptyset, t)$ has the value $d(\emptyset, s, t)$, and so must produce an element with the descriptor D_3. In the remaining case it follows inductively (given that s and t have the respective descriptors D_1, D_2) that $f(x, h(s, t))$ has the descriptor D_3 for every $x \in s$, so that $g(f(x, h(s, t)), s, t)$ has the descriptor D', and so

$$(*) \qquad \left\{ g\Big(f(x, h(s, t)), s, t \Big) \ : \ x \in s \ \middle| \ P\Big(x, f(x, h(s, t)), s, t \Big) \right\}$$

has the descriptor $\{D'\}$. Therefore the right side of the recursive definition seen above has the descriptor D_3, and it follows inductively that f has the descriptor $\left\{\left[[D_1, D_2], D_3\right]\right\}$.

If s has the descriptor $\mathcal{F}in$, then the set $(*)$ will have this descriptor also, and so if d has the descriptor $\left\{\left[[\mathcal{F}in, \mathcal{F}in], D_2\right], \mathcal{F}in\right]\right\}$, f will have the descriptor $\left\{\left[[\mathcal{F}in, D_2], \mathcal{F}in\right]\right\}$. On the other hand, if d is a monadic operator like **arb** (which is postulated to satisfy $\mathbf{arb}(\emptyset) = \emptyset \ \& \ \big(X \neq \emptyset \rightarrow \mathbf{arb}(X) \in X)\big)$, and so has the descriptor $\{[\{D_3\}, D_3]\}$ (where the null set must have the descriptor D_3), then g must have the descriptor $\left\{\left[[\mathcal{F}in, \mathcal{F}in], D_2\right], \mathcal{F}in\right]\right\}$, and s the descriptors $\{\mathcal{F}in\}$ and $\mathcal{F}in$, for $f(s, t)$ to have the descriptor $\mathcal{F}in$. In this case f has once again the descriptor $\left\{\left[[\mathcal{F}in, D_2], \mathcal{F}in\right]\right\}$.

5   **THEORY** Application

Ref incorporates a technical notion of 'theory' designed, for large-scale proof-development, to play a role similar to the notion of object class in large-scale programming. As discussed in [18], such a mechanism can be very useful for 'proof-engineering'.

The theories we allow, like procedures in a programming language, have lists of formal parameters. Each 'theory' requires its parameters to meet a set of assumptions. When 'applied' to a list of actual parameters that have been shown to meet the assumptions, a theory will instantiate several additional 'output' set, predicate, and function symbols, and then supply a list of theorems initially proved explicitly (relative to the formal parameters) by the user inside the theory itself. These theorems will generally involve the new symbols.

Again from [18], we borrow the following example of a familiar theory:

THEORY equiv_classes(s, Eq)
$$\Big\langle \forall x \in \mathsf{s} \,\big|\, \mathsf{Eq}(x, x) \Big\rangle$$
$$\Big\langle \forall x \in \mathsf{s}, y \in \mathsf{s}, z \in \mathsf{s} \,\big|\, \mathsf{Eq}(x, y) \ \longrightarrow \ \big(\mathsf{Eq}(y, z) \ \leftrightarrow \ \mathsf{Eq}(z, x)\big) \Big\rangle$$
$$\Longrightarrow (\mathsf{Eqc}_\Theta, \mathsf{r}_\Theta) \text{ -- 'quotient'-set and globalized 'canonical embedding'}$$

$$\langle \forall x \in s,\, y \in s \mid \mathsf{Eq}(x,y) \;\leftrightarrow\; \mathsf{Eq}(y,x)\rangle$$
$$\langle \forall x \in s,\, y \in s \mid \mathsf{Eq}(x,y) \;\leftrightarrow\; \mathsf{r}_\Theta(x) = \mathsf{r}_\Theta(y)\rangle$$
$$\langle \forall b \in \mathsf{Eqc}_\Theta \mid \mathbf{arb}(b) \in s \;\&\; \mathsf{r}_\Theta(\mathbf{arb}(b)) = b\rangle$$
$$\langle \forall x \in s \mid \mathsf{r}_\Theta(x) \in \mathsf{Eqc}_\Theta\rangle$$
$$\langle \forall x \in s \mid \mathsf{Eq}(x, \mathbf{arb}(\mathsf{r}_\Theta(x)))\rangle$$

END equiv_classes.

As an illustration of the usefulness of the **THEORY** construct, let us exploit the theory just seen in order to define the set $\mathbb{R}$ of all real numbers. Since the apparent simplicity of the reals as Dedekind cuts is marred by problems concerning the treatment of negative reals, we opted for Cantor's approach based on rational Cauchy sequences. Thus our construction of the reals runs as follows:

-- The set of rational sequences

DEF 46. $\mathsf{Seq}_{Q} =_{\mathrm{Def}} \{f : f \subseteq \mathbb{N} \times \mathbb{Q} \mid \mathrm{domain}(f) = \mathbb{N} \;\&\; \mathsf{Svm}(f)\}$

-- The constant 0 rational sequence

DEF 47. $\mathbf{0}_{QS} =_{\mathrm{Def}} \mathbb{N} \times \{\mathbf{0}_{Q}\}$

-- The constant 1 rational sequence

DEF 48. $\mathbf{1}_{QS} =_{\mathrm{Def}} \mathbb{N} \times \{\mathbf{1}_{Q}\}$

-- Pointwise sum of rational sequences

DEF 49. $F +_{QS} G =_{\mathrm{Def}} \{[p^{[1]}, p^{[2]} +_{Q} G\lceil p^{[1]}] : p \in F\}$

-- Pointwise additive inverse of rational sequence

DEF 50. $\mathsf{Rev}_{QS}(F) =_{\mathrm{Def}} \{[p^{[1]}, \mathsf{Rev}_{Q}(p^{[2]})] : p \in F\}$

-- Pointwise absolute value of rational sequence

DEF 51. $|F|_{QS} =_{\mathrm{Def}} \left\{\left[p^{[1]}, \left|p^{[2]}\right|_{Q}\right] : p \in F\right\}$

-- Pointwise difference of rational sequences

DEF 52. $F -_{QS} G =_{\mathrm{Def}} F +_{QS} \mathsf{Rev}_{QS}(G)$

-- Product of rational sequences

DEF 53. $F *_{QS} G =_{\mathrm{Def}} \{[p^{[1]}, p^{[2]} *_{Q} G\lceil p^{[1]}] : p \in F\}$

-- Pointwise reciprocal of rational sequence

DEF 54. $\mathsf{Recip}_{QS}(F) =_{\mathrm{Def}} \mathsf{Shifted\_seq}\big(\left\{\left[i, \mathsf{Recip}_{Q}(F\lceil i)\right] : i \in \mathbb{N}\right\},$
$$\mathbf{arb}\{h \in \mathbb{N} \mid \langle \forall i \in \mathbb{N} \backslash h \mid F\lceil i \neq \mathbf{0}_{Q}\rangle\}\big)$$

-- Pointwise quotient of rational sequences

DEF 55. $F /_{QS} G =_{\mathrm{Def}} F *_{QS} \mathsf{Recip}_{QS}(G)$

-- Rational Cauchy sequences

DEF 56. $\mathsf{Cau}_{Q} =_{\mathrm{Def}} \Big\{f : f \in \mathsf{Seq}_{Q} \mid \langle \forall \varepsilon \in \mathbb{Q} \mid \varepsilon >_{Q} \mathbf{0}_{Q} \rightarrow$
$$\mathscr{F}in(\{i \cap j : i \in \mathbb{N}, j \in \mathbb{N} \mid |f\lceil i -_{Q} f\lceil j|_{Q} >_{Q} \varepsilon\})\rangle\Big\}$$

-- Equivalence of rational sequences

DEF 57. $F \approx_{QS} G \;\leftrightarrow_{\mathrm{Def}}\; \langle \forall \varepsilon \in \mathbb{Q} \mid \varepsilon >_{Q} \mathbf{0}_{Q} \rightarrow$
$$\mathscr{F}in(\{x : x \in \mathrm{domain}(F) \mid |F\lceil x -_{Q} G\lceil x|_{Q} >_{Q} \varepsilon\})\rangle$$

THEOREM 465: $F \in \mathsf{Cau}_{Q} \rightarrow F \approx_{QS} F$. PROOF: $\cdots$

THEOREM 466: $F, G, H \in \mathsf{Cau}_{Q} \rightarrow (F \approx_{QS} G \rightarrow (G \approx_{QS} H \leftrightarrow H \approx_{QS} F))$. PROOF: $\cdots$

-- Now that we know that $\approx_{QS}$ is an equivalence relationship, we can apply the equiv_classes theory to it, to derive

APPLY $\langle$Eqc$_\Theta$: $\mathbb{R}$, f$_\Theta$: Cau_to_Re$\rangle$ equiv_classes$\big(R(f, g) \mapsto f \approx_{QS} g, s \mapsto$ Cau$_Q\big)$
$\Longrightarrow$

THEOREM 467: $\langle \forall f \in$ Cau$_Q$, g $\in$ Cau$_Q \mid f \approx_{QS} g \leftrightarrow$ Cau_to_Re$(f) =$ Cau_to_Re$(g)\rangle$ &
$\langle \forall r \in \mathbb{R} \mid \mathbf{arb}(r) \in$ Cau$_Q$ & Cau_to_Re$(\mathbf{arb}(r)) = r\rangle$ &
$\langle \forall f \in$ Cau$_Q \mid$ Cau_to_Re$(f) \in \mathbb{R}\rangle$ & $\langle \forall f \in$ Cau$_Q \mid f \approx_{QS} \mathbf{arb}\big(Cau_to_Re(f)\big)\rangle$.

Let us observe, as an incidental remark, that in spite of its relative length this list of statements works better than Dedekind's approach, because it allows us to 'lift' laws already proved for rational numbers into corresponding laws for rational Cauchy sequences, and thereby into laws concerning the reals (which are viewed here as the $\approx_{QS}$–classes of such sequences).

Use of External Provers. The Ref proof verifier has the ability to accept proofs generated by various external provers (such as Otter [17], as exemplified in [9, p. 193]). This is done by a syntactic extension of the normal Ref APPLY directive, i.e. external provers are regarded as sources of variant Ref THEORYs. When such provers are being used, the normal keyword APPLY used to invoke a THEORY is changed to "APPLY_provername", where "provername" names the external prover in question. In this case, the normal Ref THEORY declaration is expanded to list Ref-syntax translations of all the theorems being drawn from the external prover, and of all the external symbol definitions on which these depend. An external file, also named in the modified Ref APPLY directive, must be provided as certification of each such THEORY. Ref examines this file to establish that it is a valid proof, by the external prover named, of all the theorems which the THEORY claims.

6   Context of an Inference Step

Until a proof is complete and acceptable to the Ref verifier, it is undesirable to let efficiency concerns interfere with one's focus on the logic of the proof. Once an initial version of the proof has been accepted by Ref, one can speed up its processing by supplying *contexts* (see below) for the most time-consuming proof steps. Ref allows one to optimize proof steps by automated context discovery.

Statement assertions and parts of compounds connected by the conjunction sign & can be labeled for explicit subsequent reference within a proof by appending a reserved notation of the form Stat*nnn*: to them, where *nnn* designates any integer. These are the labels used in hints of statements of the form

$$\langle e_1, ..., e_m \rangle \hookrightarrow \text{Stat}nnn \Longrightarrow \cdots$$

The context of a hint defines the collection of preceding statements, within the proof in which the hint appears, which the inference mechanism invoked by the hint should use in deducing the assertion to which the hint is attached. Since the efficiency of an inference mechanism often degrades very rapidly (e.g. exponentially or worse) with the size of the context with which it is working, appropriate restriction of context can be crucial to successful completion of an inference. Inferences which the verifier cannot complete within a reasonable amount of time are abandoned with a diagnostic message "Abandoned...", or with the more specific message "Failure..." if the inference method is able to certify that the inference finally attempted is impossible. Hint keywords like ELEM, EQUAL, SIMPLF, and ALGEBRA can be supplied with context indications by prefixing them (in the cases of ELEM and Discharge) or suffixing them (in all other cases) with a statement label, or a comma-separated list of such labels, as in the examples

$$\langle \mathsf{Stat3} \rangle \mathsf{ELEM} \Longrightarrow s \notin \{ \, x \subseteq o \mid \mathcal{O}(x) \& P(x) \, \}$$

and

$$\langle \mathsf{Stat3, Stat4, Stat9} \rangle \mathsf{ELEM} \Longrightarrow s \notin \{ \, x \subseteq o \mid \mathcal{O}(x) \& P(x) \, \} \, .$$

The first form of prefix defines the context of an inference to be the collection of all statements in the proof, back to the point of last previous occurrence of the statement label in the proof (but not within ranges of the proof that are already closed in virtue of the fact that they are included between a preceding Discharge statement and its matching Suppose statement—see below). The second form of prefix defines the context of an inference to be the collection of statements explicitly named in the prefix. If no context is specified for an inference, then its context is understood to be the collection of all preceding statements in the same proof (not including statements enclosed within previously closed Suppose/Discharge ranges). This unrestricted default context is workable for simple enough inferences in short enough proofs.

The Ref Proof Step Optimizer. Ref's automated proof optimizer attempts to determine, for each line L in a proof, a close-to-minimal subset of the set of all prior lines in the proof which is large enough to serve as a context for the proof of L, i.e. large enough to be inconsistent with the negation $\neg L$ of L. To this end, it collects a list of prior statements, called 'critical', which it believes to be necessary for the desired inconsistency. Initially this list of critical statements consists of all the statements preceding L. A first binary search over ranges of statements shortens this to the smallest range R of statements preceding L which is large enough to be inconsistent with $\neg L$. The first statement F in this range is added to an (initially empty) list C. This reflects the fact that if F is removed from R, the set $R \cup \{\neg L\}$ of statements is no longer inconsistent.

Let R' be R after F is removed. Plainly $C \cup R' \cup \{\neg L\}$ is inconsistent. But R' may be larger than it need be to guarantee this property. So a second binary search is made, to shorten R' to the smallest range R'' of statements which is large enough for $C \cup R'' \cup \{\neg L\}$ to be inconsistent. The first statement of R'' is then moved from R'' to C. This operation is repeated as often as needed to produce a final list C of critical statements such that $C \cup \{\neg L\}$ is inconsistent. This list C of statements is returned by the proof optimizer as the context to be used in proving L.

The code described in the preceding paragraphs is organized using a

procedure test_range(critical_list,range_tup,statement)

which sets up the inconsistency tests described and then calls Ref's underlying ELEM procedure.

To mark a proof for invocation of the automated analysis just described, one simply changes the normal "$\Longrightarrow$" mark of its initial Suppose_not to "$\Rrightarrow$". To mark a single step of a proof for application of this analysis, one changes its "$\Longrightarrow$" mark to "$\Rrightarrow$". The first such mark encountered in a proof (if any) turns off the 'analyze by default' option if this has been set by marking the initial Suppose_not.

Here are a few illustrative examples of the output produced by Ref's automated proof optimizer:

The lines of context needed to prove citation of theorem T116 in line 9, namely: (domain(f) $\subseteq$ $\mathbb{N}$) & (range(f) $\subseteq$ $\mathbb{Q}$) are T116 plus [1, 5]
The lines of context needed to prove citation of theorem T220 in line 10, namely: $g \subseteq (\mathbb{N} \times \mathbb{Q})$ are T220 plus [1, 7, 9]
The lines of context needed to prove citation of theorem T85 in line 12, namely: domain($f \circ h$) = domain(h) are T85 plus [1, 7]

7 Case-Study on Bisimulations: Towards a Theory of Labeled Graphs

Although one has several choices in constructing the basic notions of analysis, the hierarchic pattern into such notions is more or less established [3]. The situation is somewhat less standard-prone in discrete mathematics, because in limited frameworks one can very well do with *ad hoc* notions (compare, e.g., the definition of ordered tree in [16] with various others present in texts on algorithms). This state of affairs is quite acceptable as long as the different views are coherent: one can contend that mathematical insight benefits from multiple views on the same notion rather than from formalistic scruples. The issue looks different if we want to rely on a verifier, e.g. for the reasons put forward in [15]. As a contribution towards standardization, let us see how we began to set up a scenario on (bi)simulations (cf. [10,14]) in Ref.

A study of this kind splits naturally into one strictly mathematical layer (where the degree of abstraction reaches infinite sets and even proper classes) and one layer where the issues are more algorithmic (and, accordingly, are more concerned with combinatorics and finiteness). In the case at hand, the end-product in the mathematical layer can take the form of a theory such as the following (where trans_reflCl(R,V) designates the transitive-reflexive closure of a *map*, i.e. of a set R of pairs, on a given domain V):

THEORY labeledGraph(vertices, edges, tags)
    edges $\subseteq$ vertices $\times$ vertices
    vertices $\subseteq$ $\bigcup$ tags

$\langle \forall x \in \mathsf{tags}, y \in \mathsf{tags} \mid x \neq y \rightarrow x \cap y = \emptyset \rangle$
$\Longrightarrow (\mathsf{tag}_\Theta, \mathsf{block}_\Theta, \mathsf{SameTag}_\Theta, \mathsf{bisim}_\Theta, \beta_\Theta, \mathsf{sim}_\Theta, \sigma_\Theta)$
$\quad \langle \forall e \in \mathsf{edges} \mid e = \left[e^{[1]}, e^{[2]} \right] \ \& \ e^{[1]}, e^{[2]} \in \mathsf{vertices} \rangle$
$\quad \langle \forall v, \exists y \mid v \in \mathsf{vertices} \rightarrow y \in \mathsf{tags} \ \& \ v \in y \rangle$
$\quad \langle \forall x \in \mathsf{tags}, y \in \mathsf{tags} \mid x \neq y \rightarrow x \cap y = \emptyset \rangle$
$\quad \langle \forall v \in \mathsf{vertices} \mid \mathsf{tag}_\Theta(v) \in \mathsf{tags} \ \& \ v \in \mathsf{tag}_\Theta(v) \rangle$
$\quad \langle \forall v \mid \mathsf{block}_\Theta(v) = \textbf{if} \ v \in \mathsf{vertices} \ \textbf{then} \ \{w \in \mathsf{tag}_\Theta(v) \mid w \in \mathsf{vertices}\} \ \textbf{else} \ \emptyset \ \textbf{fi} \rangle$
$\quad \langle \forall v, w \mid \mathsf{SameTag}_\Theta(v, w) \leftrightarrow \mathsf{block}_\Theta(v) = \mathsf{block}_\Theta(w) \rangle$
$\quad \langle \forall v \mid \mathsf{SameTag}_\Theta(v, v) \rangle$
$\quad \langle \forall q, r, x \mid q \subseteq r \cup r^\leftarrow \ \& \ \mathsf{Is\_map}(r) \ \& \ \langle \forall e \in r \mid \mathsf{SameTag}_\Theta(e^{[1]}, e^{[2]}) \rangle \ \&$
$\qquad x \in \mathsf{domain}(q) \longrightarrow \mathsf{SameTag}_\Theta(x, q \restriction x) \rangle$
$\quad \langle \forall q, r, x, y \mid q \subseteq r \cup r^\leftarrow \ \& \ \mathsf{Is\_map}(r) \ \& \ \langle \forall e \in r \mid \mathsf{SameTag}_\Theta(e^{[1]}, e^{[2]}) \rangle \ \&$
$\qquad \mathsf{Straight}(q, x, y) \longrightarrow \mathsf{SameTag}_\Theta(x, y) \rangle$
$\quad \langle \forall r, x, y, v \mid \mathsf{Is\_map}(r) \ \& \ \langle \forall e \in r \mid \mathsf{SameTag}_\Theta(e^{[1]}, e^{[2]}) \rangle \ \&$
$\qquad [x, y] \in \mathsf{trans\_reflCl}(r \cup r^\leftarrow, v) \longrightarrow \mathsf{SameTag}_\Theta(x, y) \rangle$
$\mathsf{bisim}_\Theta =_{\mathrm{Def}} \{b \subseteq \{[v, w] : v \in \mathsf{vertices}, w \in \mathsf{vertices} \mid \mathsf{SameTag}_\Theta(v, w)\} \mid$
$\qquad \mathsf{edges} \circ b^\leftarrow = b^\leftarrow \circ \mathsf{edges} \ \& \ \mathsf{edges} \circ b = b \circ \mathsf{edges}\}$
$\beta_\Theta =_{\mathrm{Def}} \bigcup \mathsf{bisim}_\Theta$
$\mathsf{sim}_\Theta =_{\mathrm{Def}} \{b \subseteq \{[v, w] : v \in \mathsf{vertices}, w \in \mathsf{vertices} \mid \mathsf{SameTag}_\Theta(v, w)\} \mid$
$\qquad \mathsf{edges} \circ b^\leftarrow = b^\leftarrow \circ \mathsf{edges}\}$
$\sigma_\Theta =_{\mathrm{Def}} \bigcup \mathsf{sim}_\Theta \cap \bigcup \mathsf{sim}_\Theta^\leftarrow$
$\quad \langle \forall x \mid \emptyset \in \mathsf{bisim}_\Theta \ \& \ \mathsf{bisim}_\Theta \subseteq \mathsf{sim}_\Theta \ \& \ (x \in \mathsf{bisim}_\Theta \rightarrow x \subseteq \mathsf{vertices} \times \mathsf{vertices}) \rangle$
$\quad \langle \forall b \in \mathsf{bisim}_\Theta \mid b \subseteq \beta_\Theta \rangle$
$\quad \langle \forall b \in \mathsf{bisim}_\Theta \mid b^\leftarrow \in \mathsf{bisim}_\Theta \rangle$
$\beta_\Theta \subseteq \sigma_\Theta$
$\quad \langle \forall b \in \mathsf{sim}_\Theta \mid \mathsf{trans\_reflCl}(b, \mathsf{vertices}) \in \mathsf{sim}_\Theta \rangle$
$\quad \langle \forall b \in \mathsf{bisim}_\Theta \mid \mathsf{trans\_reflCl}(b \cup b^\leftarrow, \mathsf{vertices}) \in \mathsf{bisim}_\Theta \rangle$
$\quad \langle \forall r \in \mathsf{sim}_\Theta, s \in \mathsf{sim}_\Theta \mid r \cup s \in \mathsf{sim}_\Theta \rangle$
$\quad \langle \forall r \in \mathsf{bisim}_\Theta, s \in \mathsf{bisim}_\Theta \mid r \cup s \in \mathsf{bisim}_\Theta \rangle$
$\quad \mathsf{Transitive}(\beta_\Theta) \ \& \ \mathsf{Symmetric}(\beta_\Theta) \ \& \ \mathsf{Reflexive}(\beta_\Theta, \mathsf{vertices})$
$\quad \mathsf{Transitive}(\sigma_\Theta) \ \& \ \mathsf{Symmetric}(\sigma_\Theta) \ \& \ \mathsf{Reflexive}(\sigma_\Theta, \mathsf{vertices})$
END labeledGraph

In the development of such a theory, a modular subdivision of the work will, as opposed to a direct approach, give us two advantages: on the one hand, preliminary definitions and intermediate theories will, if properly designed, bear an autonomous value and can be reused in a variety of situations; on the other hand, they will make the overall task more affordable and manageable.

One readily sees that the notion of graph underlying the above theory can be tackled at a higher degree of abstraction where the class of edges is not taken to be necessarily a set:

THEORY $\mathsf{taggedGraph}\big(\mathsf{Is\_vertex}(V), \mathsf{Is\_edge}(E), \mathsf{Is\_tag}(X)\big)$
$\quad \langle \forall e \mid \mathsf{Is\_edge}(e) \rightarrow e = \left[e^{[1]}, e^{[2]} \right] \ \& \ \mathsf{Is\_vertex}(e^{[1]}) \ \& \ \mathsf{Is\_vertex}(e^{[2]}) \rangle$
$\quad \langle \forall v, \exists y \mid \mathsf{Is\_vertex}(v) \rightarrow \mathsf{Is\_tag}(y) \ \& \ v \in y \rangle$
$\quad \langle \forall x, y \mid \mathsf{Is\_tag}(x) \ \& \ \mathsf{Is\_tag}(y) \ \& \ x \neq y \rightarrow x \cap y = \emptyset \rangle$

$\Longrightarrow (\mathsf{tag}_\Theta, \mathsf{block}_\Theta, \mathsf{SameTag}_\Theta)$
$\langle \forall v \mid \mathsf{Is\_vertex}(v) \rightarrow \mathsf{Is\_tag}(\mathsf{tag}_\Theta(v)) \ \& \ v \in \mathsf{tag}_\Theta(v)\rangle$
$\langle \forall v \mid \mathsf{block}_\Theta(v) = \mathbf{if} \ \mathsf{Is\_vertex}(v) \ \mathbf{then} \ \{w \in \mathsf{tag}_\Theta(v) \mid \mathsf{Is\_vertex}(w)\} \ \mathbf{else} \ \emptyset \ \mathbf{fi}\rangle$
$\langle \forall v, w \mid \mathsf{SameTag}_\Theta(v, w) \leftrightarrow \mathsf{block}_\Theta(v) = \mathsf{block}_\Theta(w)\rangle$
$\langle \forall x \mid \mathsf{SameTag}_\Theta(x, x)\rangle$
$\langle \forall x, y, z \mid \mathsf{SameTag}_\Theta(x, y) \ \& \ \mathsf{SameTag}_\Theta(y, z) \rightarrow \mathsf{SameTag}_\Theta(z, x)\rangle$
$\langle \forall q, r, x \mid q \subseteq r \cup r^\leftarrow \ \& \ \mathsf{Is\_map}(r) \ \& \ \langle \forall e \in r \mid \mathsf{SameTag}_\Theta(e^{[1]}, e^{[2]})\rangle \ \&$
$\qquad x \in \mathsf{domain}(q) \rightarrow \mathsf{SameTag}_\Theta(x, q \restriction x)\rangle$
$\langle \forall q, r, x, y \mid q \subseteq r \cup r^\leftarrow \ \& \ \mathsf{Is\_map}(r) \ \& \ \langle \forall e \in r \mid \mathsf{SameTag}_\Theta(e^{[1]}, e^{[2]})\rangle \ \&$
$\qquad \mathsf{Straight}(q, x, y) \rightarrow \mathsf{SameTag}_\Theta(x, y)\rangle$
$\langle \forall r, x, y, v \mid \mathsf{Is\_map}(r) \ \& \ \langle \forall e \in r \mid \mathsf{SameTag}_\Theta(e^{[1]}, e^{[2]})\rangle \ \&$
$\qquad [x, y] \in \mathsf{trans\_reflCl}(r \cup r^\leftarrow, v) \rightarrow \mathsf{SameTag}_\Theta(x, y)\rangle$
END taggedGraph

As can be seen by comparing the two specifications, some results are imported from this more general taggedGraph theory into the labeledGraph theory; for its part, the more abstract notion of graph ensures a higher degree of reusability.

Working in the opposite direction of an increasing concreteness, we designed the following narrow-scope theory, where the set of edges is taken to be finite and cycle-free, so that a height function for its vertices can be defined within the theory by means of a general form of recursion available in connection with any well-founded relation [18]:

THEORY acyclicFiniteGraph(vertices, edges)
$\quad \mathsf{edges} \subseteq \mathsf{vertices} \times \mathsf{vertices}$
$\quad \mathcal{F}in(\mathsf{edges})$
$\quad \mathsf{Acyclic}(\mathsf{edges})$
$\Longrightarrow (\mathsf{height}_\Theta)$
$\quad \langle \forall t \subseteq \mathsf{vertices} \mid t \neq \emptyset \rightarrow \langle \exists w \in t, \forall v \in t \mid [v, w] \notin \mathsf{transCl}(\mathsf{edges})\rangle\rangle$
$\quad \langle \forall x \in \mathsf{vertices} \mid \mathsf{height}_\Theta(x) = \bigcup \{\mathsf{next}(\mathsf{height}_\Theta(y)) : y \in \mathsf{vertices} \mid [y, x] \in \mathsf{edges}\}\rangle$
END acyclicFiniteGraph

But how were acyclicity and the transitive closure operation defined in the first place? Here is a viable approach, where the presupposed $\mathsf{Svm}(\cdot)$ notion of single-valued map gets refined into notions of various kinds of simple path:

$\qquad$ -- permutation of a finite set
DEF 943. $\mathsf{Is\_perm}(P) \ \leftrightarrow_{\mathrm{Def}} \ \mathsf{Svm}(P) \ \& \ \mathcal{F}in(P) \ \& \ \mathsf{range}(P) \supseteq \mathsf{domain}(P)$

$\qquad$ -- simple cyclic permutation
DEF 944. $\mathsf{Scycle}(C) \ \leftrightarrow_{\mathrm{Def}} \ \{p \subseteq C \mid p \neq \emptyset \ \& \ \mathsf{Is\_perm}(p)\} = \{C\}$

$\qquad$ -- acyclicity property of a map
DEF 945. $\mathsf{Acyclic}(A) \ \leftrightarrow_{\mathrm{Def}} \ \{c \subseteq A \mid \mathsf{Is\_perm}(c)\} = \{\emptyset\}$

$\qquad$ -- cycle-free path connecting two nodes, or empty path
DEF 946. $\mathsf{Straight}(P, X, Y) \ \leftrightarrow_{\mathrm{Def}} \ \mathsf{Scycle}(P \cup \{[Y, X]\}) \ \& \ [Y, X] \notin P$

-- prefixed simple path in a map (this DEF exploits built-in $\in$-recursion)

DEF 947. $\mathsf{sgm}(\mathsf{N},\mathsf{P},\mathsf{S}) =_{\mathrm{Def}} \mathsf{S} \cup \{[\mathsf{x},\mathsf{P}\!\restriction\!\mathsf{x}] : i \in \mathsf{N}, \mathsf{x} \in \mathsf{range}(\mathsf{sgm}(i,\mathsf{P},\mathsf{S})) \cap \mathsf{domain}(\mathsf{P})\}$

-- restriction of a multi-valued map

DEF 948. $\mathsf{on}(\mathsf{R},\mathsf{N}) =_{\mathrm{Def}} \mathsf{R} \cap (\mathsf{N} \times \mathsf{N})$

-- transitive closure of a map

DEF 949. $\mathsf{transCl}(\mathsf{R}) =_{\mathrm{Def}} \{[\mathsf{x},\mathsf{y}] : \mathsf{q} \subseteq \mathsf{R}, \mathsf{x} \in \mathsf{domain}(\mathsf{q}), \mathsf{y} \in \mathsf{range}(\mathsf{q}) \,|$
$(\mathsf{x} = \mathsf{y} \,\&\, \mathsf{Scycle}(\mathsf{q})) \vee (\mathsf{x} \neq \mathsf{y} \,\&\, \mathsf{Straight}(\mathsf{q},\mathsf{x},\mathsf{y}))\}$

-- transitive-reflexive closure of a map

DEF 950. $\mathsf{trans\_reflCl}(\mathsf{R},\mathsf{N}) =_{\mathrm{Def}} \mathsf{transCl}(\mathsf{on}(\mathsf{R},\mathsf{N})) \cup \iota_\mathsf{N}$

8   Conclusions and Future Work

As said at the outset, we aim at exploitations of Ref in the realm of program correctness verification. A promising fact is that many algorithms can be specified, very naturally and in compact, high-level terms, by means of an executable language grounded on set theory. For example—laying bisimulations and stable partitioning algorithms momentarily aside to save space—, we can specify the construction of a *spanning tree* for a finite rooted graph as simply as by the invocation

$$\mathsf{dfst}(e, r, [\emptyset, \{r\}]),$$

where e is the set of all edges of the graph, r is a designated node, and the procedure dfst is as follows:

```
procedure dfst(graph, node, tree_and_visited);     -- depth first spanning tree
   return  if    ( avail := graph{node} - tree_and_visited(2) ) = ∅
           then  tree_and_visited
           else  dfst( graph, node,
                    dfst(graph, downto := arb(avail),
                       withall(tree_and_visited, [ [node,downto], downto ])) )
           end if;
end dfst;
procedure withall(tup_of_sets, tup_of_elts);        -- inserts elements into sets
   return  [ set with tup_of_elts(j):  set = tup_of_sets(j) ];
end withall;
```

It should be intuitively clear, indeed, that the said invocation will produce a pair $[e', v]$ where the set $e' \subseteq e$ of edges forms a tree rooted in r and v consists of all vertices reachable in the input graph from r, whose set coincides also with the set of all vertices in the tree. Making these claims rigorous amounts to developing a set-theoretic proof, which one would like to do with the assistance of Ref. Although Ref does not, up until today, encompass the programming notation exemplified by the above procedure, it is easy to conceive an integration of such notation with Ref's logical notation—the one where one characterizes rooted graphs and spanning trees—, thanks to the set-theoretic background common to both languages. After such an integration, proving that a procedure behaves as desired could be done, in full, under the surveillance of our automated verifier.

References

1. L. Aiello and R.W. Weyhrauch. Using meta-theoretic reasoning to do algebra. In [2] pp. 1-13.
2. W. Bibel and R. Kowalski, eds. *Proceedings of the 5th conference on Automated Deduction*, LNCS 87, Springer-Verlag, 1980.
3. D.S. Bridges. *Foundations of Real and Abstract Analysis*, vol. 174 of *Graduate Texts in Mathematics*. Springer-Verlag, 1997.
4. D. Cantone. A fast saturation strategy for set-theoretic tableaux. In D. Galmiche, ed., *Proc. of the International Conference on Automated Reasoning with Analytic Tableaux and Related Methods*, volume 1227 of *LNAI*, pp. 122-137. Springer-Verlag, 1997.
5. D. Cantone and C.G. Zarba. A new fast tableau-based decision procedure for an unquantified fragment of set theory. In R. Caferra, G. Salzer (Eds.), *Proc. of the International Workshop on First-Order Theorem Proving (FTP'98)*, LNAI 1761, pp. 126-136. Springer-Verlag, 2000.
6. D. Cantone, E.G. Omodeo, and A. Policriti. The automation of syllogistic. II. Optimization and complexity issues. *J. Automated Reasoning*, 6(2):173-187, 1990.
7. D. Cantone, E.G. Omodeo, and A. Policriti. *Set Theory for Computing. From Decision Procedures to Declarative Programming with Sets*. Monographs in Computer Science. Springer-Verlag, New York, 2001.
8. D. Cantone, A. Formisano, E.G. Omodeo, and J.T. Schwartz. Various commonly occurring decidable extensions of multi-level syllogistic. In S. Ranise and C. Tinelli, eds., *Pragmatics of Decision Procedures in Automated Reasoning 2003*, PDPAR'03 (CADE19), Miami, USA, July 29, 2003.
9. D. Cantone, E.G. Omodeo, J.T. Schwartz, and P. Ursino. Notes from the logbook of a proof-checker's project. In N. Dershowitz ed., *International symposium on verification (Theory and Practice)* celebrating Zohar Manna's 1000000_2-th birthday. LNCS 2772, pp. 182-207. Springer-Verlag, 2003.
10. A. Dovier, C. Piazza, and A. Policriti. An Efficient Algorithm for Computing Bisimulation Equivalence. *Theoretical Computer Science*, 311(1-3):221-256, 2004.
11. A. Ferro, E.G. Omodeo, and J.T. Schwartz. Decision procedures for some fragments of set theory. In [2] pp. 88-96.
12. A. Ferro, E.G. Omodeo, and J.T. Schwartz. Decision procedures for elementary sublanguages of set theory. I. Multi-level syllogistic and some extensions. *Comm. Pure Applied Math.*, 33(5):599-608, 1980.
13. M.C. Fitting. *First-Order Logic and Automated Theorem Proving*. Graduate Texts in Computer Science. Springer-Verlag, Berlin, 2nd edition, 1996. 1st ed., 1990.
14. R. Gentilini, C. Piazza, and A. Policriti. From Bisimulation to Simulation. Coarsest Partition Problems. *Journal of Automated Reasoning*, 31(1):73-103, 2003.
15. J.-P. Keller and R. Paige. Program derivation with verified transformations - A case study. *Comm. Pure Appl. Math.*, 48(9-10):1053-1113, 1995. *Special issue in honor of J.T. Schwartz*.
16. J. W. Lloyd. *Foundation of Logic Programming*. Springer, 1987. Second edition.
17. W.W. McCune. Otter 2.0 User Guide. Technical Report ANL-90/9, Argonne National Laboratory, Argonne, Illinois, 1990.
18. E.G. Omodeo and J.T. Schwartz. A 'Theory' mechanism for a proof-verifier based on first-order set theory. In A. Kakas and F. Sadri (Eds.), *Computational Logic: Logic Programming and beyond, Essays in honour of Robert Kowalski*, part II, LNAI 2408, pp. 214-230. Springer-Verlag, 2002.

Appendix: Multi-level Syllogistic

MLSS (*multilevel syllogistic with singleton*) is the unquantified language of set theory consisting of a denumerable infinity u, v, w, x, y, z, ... of set variables, the 'null set' constant $\emptyset$, the set operators $\cdot \cap \cdot$, $\cdot \setminus \cdot$, $\cdot \cup \cdot$, $\{\cdot, \ldots, \cdot\}$, the set predicates $\cdot \in \cdot$, $\cdot = \cdot$, $\cdot \subseteq \cdot$, and propositional connectives.

The semantics of **MLSS** is based upon the von Neumann cumulative hierarchy $\mathcal{V}$ defined as follows (where $\mathcal{O}$ and $\mathcal{P}(X)$ designate the class of all ordinals and the power-set of X):

$$\mathcal{V}_\alpha =_{\mathrm{Def}} \bigcup_{\mu < \alpha} \mathcal{P}(\mathcal{V}_\mu) \,, \text{ for each ordinal } \alpha \,;$$
$$\mathcal{V} =_{\mathrm{Def}} \bigcup_{\mathcal{O}(\alpha)} \mathcal{V}_\alpha \,.$$

An *assignment* $\mathcal{M}$ over a collection of variables V is any map from V into $\mathcal{V}$. Let φ be an **MLSS**-formula over a collection V of variables, and let $\mathcal{M}$ be an assignment over V. By $\varphi^{\mathcal{M}}$ we denote the truth-value of φ obtained by interpreting each variable $x \in V$ with the set $x^{\mathcal{M}}$ and the set operators and propositional connectives according to their standard meanings. Such a φ is said to be *satisfiable* if it has a *model*, namely, an assignment $\mathcal{M}$ making $\varphi^{\mathcal{M}}$ true.

The satisfiability problem for **MLSS** is the problem of determining whether or not any given **MLSS**-formula φ is satisfiable. It was first solved in [11]. Subsequently, it was shown that the satisfiability problem for conjunctions of 'flat' **MLSS**-literals of the forms

$$x = y, \quad x \neq y, \quad x \in y, \quad x \notin y, \quad x = y \cup z, \quad x = y \setminus z, \quad x = \{y\}, \tag{1}$$

to be called *normalized* **MLSS**-conjunctions, is *NP*-complete (cf. [6]); more recently, its decision procedure was optimized in [5,7] by means of semantic tableaux. For the reader's convenience, we sketch a decision procedure for normalized **MLSS**-conjunctions based on semantic tableaux.

Table 1 lists the rules of a tableau calculus for **MLSS**. Notice that the rules (2), (5), and (9) cause branch splits.

Next we define **MLSS**-tableaux (for general notions on tableaux, the reader is referred to [13]).

Let S be a finite collection of flat **MLSS**-literals of the form (1). An INITIAL **MLSS**-TABLEAU for S is a one-branch tree whose nodes are labeled by the literals in S.

An **MLSS**-TABLEAU for S is a tableau labeled with **MLSS**-literals which can be constructed from the initial tableau for S by a finite number of applications of the rules (1)–(11) of Table 1.

Let $\mathcal{T}$ be an **MLSS**-tableau for S. A branch ϑ of $\mathcal{T}$ is said to be

- STRICT, if no rule has been applied more than once on ϑ to the same literal occurrences;
- SATURATED, if each of the tableaux rules (1)–(11) has been applied at least once on each instance of its premises on ϑ;
- CLOSED, if either ϑ contains a set of literals of the form $x \in x_1$, $x_1 \in x_2$, ..., $x_{n-1} \in x_n$, $x_n \in x$, for some variables $x, x_1, \ldots, x_n$ with $n \geq 0$, or it contains a pair of complementary literals X, $\neg X$;
- OPEN, if it is not closed;

Table 1. Tableaux rules for **MLSS**

$$
\frac{\begin{array}{c} x = y_1 \cup y_2 \\ z \in y_i \end{array}}{z \in x} \ (1)
\qquad
\frac{\begin{array}{c} x = y_1 \cup y_2 \\ z \in x \end{array}}{z \in y_1 \mid z \in y_2} \ (2)
\qquad
\frac{\begin{array}{c} x = y_1 \setminus y_2 \\ z \in x \end{array}}{\begin{array}{c} z \in y_1 \\ z \notin y_2 \end{array}} \ (3)
\qquad
\frac{\begin{array}{c} x = y_1 \setminus y_2 \\ z \in y_1 \\ z \notin y_2 \end{array}}{z \in x} \ (4)
$$

$$
\frac{\begin{array}{c} x = y_1 \setminus y_2 \\ z \in y_1 \end{array}}{z \in y_2 \mid z \notin y_2} \ (5)
\qquad
\frac{x = \{y\}}{y \in x} \ (6)
\qquad
\frac{\begin{array}{c} x = \{y\} \\ z \in x \end{array}}{z = y} \ (7)
\qquad
\frac{\begin{array}{c} y_1 \in x \\ y_2 \notin x \end{array}}{y_1 \neq y_2} \ (8)
$$

$$
\frac{x \neq y}{\begin{array}{c} w \in x \mid w \notin x \\ w \notin y \mid w \in y \end{array}} \ (9)^a
\qquad
\frac{\begin{array}{c} x = y \\ \phi \end{array}}{\phi_y^x} \ (10)^b
\qquad
\frac{\begin{array}{c} y = x \\ \phi \end{array}}{\phi_y^x} \ (11)^b
$$

---

[a] w must be a new variable not occurring on the branch to which the rule is applied.
[b] By ϕ_y^x we denote the formula resulting by substituting in ϕ each occurrence of x with y.

- SATISFIABLE, if there exists a set model for the literals occurring on ϑ.

A tableau $\mathcal{T}$ is said to be

- STRICT, or SATURATED, or CLOSED, if such are all of its branches;
- SATISFIABLE, or OPEN, if such is at least one of its branches.

Notice that according to the above definition, any closed branch, and therefore any closed **MLSS**-tableau, is unsatisfiable.

The system of rules (1)–(11) is plainly *sound*, namely any **MLSS**-tableau for a satisfiable normalized **MLSS**-conjunction must be satisfiable, and therefore must be open.

In addition, the tableau calculus in Table 1 is *complete*, namely any unsatisfiable normalized **MLSS**-conjunction has a closed **MLSS**-tableau. What is important for our decidability purposes is that completeness is not disrupted even when the tableau rules are subject to the following restrictions, which guarantee termination:

R1. all applications of tableau rules are strict;
R2. rule (9) is applied only to literals of the form $x \neq y$, with x and y occurring in the initial collection of **MLSS**-literals.

It can easily be seen that starting with an initial collection $\mathcal{S}$ of flat **MLSS**-literals, any tableau construction rule subject to the above restrictions R1 and R2 must terminate in a finite number of steps, generating a saturated tableau $\mathcal{T}_\mathcal{S}$ for $\mathcal{S}$. Then the decidability of **MLSS** follows from the fact that $\mathcal{S}$ is satisfiable if and only if the tableau $\mathcal{T}_\mathcal{S}$ is open.

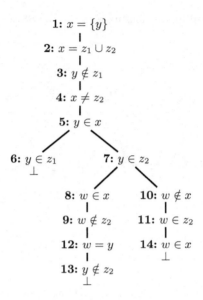

Fig. 1. A closed **MLSS**-tableau

From the soundness of rules (1)–(11), one only needs to check that if $\mathcal{T}_\mathcal{S}$ is open then $\mathcal{S}$ is satisfiable. Thus, let us assume that $\mathcal{T}_\mathcal{S}$ is open and let ϑ be an open (saturated) branch of $\mathcal{T}_\mathcal{S}$. Let

$V_\mathcal{S}$ be the collection of variables occurring in $\mathcal{S}$;

T be the collection of variables occurring on ϑ other than $V_\mathcal{S}$;

$\sim_\mathcal{S}$ be the equivalence relation induced on $V_\mathcal{S} \cup T$ by equality literals $x = y$ in ϑ;

T' be the set $\{t \in T : t \not\sim_\mathcal{S} x$, for all $x \in V_\mathcal{S}\}$;

V' be the set $(V_\mathcal{S} \cup T) \setminus T'$;

$\widehat{\in}_\vartheta$ be the dyadic relation on $V' \cup T'$ defined as follows:
$$x \widehat{\in}_\vartheta y \quad \text{iff} \quad \text{the literal } x \in y \text{ is in } \vartheta.$$

In addition, for each $t \in T'$, let $\boldsymbol{u}_t$ be an assigned set.

Since the branch ϑ is not closed, the relation $\widehat{\in}_\vartheta$ is acyclic. Therefore we can recursively define the following assignment, called the *realization* of the branch ϑ relative to $\mathcal{S}$ and the sets $\boldsymbol{u}_t$, for $t \in T'$:

$$R_\vartheta x = \{R_\vartheta y \mid y \widehat{\in}_\vartheta x\}, \qquad \text{if } x \in V'$$
$$R_\vartheta t = \boldsymbol{u}_t, \qquad \text{if } t \in T'.$$

It can be checked that if the sets $\boldsymbol{u}_t$ satisfy the conditions

(a) $\boldsymbol{u}_{t_1} \neq \boldsymbol{u}_{t_2}$, for every pair of distinct $t_1, t_2 \in T'$,

(b) $\boldsymbol{u}_t \neq R_\vartheta x$, for all $t \in T'$ and $x \in V'$, then the realization R_ϑ is a model for ϑ, and in turn for $\mathcal{S}$. Since conditions (a) and (b) can always be enforced, for instance by choosing $|T'|$ distinct sets u_t of large enough cardinalities, we have the completeness of our tableau calculus.

It is also interesting to note that the realization R_ϑ can be used on open non-saturated branches to guide the saturation process, as discussed in [4].

Figure 1 contains a closed **MLSS**-tableau for the collection

$$\mathcal{S} = \{x = \{y\},\ x = z \cup w,\ y \notin z,\ x \neq w\}$$

of flat **MLSS**-literals.

Notice that in the above **MLSS**-tableau

- literals 1–4 form the initial tableau for $\mathcal{S}$;
- literal 5 has been added by rule (6);
- literals 6 and 7 have been added by rule (2);
- literals 8–11 have been added by rule (9);
- literal 12 has been added by rule (7);
- literal 13 has been added by rule (10);
- literal 14 has been added by rule (1).

About Implicit and Explicit Shape Representation

Fiora Pirri

Dipartimento di Informatica e Sistemistica,
via Salaria 113, 00198 Roma, Italy

Abstract. We present a composite analysis of shapes based on form and features. We discuss how form and features are two facets of object representation and how similarity measures are used to understand the relation between two objects' images. We present a novel approach to approximate a shape that can still make use of Procrustes distance, leading to a relaxed notion of similarity measure. We introduce also a study on the similarity measures for non-parametric kernel densities. Finally we briefly discuss how these distance measures can be combined and represented into a Bayesian network, to learn the parameters of the defined similarity function.

To Gigina

1   Introduction

The human inner models of visual perception have been represented in several forms over the history of figurative art and attained a huge amount of structures modeling the body of symbolic features and traits. These structures, together with the way human beings perceive their representation, can provide a deep insight to automatic visual recognition. We argue that these models, accounting for the human representation of visual perception, can integrate those approaches to human object recognition (see e.g. [20,35,19]) inspired by the biological and neurophysiological aspects of human and animal perception.

Consider, for example, the paintings in Figure 1. If you look at the left drawing, few primitive traits are sufficient to denote a face, even a known face (Michelangelo's David); on the other hand in the second one a rich representation of shadows and lights of the lower part of a woman body can be misleading. In the third one, a particular of Seraut pointillism anticipates how shape emerges from pixelization. In this paper we argue that there are at least two possible aspects of shape representation, paradigmatically connoted by the traits of the David, by the shades in the Leonardo's studio of drapery and by the Seraut pointillism, and inspired by the human conception of representation. These two aspects can be investigated for recognition purposes, and they are the implicit and explicit representation of shapes. Where the term "explicit" is used to account for the main traits of what, in the human representation, connotes a principal source of information regarding a category, sort of visual synecdoche, e.g. the pupil for the

O. Stock and M. Schaerf (Eds.): Aiello Festschrift, LNAI 4155, pp. 141–158, 2006.

Fig. 1.

eye, a nostril for the nose, etc. The term "implicit", on the other hand, accounts for the shape rendered by shadows and lights, scale pixelization, and volumetric impression, sort of visual metonymy in which specific features account for the whole shape. This view of recognition is in between a mereological approach to shape (e.g. [39] for the gestaltic view, and [44] for the geometrical view) based principally on relations holding between parts, and the computational or biologically inspired approach, in which the shape is a signal to be interpreted analytically.

2   Primitives of Perception

A clear understanding of the concept of primitives in visual representation would solve most of the thorniest tasks of recognition. In a recent paper discussing the primitives of perception Chen [11] emphasises that "physically or computationally simple does not necessarily mean psychologically simple or perceptual primitive". In his seminal paper on *visual perceptual organization* [34] Pentland has pointed out that perception is successful because of an inner structuring of our environment and because of the human ability to identify the connections between these environmental regularities and primitive elements of cognition. The model-based approaches to perception have been strongly influenced by this view (see e.g. [15,46,38]). Among the model-based approaches, the constructive approach, known as recognition by components (RBC), was pioneered by [33], [43], [34], and especially by [7], and finally by [46,40]. Further [28] have introduced the concept of vantage points in the representation of an object components introducing the notion of aspect graph, and in [13] they suggested a 3D modeling of an object via a hierarchical aspect representation based on the projected surfaces of the primitives. Similarly in [36,37] objects categories are represented through their common parts, which are recognized according to a decomposition into primitives and recomposition is achieved according to an algebra of figures and a Bayes aspect graph. Recently (see e.g

[3,45,9,16,17]), in the stream of object representation approaches based on categories gathering similar parts, the problem has been faced in new terms considering features and appearances of parts, so as to overcome all the occlusion and vantage point problems raised in the model based approaches. In particular Perona and colleagues (see [10,16,17]) have proposed modeling objects as a constellation of parts proposing a successful method to learn object categories from cluttered data, with unsupervised labeling, in so relieving from the burden of manually labeling the images. In [17] features are found using the detector described in [24,25], and features are represented in an appearance space, where each part composing an object has a Gaussian density. Analogously both shapes and relative scales are represented by a joint Gaussian and thus the recognition model is based on maximum likelihood estimation of the parameters composition.

This paper is organized as follows, after few words of preliminaries, in Section 4 we introduce the explicit analysis and discuss in Section 5 the transformations of data representation in order to easily apply Procrustes methodology to deal with a relaxed notion of similarity. In Section 6 we discuss the distance measures for non parametric kernel densities and, finally, before concluding, in Section 7 we briefly hint about a combined distance learning methodology.

3   Preliminaries

We assume that an object is any element in the image that can be specifically named, e.g. a *cat*, a *table*, even if it is partially occluded. Note that we are not interested, here, in describing how an object is isolated from the background, or how features are extracted. For the sake of completeness we outline the segmentation methodologies used, from which the contour is obtained. In summary, for the extraction of the contour of a shape from a cluttered image we have been using two different segmentation methodologies depending on the type of available image. If the 3d information is available (e.g. the images are acquired from robot perception) then segmentation is attained by a k-mean clustering of the 3d image map, initialized with the histogram of the 3d information. Further, from the convex-hull of each segment a shape is obtained by combining the region of interest with the texture features. See the upper images of Figure 2. Finally with a Canny edge detector the contour is obtained. On the other hand, if the 3d information is not available, the segmentation methodology used is based on the assumption that the interesting object lies in the center of the image. On this basis an ellipse, whose dimension and orientation is subject to the gradient of the area, is drawn and from its texture a multivariate Gaussian mixture is obtained. From the back-projected image of the density (see the first picture of the third row in Figure 2) the contour is easily obtained. For general methods for features extraction (see e.g. [30]). For the purpose of illustrating our methodology on recognition we assume that the representation of an object is modeled by a pair $\langle E, I \rangle$ where E is the shape representation as a logical matrix, which we call the *drawing* and I is the feature representation of the points bound by the

Fig. 2. The preliminary analysis leading to the shape extraction. The upper images are obtained by K-mean clustering on a 3D image (by Bumblebee, @PtGray, on board of the robot, from which the images have been taken), conjugated with gradient and texture segmentation on the RGB image. The lower images illustrate different phases of feature extraction. Second row: the second image illustrates the gradient features of the image, the third image shows the ellipse from which textures are sampled. The segmented image, first image in the third row, is obtained by backprojecting the pdf of the multivariate Gaussian mixture computed from the sampled textures. The last image is obtained by edge detecting the segmented one, with suitable region growing to smooth the contours.

E contour, which we call the *painting*. Some examples of these pairs are given in Figure 3.

In the following, by $\mathbf{H} : M \times N \times K$ we denote a matrix $\mathbf{H}$ of dimension $M \times N \times K$. An image (or *figure*) is a matrix $\mathbf{I} : M \times N \times K$ such that M, N are the location dimension and K is the feature dimension. Each element (x, y) in the location dimension indicates the location of the pixel whose values belongs to the feature space. If the features space is the color space then it can either be boolean or ranging over intensities (in which cases the feature dimension is one) or over different colour representations (e.g. RGB, YIQ, HSV, YCbCr etc.). Namely $v : \mathbf{I} \mapsto V$ with V the feature space. By shape (or *drawing*) we mean an image $\mathbf{E}$ such that $V = \{0, 1\}$, and the shape is defined to be the location $\{x_i, y_i\}_{i=1...N}$ in $\mathbf{E}$ of the points set having value 1. By the painting we mean an instance $\{f_{1i}, \ldots, f_{4i}\}_{i=1..NM}$ of the feature space of the image $\mathbf{I} : M \times N \times 3$.

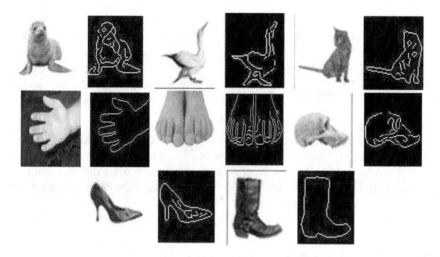

Fig. 3. Examples of pairs $\langle E, I \rangle$, on the left the *painting* and on the right the *drawing*, that we consider in our analysis for the implicit and explicit description of shapes

4   Explicit Analysis

Shape recognition relies on the use of inner products or distance measures between patterns. A rich literature is devoted to understanding the distance between generic shapes, to cite some consider the analysis of moments, shock analysis, skeleton analysis, Hausdorff distance, Procrustes methods (see [42, 29, 27, 32, 41]) and more, see also [6]. However for recognition we need to smooth the notion of distance and distill from it a notion of similarity that can suit the categories of interest, i.e. those learned or memorized. In the next paragraph we recall the notion of Procrustes distance.

4.1   Procrustes Distance

In general, two shapes are considered *congruent* if they differ by a rigid body transformation. We first consider the form, that refers to the figure with location and orientation removed. More specifically two figures $\mathbf{A} : N \times K$ and $\mathbf{B} : N \times K$ have the same form if:

$$\mathbf{B} = \mathbf{A}\Gamma + \mathbf{1_N}\gamma^\top \tag{1}$$

where $\Gamma : K \times K$ is a rotation $|\Gamma|=1$, $\mathbf{1_N}$ is a vector of ones and $\gamma : K \times 1$ is a translation.

On the other hand the *shape* refers to the equivalence class of figures having the same form suitably scaled, so that two figures $\mathbf{A}$ and $\mathbf{B}$ have the same shape, i.e. they belong to the same equivalence class $[\mathbf{A}] = [\mathbf{B}]$, if:

$$\mathbf{B} = \beta\mathbf{A}\Gamma + \mathbf{1_N}\gamma^\top \tag{2}$$

with $\beta > 0$ a scalar. In Procrustes analysis the transformations are found by premultiplying the shape $\mathbf{A}$ by an $(N-1) \times N$ Helmert matrix $\mathbf{H}$. Where $\mathbf{H}$ has orthonormal rows, each orthogonal to the unit vector $\mathbf{1}_N/\sqrt{N}$. The rows of the matrix $\mathbf{A}_\mathbf{H} = \mathbf{HA}$ are the coordinates of the derived landmarks. The centered landmarks, i.e. with the location removed, are obtained by the derived landmarks as

$$\mathbf{A}_C = \mathbf{H}^\top \mathbf{A}_\mathbf{H} \tag{3}$$

The matrix of the derived landmarks is said to be in preform space $\mathbb{R}^{(N-1)K}$, while the original figure is in figure space $\mathbb{R}^{(N)K}$ (Goodall [18] notes that any statistical model for the matrix in the preform space can be derived from the figure matrix in the figure space). The derived landmarks are centered and scaled by:

$$\mathbf{Z}_A = \mathbf{H}^\top \frac{\mathbf{HA}}{||\mathbf{HA}||} \tag{4}$$

The Procrustes distance between $\mathbf{A}$ and $\mathbf{B}$ is thus

$$d(\mathbf{A},\mathbf{B}) = \inf_{\Gamma,\beta} ||\mathbf{Z}_B - \beta\mathbf{Z}_A\Gamma|| \tag{5}$$

Where $\Gamma = \mathbf{UV}^\top$, with $\mathbf{U}$ and $\mathbf{V}$ obtained by the singular values decomposition of $\mathbf{Z}_A^\top\mathbf{Z}_B$ and:

$$\beta = \sum_i^k \lambda_i, \quad \lambda_1 \geq \lambda_2 \geq \ldots \geq \lambda_k, \text{ the singular values}$$

Different distance representations are given by Bookstein and Kendall [8,26].

5   Approximating PA

Procrustes analysis of shapes requires shapes to have the same form, in the sense of equation (1). Under this perspective a major issue in comparing two shapes using the Procrustes methodology is data representation. Given figure $\mathbf{A} : K \times N$, and figure $\mathbf{B} : M \times P$, their shapes are given by the 2D landmark representation $A = \{x_i, y_i\}_{i=1..KN}$ and $B = \{x_j, y_j\}_{j=1..MP}$ obtained by choosing a set of landmarks in $\mathbb{R}^2$. Usually the landmarks are chosen from those pixels having value 1. Since Procrustes distance is based on least squares, it asks for an exact correspondence between the two shapes instances to be compared, that is, points set $A = \{x_i, y_i\}_{i=1..N}$ is aligned to points set $B = \{x_j, y_j\}_{j=1..K}$, w.r.t. a transformation group $\mathcal{T}$ if the distance cannot be decreased by applying to B a transformation from $\mathcal{T}$ (see [14]). Furthermore the size of the two sets of landmarks should be the same. However, it is easy to see that, given a figure $\mathbf{A} : K \times N$, and its shape as points set $A = \{x_i, y_i\}_{i=1..M}$, there are 2^M possible instances of the points set of $\mathbf{A}$, each constituting an approximate representation according to some linear transformation. Hence the Procrustes distance between two shapes varies according to the choice of the instances, i.e. of the

Fig. 4. Two similar cats with slit change in the head and tail position. The non continuous traits show the approximation obtained from the critical points.

point set. Because we want to use Procrustes distance for the general problem of establishing the similarity between two shapes, we need to precise this notion better.

Consider Figure 4, the two shapes (say **A** and **B**) are different under rigid transformations, yet similar. Our cognitive understanding of similarity relies on the fact that the two shapes represent two cats in similar position. And indeed similarity is far from being a precise measure, as it can only be approximated. Under which conditions two shapes can be considered similar, given that we do not know the class (or category) they belong to, is still an unsolved question.

Our hypothesis is the following. Let τ be a threshold value for the Procrustes distance between two shapes (see equation 5), e.g. we take $\tau = 0.59$. Let $\mathbb{P}$ be the set of shapes labeled by a specific category (e.g cats), then we say that two instances $A = \{x_i, y_i\}_{i=1..M}$ and $B = \{x_j, y_j\}_{j=1..N}$ of shapes **A** and **B**, respectively, are similar if:

i. there exist approximations $t_1(A)$ of the instance A of **A** and $t_2(B)$ of the instance B of **B** such that $d(\Sigma_X, \Sigma_{t_i(X)}) = 0$, $i = 1, 2$, $X \in \{A, B\}$, where d is the Procrustes distance, Σ_Y is the empirical variance-covariance matrix of Y, and $t_i(X) = \{x_h, y_h\}_{h=1..K}$ with K as required in the next item.

ii. There exist instances $C_1 = \{x_i, y_i\}_{i=1..M}, C_2 = \{x_j, y_j\}_{j=1..N}$ in some predefined category $\mathbb{P}$ such that Y and C_i have the same size and $d(Y, C_i) \leq \tau$, $i = 1, 2$, $Y \in \{t_1(A), t_2(B)\}$, where d is the Procrustes distance.

The first condition simply says that whenever a transformation is applied to the input data the transformation has to be statistically consistent, i.e. the new set of points will preserve the empirical variance-covariance. This is an obvious requirement, otherwise similarity would be trivially satisfied: take a single point $\mathbf{x} = (x_0, y_0)$, from each of the two shapes, they are obviously similar.

And the second condition says that the concept of similarity for a shape can be established only if a category for that shape is defined. Under the above conditions the rigid transformation is weakened, because the least square is carried on a subspace $(N - K)2$ of the original figure space $(N)2$.

By an approximation $Y = t(X)$ of an instance $X = \{x_i, y_i\}_{i=1..N}$ of a shape we mean a linear transformation

$$Y = (X^\top W_1)H_1^\top + \ldots + (X^\top W_n)H_n^\top \tag{6}$$

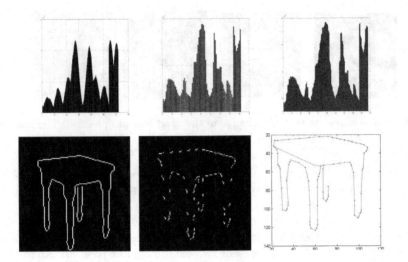

Fig. 5. In the images on the top the probability of each points set on the contour of shape **A** depicted below, minima and maxima corresponds to the critical points. In the images below, the first is the edge detection, the second is the points set obtained by approximation, the third is its polygonal reconstruction.

such that $Y = \{x_j, y_j\}_{j=1..K}$, with $Y \subset X$, preserving the covariance under rigid transformations:

$$\Sigma_Y = \beta \Sigma_X \Gamma + \mathbf{1_N} \gamma^\top$$

Here $W_i : N \times 2$ is the selection matrix, of all zeroes but a 1 in the first column of row i; and $H_j : 2 \times K$ is the positioning matrix, of all zeroes but the column j, made of ones. The transformation shall select from the shape the salient points, those that characterize the shape. To understand how an approximation satisfying these constraint is built, consider the complete instance $X = \{x_i, y_i\}_{i=1..N}$ of the initial shape **A**. The critical points of a shape are the local minima, local maxima and the saddle points, easily obtainable by the Hessian determinant, only if for each $\{x_i, y_i\} = p \in X$ the following condition is met: if q is contiguous to p then $q = \{x_{i+1}, y_{i+1}\}$ or $q = \{x_{i-1}, y_{i-1}\}$, hence from the matrix representation (which otherwise is described column wise: $(x_1, y_1), (x_2, y_2), \ldots, (x_{MK}, y_{MK})$ are sorted according to the column indexing), the contour path $t(X)$ is obtained as follows.

Following a clock-wise direction, add to the current point $(x_i, y_i) \in t(X_i)$ the matrix $Q : 8 \times 2$:

$$\mathbf{1_8}(x_i, y_i) + \begin{bmatrix} -1 & -1 & 0 & 1 & 1 & 1 & 0 & -1 \\ 0 & 1 & 1 & 1 & 0 & -1 & -1 & -1 \end{bmatrix}^\top$$

Let $t(X_k)$ be the contour reconstructed up to point (x_q, y_q), $k \leq q$, then if the path reaches a dead end it must follow its steps back by jumping out of the loop, finding a point $p' \notin t(X_k)$ with minimal distance:

$$p' = min_\delta \delta(p, p'), p \in t(X_k), p' \notin t(X_k)$$

Once the contour of the shape is defined as a continuous path of adjacent points then critical points $\mathbf{Cp}$ of $t(X)$ can be easily determined, according to the rules of the Hessian. Note that the critical points corresponds also to the local minima and maxima of the probability $p_X(x_i, y_i) = \mathcal{N}(\mu_X, \Sigma_X)$ (see Figure 5).

Given the critical points $\mathbf{Cp}$ an approximation $Y = \{x_j, y_j\}_{j=1..K}$ satisfying the condition (ii.) above can be found considering the neighbourhood of each $\{x_k, y_k\} \in \mathbf{Cp}$ up to a specific distance δ, less than a threshold ϵ. Note that the distance has to be defined on the contour, let $\mathbf{x}_j = (x_j, y_j)$:

$$\mathbf{n}(\mathbf{x}_i) = \{\mathbf{y} \in t(X) \,|\delta(\mathbf{x}_i, \mathbf{y}) < \epsilon \wedge$$
$$\forall \mathbf{z}.\mathbf{n}(\mathbf{y}, \mathbf{z}) \wedge \mathbf{n}(\mathbf{z}, \mathbf{x}_i) {\to} \delta(\mathbf{z}, \mathbf{x}_i) \le \epsilon\}$$

Thus starting with $\mathbf{Cp}$ the approximation $Y = \{x_j, y_j\}_{j=1..K}$ can be iteratively constructed using the linear transformation (6) by adding further neighbours, at each step, according to a clockwise, left-handed direction, as far as condition (ii) is satisfied and the approximation Y meets the approximation Z of the shape $\mathbf{B}$, with which the similarity has to be established; that is, the two approximations need to have the same size. We can thus express the notion of *pre-similarity*, in which $\mathbf{B}$ is the reference category of $\mathbf{A}$ (to comply with definition (ii.)), as follows. Let $\mathbf{A}$ and $\mathbf{B}$ be two shapes and $A = \{x_i, y_i\}_{i=1..N}, B = \{x_i, y_i\}_{i=1..M}$ any instances of them. We say that $\mathbf{A}$ and $\mathbf{B}$ are pre-similar according to linear transformations t_1 and t_2, and we denote it by $\sim_p$ according to the following definition:

$$\mathbf{A}(t_1) \sim_p \mathbf{B}(t_2) \equiv d(\Sigma_{t_1(A)}, \Sigma_A) = 0 \wedge d(\Sigma_{t_2(B)}, \Sigma_B) = 0 \wedge |t_1(A)| = |t_2(B)|$$

Here $|.|$ is used to denote the size and Σ_Y is the empirical variance covariance matrix of Y. Now, given that two shapes are pre-similar, then they are similar with respect to the approximation (denoted by $\sim_A$) according to:

$$\mathbf{A} \sim_A \mathbf{B} \equiv \mathbf{A}(t_1) \sim_p \mathbf{B}(t_2) \wedge$$
$$\mathbf{B}(t_2) = \beta \mathbf{A}(t_1)\Gamma + \mathbf{1_N}\gamma^\top \tag{7}$$

Here $\Gamma : K \times K$ is a rotation $|\Gamma|=1$, $\mathbf{1_N}$ is a vector of ones, $\beta > 0$ is a scalar and $\gamma : K \times 1$ is a translation, t_1 and t_2 indicate the linear transformation for the approximation.

Finally, from the approximation a polygonal shape can be reconstructed, see Figure 5, connecting the points set obtained with a line. Given the polygonal approximation several distances on polygons, such as for example the turning function Θ_A of [4], the Hausdorff distance [21], or the Frechet distance (e.g. see [2]) can be used to check the distance between the two shapes.

The experiments are described in Section 8.

6   Implicit Analysis

For the implicit component of recognition we consider a non-parametric density estimation of the joint location-feature space. Let $I = \{x_i, y_i, k_i, \nabla k_i\}_{i=1..N}$ be the sample points in the implicit model of the image, where $(x_i, y_i) = \mathbf{x}$ denotes the 2D coordinates of the i-th point, k_i is the intensity of $\mathbf{x}$, normalized, and ∇k_i is the gradient. Note that I might have been earlier smoothed by a Gaussian filter. We consider a Gaussian kernel $\mathcal{K} = (1/2\sqrt{2\pi}) \exp\left(-1/2u^2\right)$, and the non-parametric density as a product of kernels as follows:

$$\hat{f}_h(\mathbf{z}) = \frac{1}{n} \sum_{i=1}^{n} \left\{ \prod_{j=1}^{4} h^{-1} \mathcal{K}(\frac{\mathbf{z}_j - X_{ij}}{h_j}) \right\}$$

Here each bandwidth is obtained from the median $\overline{m}$ as follows:

$$h_j = \frac{1}{w}\overline{m}(abs(X_{ij} - \overline{m}(X_{ij})))$$

where w depends on the standard normal distribution, at the points of the r, g, b components of I. Densities of some of the objects in Figure 9 are illustrated in Figure 6.

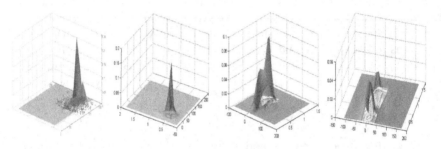

Fig. 6. Non-parametric density estimations with Gaussian Kernels. The first two kernels are from the r and g component of the first two dogs in Figure 9, and the second and third kernels are from the first two sofas in Figure 13.

Now, given two implicit models M_1 and M_2 of two images I_1 and I_2 we establish their *implicit* similarity according to a information-theoretic distance between their pdfs $\hat{f}_1$ and $\hat{f}_2$. We have, indeed, considered some of the distances reported in [1]) and a recently introduced one (see [47]). The distances performance, obtained by our experiments, is illustrated in Figure 7, note that we have grouped the distances consistently with their module and have computed in particular that of [47] according to our defined bandwidth. It is clear from the graphs in Figure 7 that those performing better are the one of [47] and the Bhattacharyya one. However Yang et Al. distance is particularly interesting for non-parametric estimation because it does not require a previous computation of the pdf, hence it is certainly the most suitable, beside being quite accurate, as emerges from our experiments. The advantage of a non-parametric estimation is due both to the fact that the number of components needs not to be known, and to the greater adaptability to object represented by composite features. However the difficulty with non-parametric densities is the need to memorizing the

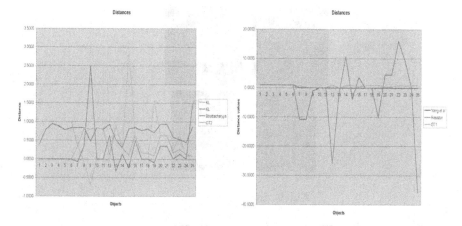

Fig. 7. Two groups of information theoretic distances over the estimated pdf of the objects illustrated in Figure 9. Ground truth is in light green.

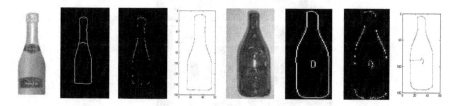

Fig. 8. Similarity=0.27

whole shape, while with parametric models, such as Gaussian mixtures, only the parameters need to be allocated.

Recently Jebara and Kondor (see [22,23]) have introduce a new class of kernels that can be used with parametric models. Their general *Probability Product Kernel K* is:

$$K_\rho(p, p') = \int p(x)^\rho p'(x)^\rho dx$$

With a close form solution for the exponential family distributions (see [5]). We have thus considered also Gaussian mixtures and the expected likelihood kernel proposed by Lyu in [31]. From these last experiments the results seem to be less stable than the distances studied for non parametric densities. The experiments for the distances on implicit shapes are reported in Section 8.

7 Combining Distances and Causal Relationships

In the previous sections we have described two methods for shape approximate representation, namely the explicit and implicit ones, leading to two distance

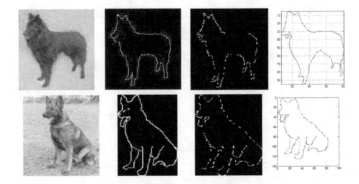

Fig. 9. Similarity=0.43

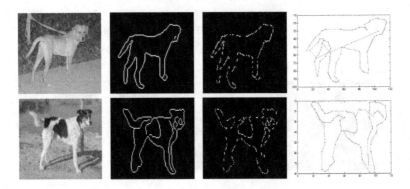

Fig. 10. Similarity=0.58

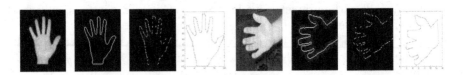

Fig. 11. Similarity=0.47

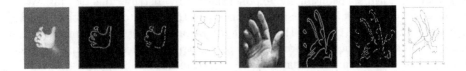

Fig. 12. Similarity=0.76

measures d_e, which is the distance between the polygons obtained by the transformation associated with the Procrustes analysis, and the distance d_i, which is the distance between the kernels representing the set of selected features of the shapes. As addressed in [16,17] each recovered part can constitute an hypothesis of being the component of a structured object, and an hypothesis can be certainly reinforced by the presence of some component which might be clearly identified, such as a leg, or a ear, or a foot. This is a crucial aspect to be considered for learning the combined similarity measure. We could suppose that the two distances, d_e and d_i can be linearly combined to form the distance $d = \alpha d_e + \beta d_i$, but α and β are unknown. Because the probability that an object A belongs to a category C depends on the distance between A and the elements of C one can take α and β to be the likelihood ratio between the successful matches against all the comparisons. For example in our experiments we have empirically established a threshold on the basis of the correct matches against the whole amount of tests. So the probability that A belongs to a category C given the background knowledge $\mathcal{D}$ is:

$$P(A = C|\mathcal{D}) = \frac{\sum_{c \in \mathcal{C}} (\alpha d_e(A, c) + \beta d_i(A, c))U}{\sum_{d \in \mathcal{D}} (\alpha d_e(A, d) + \beta d_i(A, c))} \tag{8}$$

Here $U = I(\alpha d_e + \beta d_i > \tau)$ is an indicator. To this end it is reasonable to configure a model on the basis of structural relationships among the elements, or on the basis of hierarchical relations. In this way, for example, exploiting independence between certain elements, as in Bayesian networks, it would be both possible to determine the parameters for the distance without any concern about mutual information, and to add nodes to the network as soon as a certain distance is learned. This would be facilitated if, on the basis of an excluded middle principle peculiar to objects, it is assumed that an object cannot belong to two categories at the same time. Therefore if B and D are independent categories then the estimate of the parameters for A would be:

$$\begin{aligned} P(A = C|B, D) &= \frac{P(A = C|B)P(D|B, A = C)}{P(D|B)} \\ &= \frac{P(A = C|B)P(D|A = C)}{P(D)} \\ &= P(A = C|B)P(A = C|D) \end{aligned} \tag{9}$$

Therefore if one assume that there exists a network structure $\mathbf{H}$, with parameters θ (which in our case are the α_i and β_i) and Pa are the parents of the element of the current category of interest for the observation at hand, then for $c_i \in C$:

$$P(A = C|\mathbf{H}, \theta) = \prod_{i=1}^{n} P(A = c_i|Pa(c_i), \theta_i, \mathbf{H})p(\theta_i|\mathbf{H})$$

Which amount, indeed, to learning the parameters, i.e. to estimate $P(\theta_i|\mathbf{H})$, by maximum likelihood estimation. It is easy to see that even under a different setting, the problem of capturing the underlying structure of similarity between objects and their parts, at the end reduce to the approach of [17] and other analogous approaches, in which the model is based on fitting Gaussian mixtures and hypotheses are mixture hidden variables recovered via the EM.

Fig. 13. Similarity measures of non-parametric kernel densities, from 1 (left) to 4(right):

	KL	B	Y
1 − 2	0.36	0.84	0.67
1 − 3	NaN	0.79	0.04
3 − 4	0.63	0.93	0.03

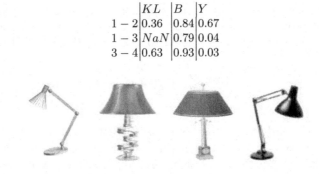

Fig. 14. Similarity measures of non-parametric kernel densities, from 1 (left) to 4(right):

	KL	B	Y
1 − 2	0.81	0.59	0.008
1 − 3	0.067	0.78	0.006
1 − 4	0.19	0.81	0.002
2 − 3	0.09	0.71	0.019
2 − 4	0.4	0.78	0.004

The experiments are summarized by the graphs in Figure 17, we have considered the sensitivity at a cut point of 0.59 below this value we considered the similarity accepted and above rejected, thus the cases of dissimilar shapes are penalized. The ROC curve is illustrated in Figure 17.

8   Experiments

In this section we show the results of some of the experiments concerning the two distance measures d_e and d_i with respect to explicit shape description (d_e) and implicit shape description (d_i). In the first set of images, namely Figure 9, till Figure 12 we show two pairs of image sequences each composed of 4 images as follows. The first is the input image. The second is the edge image obtained by edge detection (here we used Canny algorithm), after a smoothing with a Gaussian filter with variance varying in [0.3 0.6] according to the entropy of the image. For instance

Fig. 15. Similarity measures of non-parametric kernel densities, from 1 (left) to 4(right):

	KL	B	Y
$1-2$	0.34	0.84	0.01
$1-3$	NaN	0.88	0.006
$1-4$	0.66	0.84	0.003
$2-3$	NaN	0.88	0.012
$2-4$	0.02	0.85	0.003
$3-4$	0.005	0.71	0.003

Fig. 16. Similarity measures of non-parametric kernel densities, from 1 (left) to 4(right):

	KL	B	Y
$1-2$	0.62	0.82	0.006
$1-3$	0.99	0.70	0.009
$1-4$	NaN	0.69	0.005
$1-5$	NaN	0.85	0.028
$2-3$	0.06	0.70	0.007
$2-4$	0.66	0.70	0.011
$2-5$	0.9	0.65	0.015

the image of the second dog in Figure 9 has entropy $e = -\sum(p\log(p)) = 6.67$ and we used a variance of 0.4 while the first dog has entropy 7.5 and we used a variance of 0.5, according to the formula $\sigma = (e10^3)/(NK)$ with N, K the size of the image. The third image is the approximation obtained as described in Section 5. The last image is the polygonal reconstruction obtained by the approximation. The value in the caption of the two sequences denote the Procrustes distance computed between the approximations. We do not report here the values of the other distances (e.g the Hausdorff, the Frechet etc. and the distance based on the turning function), computed over the polygonal reconstruction.

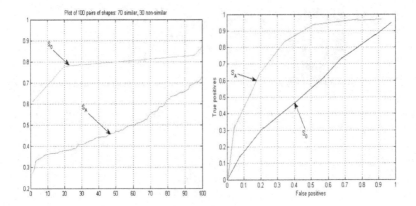

Fig. 17.

In the second set of images we report on the similarity measures for the non parametric kernel density. In particular we report on the three distances, Kullback-Leibler (KL), Bhattacharyya coefficients (B) (as presented in [12]), and Yang et Al. (Y) [47]. Note that while Bhattacharyya upperbound for similarity is 1, the similarity measure of Yang et Al. is reversed, i.e. two distributions are similar if their distance is closer to 0.

9   Conclusions

In this paper we have related the similarity between objects to their explicit and implicit representation. In other words we have discussed how the approximation of the shapes of two objects can influence their similarity. In particular we have shown how to extend Procrustes distance to cope with a notion of similarity that is not constrained to a rigid definition of form, by introducing the novel concept of instance of a shape. We have also compared different similarity measures for the nonparametric kernel density. Finally we have hinted on how these two distances could be combined to learn a more reliable similarity measure, considering the causal relations of an object with its part or categories, as represented by a Bayes network.

Acknowledgments. I have begun working on theory of perception with Gigina, and despite most of her work has been on the mechanization of reasoning, she has been interested in extending reasoning with perception. That's why I thought that dedicating this paper to her would have been a way to acknowledge her contribution also on this aspect of cognition.

References

1. S. M. Ali and S. D. Silvey. A general class of coefficients of divergence of one distribution from another. *J. Roy. Statist. Soc. Ser B*, 28:131–142, 1966.
2. H. Alt and M. Godau. Computing the fréchet distance between two polygonal curves. *Internat. J. Comput. Geom. Appl.*, 5:75–91, 1995.

3. Y. Amit and D. Geman. A computational model for visual selection. *Neural Computation*, 11(7):1691–1715, 1999.
4. E. M. Arkin, L. P. Chew, D. P. Huttenlocher, K. Kedem, and J. S. B. Mitchell. An efficiently computable metric for comparing polygonal shapes. *IEEE Trans. Pattern Anal. Mach. Intell.*, 13(3):209–216, 1991.
5. O. Barndorff-Nielse. *Information and exponential families in statistical theory.* John Wiley & Sons, 1978.
6. S. Belongie, J. Malik, and J. Puzicha. Shape matching and object recognition using shape contexts. *IEEE Trans. Pattern Anal. Mach. Intell.*, 24(4):509–522, 2002.
7. I. Biederman. Recognition by components - a theory of human image understanding. *Psychological Review*, 94(2):115–147, 1987.
8. F.L. Bookstein. Size and shape spaces for landmark data in two dimensions (with discussion) statist. sci. *Journal of the Royal Statistica Society, Series B (Methodological)*, 4:181–242, 1986.
9. E. Borenstein and S. Ullman. Class-specific, top-down segmentation. In *ECCV (2)*, pages 109–124, 2002.
10. M. C. Burl, M. Weber, and P. Perona. A probabilistic approach to object recognition using local photometry and global geometry. In *ECCV '98: Proceedings of the 5th European Conference on Computer Vision-Volume II*, pages 628–641, London, UK, 1998. Springer-Verlag.
11. L. Chen. The topological approach to perceptual organization. *Visual Cognition*, 12(4):553–637, 2005.
12. D. Comaniciu, V. Ramesh, and P. Meer. Kernel-based object tracking. *IEEE Trans. Pattern Anal. Mach. Intell*, 25(5):564–575, 2003.
13. S. Dickinson, A. Pentland, and A. Rosenfeld. 3d shape recovery using distributed aspect matching. *PAMI*, 14(2):174–198, 1992.
14. N. Duta, A. K. Jain, and M.-P. Dubuisson-Jolly. Learning 2d shape models. In *CVPR*, pages 2008–2014, 1999.
15. S. Edelman. Computational theories of object recognition. pages 296–304, 1997.
16. R. Fergus, P. Perona, and A. Zisserman. Object class recognition by unsupervised scale-invariant learning. In *CVPR (2)*, pages 264–271, 2003.
17. R. Fergus, P. Perona, and A. Zisserman. A sparse object category model for efficient learning and exhaustive recognition. In *CVPR (1)*, pages 380–387, 2005.
18. C. Goodall. Procrustes methods in the statistical analysis of shape. *Journal of the Royal Statistical Society*, 53(2):285–3399, 1991.
19. K Grill-Spector. The neural basis of object perception. *Curr. Opin. Neurobiology*, 13(2):159–166, 2003.
20. J. V. Haxby, M. I. Gobbini, M. L. Furey, A. Ishai, J. L. Schouten, and P. Pietrini. Distributed and overlapping representations of faces and objects in ventral temporal cortex. *Science*, 293(5539):2425–2430, 2001.
21. D. P. Huttenlocher, G. A. Klanderman, and W. Rucklidge. Comparing images using the hausdorff distance. *IEEE Trans. Pattern Anal. Mach. Intell.*, 15(9):850–863, 1993.
22. T. Jebara and K. Kondor. Bhattacharyya and expected likelihood kernels. *Conference on Learning Theory, COLT/KW*, 2003.
23. T. Jebara, K. Kondor, and A. Howard. Probability product kernels. *Journal of Machine Learning Research, JMLR, Special Topic in Learning Theory*, 5:819–884, July 2004.
24. T. Kadir and M. Brady. Saliency, scale and image description. *International Journal of Computer Vision*, 45(2):83–105, 2001.

25. T. Kadir, A. Zisserman, and M. Brady. An affine invariant salient region detector. In *ECCV (1)*, pages 228–241, 2004.
26. D. G. Kendall. Shape-manifolds, procrustean metrics and complex projective spaces. *Bull. Lond. Math. Soc.*, 16:811–121, 1984.
27. A. Khotanzad and Y. H. Hong. Invariant image recognition by zernike moments. *IEEE Trans. Pattern Anal. Mach. Intell.*, 12(5):489–497, 1990.
28. J. Koenderink and A. van Doorn. The internal representation of solid shape with reference to vision. *Biological Cybernetics*, 32:211 – 216, 1979.
29. F. Leymarie and M. D. Levine. Simulating the grassfire transform using an active contour model. *IEEE Trans. Pattern Anal. Mach. Intell.*, 14(1):56–75, 1992.
30. D. G. Lowe. Distinctive image features from scale-invariant keypoints. *International Journal of Computer Vision*, 60(2):91–110, 2004.
31. S. Lyu. Kernel between sets: the gaussian mixture approach. Technical Report TR1005-214, Computer Science Department, Dartmouth College, 2005.
32. K. V. Mardia. Shape statistics and image analysis. In *ACCV*, pages 297–306, 1995.
33. D. Marr and H. Nishihara. Representation and recognition of the spatial organization of three-dimensional shapes. In *Proc. R. Soc. Lond. B, vol. 200*, pages 269–294, 1978.
34. A. Pentland. Perceptual organization and the representation of natural form. *Artificial Intelligence*, 28(2):293–331, 1986.
35. P. Pietrini, M. L. Furey, E. Ricciardi, M. I. Gobbini, W. H. Wu, L. Cohen, M. Guazzelli, and J. V. Haxby. Beyond sensory images: Object-based representation in the human ventral pathway. In *Proc Natl Acad Sci USA*, volume 101, pages 5658–5653, 2004.
36. F. Pirri. The usual objects: a first draft on decomposing and reassembling familiar objects images. In *Proceedings of XXVII Annual Conference of the Cognitive Science Society*, pages 1773–1778, 2005.
37. F. Pirri and M. Romano. 2d qualitative recognition of symgeon aspects. In *Proc. KES 2003*, volume 2774 of *Lecture Notes in Computer Science*, pages 1187–1194. Springer, 2003.
38. A.R. Pope and D.G. Lowe. Learning object recognition models from images. In *ICCV93*, pages 296–301, 1993.
39. N. Rescher and P. Oppenheim. Logical analysis of gestalt concepts. *British Journal for the Philosophy of Science*, 6:89–106, 1955.
40. E. Rivlin, S. J. Dickinson, and A. Rosenfeld. Recognition by functional parts. *Computer Vision and Image Understanding: CVIU*, 62(2):164–176, 1995.
41. G. Rote. Computing the minimum hausdorff distance between two point sets on a line under translation. *Information Processing Letters*, 38:123–127, 1991.
42. T. B. Sebastian, P. N. Klein, and B. B. Kimia. Recognition of shapes by editing their shock graphs. *IEEE Trans. Pattern Anal. Mach. Intell.*, 26(5):550–571, 2004.
43. L. G. Shapiro, J. D. Moriarty, R. M. Haralick, and P. G. Mulgaonkar. Matching three-dimensional objects using a relational paradigm. *Pattern Recognition*, 17(4):385–405, 1984.
44. A. Tarsky. *Logic, Semantics, Metamathematics*, chapter Foundation of the Geometry of Solids, pages 24–29. Oxford, The Clarendon Press, 1927.
45. M. Weber, M. Welling, and P. Perona. Unsupervised learning of models for recognition. In *ECCV (1)*, pages 18–32, 2000.
46. K. Wu and M. Levin. 3D object representation using parametric geons. Technical Report CIM-93-13, CIM, 1993.
47. C. Yang, R. Duraiswami, and L. S. Davis. Efficient mean-shift tracking via a new similarity measure. In *CVPR (1)*, pages 176–183, 2005.

Agents, Equations and All That: On the Role of Agents in Understanding Complex Systems

Roberto Serra and Marco Villani

Dipartimento di Scienze Sociali, Cognitive e Quantitative
Università di Modena e Reggio Emilia
via Allegri 9, 42100 Reggio Emilia
{rserra, mvillani}@unimore.it

This paper is dedicated to Luigia Carlucci Aiello on her 60th birthday

Abstract. Differential equations and agent-based models are different formalisms which can be applied to describe the evolution of complex systems. In this paper, it is shown how differential equations can describe interactions among agents: it is pointed out that their capabilities are broader than is often assumed, and it is argued that such an approach should be preferred whenever applicable. Also discussed are the circumstances in which it is necessary to resort to agent-based models, and a rigorous approach is advocated in these cases. In particular, the relationship between the model and a theory of the processes under consideration provides both stimuli and constraints for the model. This relationship is discussed both in general terms and with reference to a specific example, which concerns a model of innovation processes.

1 Introduction

In this paper we will discuss some problems related to modelling and understanding complex systems. For the sake of definiteness we will mainly refer to social systems (although some remarks apply also to other kinds of systems, e.g. biological, artificial). This choice is largely motivated by the fact that Artificial Intelligence methods can be particularly useful in describing the interactions among different actors in a social setting.

The notion of "agent", which is widespread in Artificial Intelligence[1], has indeed contaminated several other disciplines, including Economics and the Science of Complex Systems. In the former, agent-based models have even been proposed as an alternative foundation of the discipline [2], in opposition to the so-called neoclassical approach, which is still the prevailing paradigm.

---

[1] The literature on this subject is too broad to even try a partial account; for a textbook on AI which is based on the notion of agents see [16].

O. Stock and M. Schaerf (Eds.): Aiello Festschrift, LNAI 4155, pp. 159–175, 2006.
© Springer-Verlag Berlin Heidelberg 2006

A side affect of the pervasiveness of this metaphor is that the word "agent" is used in the scientific literature with fairly different, although overlapping, meanings – which differ not only among the various disciplines, but also within each one. While this may sometimes be a source of confusion, the fuzziness of this concept has been instrumental so far in helping the development of new models and the transfer of ideas among disciplines.

Agents have been widely applied to describe the behaviour of complex systems, in particular those which are able to adapt or to learn. Agent-based models are usually contrasted to those based on differential, or difference, equations, which are often easily dismissed because they are thought to represent an over-simplified description of the real system. However, in so doing one gives up a whole set of theorems and techniques which might be valuable in order to understand the behaviour of strongly nonlinear systems.

In this paper, we will compare agent-based and equation-based approaches, showing that the latter are more expressive than usually thought. We will also stress that the supposed need for using a complicated "element" (the individual unit) should be demonstrated and not simply given a priori, since even simple elements can account for complicated behaviours.

Moreover, we will argue that the real advantages of agent-based modelling can be found in those cases where the elements must have sophisticated information processing capabilites and an internal structure, and where heterogeneity among different elements cannot be easily accounted for by choosing different parameter values.

However, a major difficulty in this kind of modelling is a lack of robust foundations: the modeller is left, in a sense, with too much freedom to choose the relevant variables and processes. Given the nonlinearity of agent-based models, they may display very different behaviours, and it is possible that one of these behaviours resembles some features observed in the social system one wants to describe. However, since observational data are often scarce, this agreement provides only a weak argument in favour of the validity of the model itself.

In order to deal with this issue, it has been proposed to relate the behaviour of the model to a theory of the social system. The relationship between theory and model is discussed with reference to a specific example, that of an agent-based model of innovation processes.

2   Agent-Based Models and Dynamical Equations

A formal dynamical model is a recipe which allows one to compute the future history of a system, from the knowledge of its present state and of the history of its interactions with the external environment, if any. The model may be deterministic or stochastic; in the latter case it is possible to compute a probability distribution of future histories, rather than a single outcome.

Let us suppose that the system is composed of a set of agents, interacting with each other and with some "external" variables, i.e. variables whose time evolution is not determined within the system itself. Such agents will belong to a single class or to a limited number of different classes. The agents which belong to the same class are similar, but not necessarily identical to each other.

Such a system is often modelled by an agent-based model (ABM), where the elements are described by suitable algorithms. The agents interact with each other and with an "environment" which they inhabit.

Note however that also a system of difference (or differential) equations can be described using an "agent" metaphor. For the sake of definiteness we will consider here a system of time-discrete difference equations of the form

$$x(t+1) = f(x(t)) \qquad (1)$$

where $x = (x_1, x_2 \ldots x_N)$. It would be straightforward to generalize the discussion below to the case of ordinary differential equations.

One can think of N agents, each one with an associated numerical variable. Agents interact in the way defined by Eq.1, which can also be given a graph representation: the N agents are associated to the nodes of the graph, and there is a direct link from node i to node k if $x_i(t)$ appears on the r.h.s of the equation for $x_k(t+1)$.

Note that the formalism of first order systems, like the one in Eq.1, can describe also higher order systems, where the values of the original state variables at time $t+1$ are influenced by previous values at time $t-1$, $t-2$, etc. In order to do so, it suffices to enlarge the set of state variables, including memory variables. For example, a second-order equation of the form

$$x(t+1) = f(x(t), x(t-1)) \qquad (2)$$

could be replaced by the equivalent system

$$y(t+1) = x(t) \qquad x(t+1) = f(x(t), y(t)) \qquad (3)$$

Sometimes agents act at time t in response to their anticipation of future situations: this can also be described by introducing "forecast" variables, where e.g. $z(t)$ represents the forecast, at time t, of the value of $x(t+1)$.

For the sake of brevity we will refer both to systems of time-discrete difference equations, like Eq. 1, as well as to their continuous-time analogue using the term "differential equations" or the shorthand "ODE" (although it is usually meant to indicate ordinary differential equations only). It is well known that continuous-time equations and their discrete analogue can have very different time behaviours, like in the case of the famous "logistic map" [21]. However, this distinction is irrelevant for the purpose of the present paper, where the important aspects are that i) interactions among agents can be described by these kind of systems and that ii) in both cases useful analytical methods and theorems are available.

From the paragraphs above one infers that ODE system formalism allows arbitrarily long memory and arbitrary forecast time windows.

Whenever an equation-based description is possible, one may take advantage of several theorems and properties which have been discovered in the long history of these studies.

Indeed, it is sometimes possible to provide an analytical solution for the evolution of the state variables in time: in this case one obtains a great deal of information about the system, because its behaviour is known at once for the whole set of parameter values, avoiding the need to resort to endless simulations to explore a large parameter space.

Even when analytical solutions which describe the whole time behaviour cannot be found, it is sometimes possible to analytically obtain useful information by studying the asymptotic dynamics of the system (the so-called "qualitative analysis" of ODE, see e.g. [21] and further references quoted there). Let us recall among these useful theorems, those concerning the stability of a fixed point using Lyapunov functions, and the Poincaré-Bendixson theorem on the existence of limit cycles.

Another approach which can sometimes be applied in cases where the complete analytical solutions cannot be found is that of finding "first integrals", i.e. conserved quantities which constrain the set of allowable solutions and allow us to draw definite conclusions about the system's possible states.

When analytical properties cannot be found, or are of limited help, it is necessary to perform an analysis of the system behaviour in different regions of parameter space. Indeed, in most models of complex systems, and in all those which describe social systems, the values of some relevant parameters are not precisely known, and it is well known that nonlinear systems can display very different behaviours for different parameter values. Therefore, an extensive study (sensitivity analysis) is necessary to draw reliable conclusions. This is required both for ODE and ABM, but in the former case the search space is more constrained than in ABM, thus making the search less arbitrary.

3   The Expressive Power of Differential Equations

So, if equations present these advantages with respect to agent-based models, why should we ever use the latter? The reason is that some properties which agents are often required to possess can be cast in the form of a dynamical system only with great difficulties and in an unnatural way.

The main weaknesses of differential equations in modelling social actors seem related to the heterogeneity among different agents and to the information processing capabilities of the agents. However, the ODE formalism is often too easily dismissed: let us therefore analyze these aspects and try to understand to what extent they can be handled using it.

Heterogeneity among different agents is very important in agent-based modelling of social and economical systems, where it can provide a way to go beyond the "representative agent" approach, typical of neoclassical economics. Yet in ODE the only natural way to differentiate agents is by using different parameter values.

However, the formalism of ODE might be stretched so as to make it possible to "choose" between alternative behaviours, depending upon some parameter values, e.g. by using Heaviside functions or interval characteristic functions.

For example, an agent described by the following Eq. 4 can switch between two different evolution functions $f(x,y)$ or $g(x,y)$ for its state variable x, depending upon the fact that parameter p is greater or less than a threshold Θ (y collectively denotes the other variables which affect the evolution of x; $H(\Theta)$ is defined as being equal to 1 if its argument is greater than or equal to 0, $H(\Theta) = 0$ if $\Theta < 0$)

$$x(t+1) = H(p-\Theta)f(x(t), y(t)) + [1 - H(p-\Theta)]g(x(t), y(t)) \qquad (4)$$

While such an equation is formally compliant with ODE, it lacks the properties of smoothness which are at the basis of many important properties and theorems on such systems, which therefore are no longer valid in this case.

Moreover, this is only a limited form of heterogeneity: for example, it would be very difficult to cast even in formally equation-like forms the process which leads the different traders in the well known SFI stock-market model [17,1] to develop their own different set of heuristic rules.

This observation leads us to another major reason why equation systems are often inadequate, i.e. that they cannot capture the sophisticated information processing capabilities of the agents which are sometimes required, which are better described by algorithms.

According to the Bohm-Jacopini theorem, any algorithm can be built using the three basic flow control structures: sequence, iteration, alternative. We have already seen that alternative can somehow be handled by Heaviside functions, so let us now turn to sequence and iteration.

Sequence is in a sense intrinsic to the difference equation model, but there it follows a unique, global clock, while agents should often be able to perform computations on time scales different from those of other agents. In this case, one might still formally consider an ODE system, where some interactions take place less frequently than others, so as to mimick the different time scales. These relatively rare interactions might then describe interactions among agents.

For example, variable x of agent A may be updated at every time step according to a rule which involves other variables from the same agent (y) and less frequently according to the variables which refer to other agents (z). Let $m(t)$ be a random function defined as follows: $m_\mu(t) = 1$ if a random number generated at time t with uniform probability between 0 and 1 has a value greater than or equal to μ, $m_\mu(t) = 0$ otherwise. Then

$$x(t+1) = f(x(t), y(t)) + m_\mu(t)g(x(t), y(t), z(t)) \qquad (5)$$

describes an agent which may have a fast internal dynamics and (if μ is sufficiently close to 1) a much slower interaction with the other agents. Another way of handling different clock frequencies makes use of periodic deterministic functions instead of random functions.

If we want to represent a situation where agent i interacts with agent k only when the latter is "ready", we could introduce a Boolean variable $b(t)$ which takes the value "1" when k is ready (e.g. when one of its variables reaches a critical threshold), and which takes the value "0" otherwise, and we could then introduce a term in the equation for i which is multiplied times $b(t)$: this term would always vanish when k is not ready.

If we were to use a continuous-time formalism, we could easily define variables with different time scales. Many studies [8] have been devoted to systems like

$$dx_i/dt = -\gamma_i x_i + f_i(x_1 \ldots x_N) \qquad (6)$$

where γ_i describes a damping term, and the f_i are nonlinear functions. By using widely different damping constants one can describe variables which evolve on different time scales.

Also in this case, like in that of the "alternative" control structure, we observe that the kind of processing allowed by ODE is wider than usually thought.

Also the "iteration" structure might be mimicked to some extent by the ODE system. Indeed, vector operations are primitives in ODE, and they allow one to perform most of the operations which in computer languages are performed by iteration structures. It is also straightforward to introduce a "counter" variable which counts how many time steps have elapsed since a particular situation occurs. If we want to perform an operation only for a finite amount of steps, this can be achieved by introducing a Heaviside function which compares the value of the counter with that of a threshold.

Here again, however, the use of Heaviside functions leads to a loss of many important properties and theorems which make equations appealing.

Finally, it should also be noted that, while the birth of new agents and new variables can be forced in ODE systems, the theoretical tools presently available are not well suited to deal with these cases, unless very specific hypotheses are introduced (as for example in some interaction network models [10,11]).

4   A Careful Approach to ABM

Therefore, in several important cases concerning social systems modelling, it seems unavoidable to abandon the ODE approach in favour of an ABM. However, we stress that this should be done only when needed, since abandoning the world of equations has a high price: theorems are no longer available, widely different behaviours of the system are possible, the space of possible alternatives to be analyzed becomes very large, and great care must be exercised in interpreting the model results.

A widespread belief is that, using a sufficient number of modifiable parameters, one can describe almost everything. While this criticism has been raised initially for empirical equation-based models, it holds a fortiori for ABM. We do not believe this criticism to be completely true, as it has been shown by working with several models of physical and biological systems, where comparison with careful experimental results is possible. In these cases the "right" behaviours did not show up, in spite of extensive parameter search, until one found out that some key processes were missing, which needed to be included in the model [4,18]. However, it is true that playing with decision rules and parameters makes it possible to obtain very different behaviours, some of which may resemble those which are actually observed. Due also to the fact that precise measurements are often lacking in social systems, so that model validation is at best weak, such a resemblance may be due to chance, rather than to a deeper reason.

A point which is worth stressing is that the use of a complicated and sophisticated model of the individual agent (the "element" of the system) is not a necessary condition to account for complex behaviours at the level of the whole system. This is apparent from the many existing systems and models where very simple elements give rise to behaviours which are "complex" in many senses:

1. complex dynamics: many systems display very different dynamical behaviours, which are often described in terms of their asymptotic attractors, i.e. fixed points, limit cycles, strange attractors [21]. Systems of three or more differential equations, as well as one-dimensional iteration maps can display all three types of attractors, as cellular automata and other lattice models can do as well

2. self-organization: the interaction of simple elements (like molecules in a fluid or in a laser, or cells in a cellular automaton, or ants in an ant colony, etc) can give rise to emerging patterns in space and/or in time with striking regularities [21,8]

3. computing properties: some classes of cellular automata have been proven to possess the property of computational universality, i.e. they are equivalent to a Turing machine [20]

A further remark in the direction of stressing the power of simplification comes from the observation that, in many cases, a model with nonlinear interactions among its elements can be substituted by an equivalent model with independent elements - which are, however, no longer the same elements as before. In order to make this somewhat obscure (and often overlooked) property clear we will mention a specific example, which has been suggested by Auyang [3] and which has the advantage of being fully amenable to mathematical treatment, namely the oscillations in a crystal.

As Einstein suggested, since atoms perform an oscillatory motion around their equilibrium positions, the crystal can be considered as a system of N harmonic oscillators which, in a first approximation, behave independently of each other. Their motion can be quantized according to the rules of quantum mechanics. A better approximation would lead us to introduce coupling between the oscillations at nearby positions, but the description in terms of N coupled oscillators can be demonstrated to be equivalent to a description which is again based on N independent oscillators, which are however no longer interpreted as individual atoms moving around their equilibrium positions (these collective oscillation modes can be quantized leading to the notion of phonons). A price which has to be paid to achieve this description is that now the independent individuals are no longer identical, as they have different oscillation frequencies. Thus the independent individual approximation can still be used, but the "individuals" should no longer be confused with the original oscillators: they are "representative individuals", modified by the interactions.

Similar remarks might apply also to economics, where the classical approach is based upon a "representative agent" whose properties (e.g. perfect rationality, infinite processing power, etc.) are highly unrealistic if applied to real human beings. But if we interpret them as idealizations of a collective interaction of real humans, then it is no longer possible to rule them out on the grounds of their unrealism. To be precise, we agree that the neoclassical paradigm has several limitations and that economics needs to be founded on more effective bases, nonetheless we simply stress here that more solid arguments need to be put forth, rather than insisting on the "unrealistic" features of the economic agent.

Another major problem is that algorithmic models like ABM may be highly arbitrary, so there is an *embarasse de richesse* and conceptual guidelines are needed to constrain the set of allowable models.

Two major considerations may help to constrain the choice of the model. The first point is that it is necessary to look for model behaviours which are robust with respect to different model perturbations, which may involve changes in parameter values and, to some extent, also in the choice of some of the functions which describe the agent. If simulation results match the (limited, unprecise) experimental data only for a limited set of parameter values, or e.g. for a very particular kind of decision rule, then suitable reasons should be found to justify this choice, before claiming that the model has anything to do with the real system it is supposed to simulate.

The second methodological consideration is that a model may be strongly constrained by its relationship with a theory of the social process which it is supposed to describe. In this way, the model should concentrate on those aspects which the theory identifies as the most relevant, dropping out or drastically simplifying a set of related, but less important aspects. This approach is similar in some sense to that of classical hard science, but with an important remark. A physical theory is composed of a set of equations and a set of rules which relate the variables which appear in the equations to some measurable quantities. A theory of a social process is often cast in qualitative terms, so formal modelling might capture some of its aspects but, in general, it is not possible to introduce all the aspects of such a theory in a single formal model.

Models may also be very useful in making the theoretical statements more precise and above all to study the unfolding of the emergent properties which are implicit in the theory, but in general they do not coincide with the theory itself.

In some cases, the theory-model relationship is short-circuited, and the formal model itself is presented as a theory of the process which it describes. While this is certainly legitimate, the differences between theory and model which have been highlighted above are often important, so an approach based on an interplay between theory and model seems more powerful.

We will illustrate this approach in the following by referring to a specific example.

5   A Theory-Based ABM of Innovation Processes

Below, we will briefly describe a model that has been developed to study innovation processes according to the approach outlined above, i.e. on the basis of a theory of the phenomenon , which has been developed in recent years by Lane and Maxfield [15,12]. The model, called I_2M (Iscom Innovation Model) has been developed within the EU_FET project Iscom; it is rather complicated and, since it is described in detail in [13,14], it will only be briefly summarized here.

The theory is fairly broad in scope and, a fortiori, it will also not be discussed, but only sketched, here; in particular, it addresses the issue of how decision makers can deal with situations of "ontological" uncertainty, which cannot be handled by probabilistic reasoning. The latter would require that different possible alternatives be identified and their probability of occurrence estimated, while in conditions of extreme uncertainty it is difficult even to identify the relevant entities and relationships which will play a role in shaping the system's future. The authors found out that the condition of managers working in rapidly changing environments can be better described by such ontological uncertainty, and instantiated their findings by careful studies of the development of new market systems (in particular, the digital telephone switching technology introduced by Rolm and the distributed control system Lonworks developed by another Silicon Valley company, Echelon).

But, if uncertainty is so high, how can decision makers cope with it? Lane and Maxfield claim that in these conditions the present situation is interpreted on the basis of past histories, which are compared and even modified to aid in making sense of the present.

The basic entities of the theory of Lane and Maxfield [15,12] are agents and artefacts. Artefacts are given meanings by the agents that interact with them, and agents play different roles. The meaning of artefacts cannot be understood without taking into account the roles which different agents can play. Thus, artefacts may be given different meanings by different agents, or by the same agent at different times.

An important principle which is called "reciprocality", essentially claims that artefacts mediate interactions between agents, and vice versa, and that both agents and artefacts are necessary for a proper understanding of the behaviour of market systems. Therefore the theory concerns the agent-artefact space and the processes which transform it.

The theory is particularly concerned with innovation, which is not just novelty but, as used by Lane and Maxfield, a modification in agent-artefact space which unleashes a cascade of further changes. The innovation may involve the introduction of a new artefact, but also a change in relationships with other agents, or even a new interpretation of an existing artefact. In order to make this notion of "new interpretation" clearer, one may consider what happened to personal computers that were introduced as "small computers" and soon became a tool for office applications, displacing typewriters and the related technology.

In a sense, this theory can be seen as a theory of the interpretation of innovation, where interpretation actually means attribution of functionalities to artefacts, and of roles to agents. According to Lane and Maxfield, a new interpretation of an artefact functionality can be put forth in the context of so-called generative relationships. We will focus below on the role of these relationships.

By interacting, a few agents come to invent and share an interpretation, based on the discovery of different perspectives and uses of existing or expected artefacts. The generative potential of a relationship may be assessed in terms of the following criteria:

- heterogeneity: the agents are different from each other, they have different features and different goals; the heterogeneity is not so intense as to prevent communication and interaction
- aligned directedness: the agents are all interested in operating in the same region (or in neighbouring regions) of agent-artefact space
- mutual directedness: the agents should be interested in interacting with each other.

Moreover, agents must be allowed to interact and to take joint actions.

Lane and Maxfield further argue that, in a situation where innovations happen at a very fast pace, predicting the future is impossible; so a better strategy would be to identify those relationships that have the potential for generativeness, and to foster them in order to effectively explore the new opportunities that can arise. It is therefore very important to be able to estimate the "generative potential" of the existing and prospective relationships.

This theory of innovation is highly sophisticated in describing the interactions between different players in innovation processes, and it cannot be entirely mapped onto a specific computer-based model. Therefore, the modelling activity aims at developing models that are based on abstractions of some key aspects of this theory, which has been stated by its authors in qualitative terms. Moreover, while the theory has been developed on the basis of case studies involving the interactions of a few agents, such a model may help in unfolding its consequences when many different agents interact.

To relate to the Lane and Maxfield theory, the model must be such that the meanings of artefacts are generated within the model itself, without resorting to an external oracle to decide *a priori* which meanings are better than others. Moreover, the roles of agents must also be generated within the model through interactions among agents and artefacts.

In the I_2M model agents can "produce" artefacts, which in turn can be used by other agents to build their own artefacts, etc. An agent can produce several artefacts for different agents (and it can sell one type of artefact to several different customers). While this model may remind us of a production network, it is intended at a fairly abstract level: the production network is one of the simplest ways to enforce "purposeful" interactions among the agents.

Each agent has a set of recipes which allows it to build new artefacts from existing ones. Agents can try to widen the set of their recipes by applying genetic operators either to their own recipes or, by cooperating with another agent, to the joint set of the recipes of both. Moreover, each agent has a store where its products are put, and from where its customers can take them.

The meaning of artefacts is just what agents do with them, while the role of agents is defined by which artefacts they produce, with whom, and for whom. The role of agents is also partly defined by the social network they are embedded into. In this network, the so-called strong ties between two agents are mediated by artefacts (i.e. two agents are linked by a strong tie if there is a customer-supplier relationship between the two). There are also weak ties between two agents ("acquaintances") which refer to the fact that agent A knows something about agent B (e.g. its products).

The value which an agent (say A) gives its relationship with another agent B is summarized in a single numerical variable (the "vote"), which is composed by the sum of two terms. The first term is increased or decreased based on the history of supplier/customer interactions between A and B, while the second term takes into account the results of previous joint cooperation in developing new projects, if any. A parameter, which can be changed in different simulations, determines the relative weight of these two terms.

A key point is the structure of artefact space. What is required is that the space has an algebraic structure, and that suitable constructors can be defined to build new artefacts by combining existing ones. For reasons discussed elsewhere [14], we have adopted the number representation and the use of mathematical operators, instead of e.g. binary strings [9], λ-calculus [6] or predicate calculus [7]. Therefore the agents are "producers" of numbers by combination of other numbers, and the recipes are defined by a sequence of operators.

Each agent is also endowed with a numerical variable (called its strength) which measures how successful it has been so far: strength increases proportionally to the number of artefacts which are sold and decreases proportionally to the number of active recipes. Note however that strength, in the present version of the model, cannot be interpreted as "money" since it is not conserved in the interactions between two agents (if A and B interact, and ΔS_A and ΔS_B represent the change in strength of the two agents due to this interaction, it may well happen that $\Delta S_A \# \Delta S_B$).

As far as innovation is concerned, let us remark that an agent can invent new recipes or eliminate old ones. In the present version of the model no new agents are generated, while agents can die because of lack of inputs or of customers.

The model is asynchronous: at each time step an agent is selected for update, and it tries to produce what its recipes allow it to do. So, for each recipe, it looks for the input artefacts and, if they are present in the stocks of its suppliers, it produces the output artefact and puts it into its stock (the supplier stocks are of course reduced). Production is assumed to be fast, i.e. it does not take multiple steps.

Besides performing the usual buy-and-sell dynamics, when its turn comes, an agent can also decide to innovate. Innovation is a two step process, the first one defines a goal, i.e. an artefact which the agent may add to the list of its product, while the second step concerns the attempt at reaching the goal.

In the goal-setting phase, and agent chooses one of the existing types of artefacts (which, recall, is a number M) and then either tries to imitate it (i.e. its goal is equal to M) or it modifies it by a jump (i.e. by multiplying M times a random number in a given range). It has been verified that imitation alone can lead to a sustainable situation where, however, innovation eventually halts.

After setting its goal, an agent tries to reach it either by using the operators of one of its recipes with different inputs, or by combining two recipes to generate a new one with genetic algorithms. In this phase, the agent can decide to cooperate with another agent, sharing the recipes of both in order to reach a common goal. The propensity of an agent to cooperate is ruled by a parameter, while the choice of the partner is usually made with a probability distribution biased in favour of those agents which have a high vote.

6   The Behaviour of the Model

The behaviours of the model can be explored, and they can raise questions which may feed back into the theory. Actually, the model results can either support the theoretical claims or, in those cases which are more interesting, they can come unexpected. Whenever this happens, it may be that the theory needs revision or improvement, or that the model needs to be better tailored for its purposes, or both. We will briefly comment below on some model behaviours, which are discussed with greater detail in [19], and on their relationship to the theory.

One of the main reasons why agent-based models are considered important in economics is the possibility of handling heterogeneous agents, overcoming the limitations of the approach based solely on the description of a "representative agent". While I_2M agents become heterogeneous because they develop different recipes, using and producing different products, it is interesting to consider a further source of heterogeneity, associated with the fact that some parameters may take different values for different agents.

We consider first those parameters which are related to the propensity and style of innovation: agents in I_2M are "natural-born innovators" that try to introduce new artefacts with a certain pace, ruled by a specific parameter. Precisely, the attempt to innovate is decided on a stochastic basis, and this parameter rules the average rate. Another important parameter which affects the way in which agents innovate is the jump range.

Different kinds of experiments can be performed by considering i) a comparison between systems which are homogeneous (i.e. all the agents have the same value for the relevant parameters) but these parameters take different values in different worlds and ii) a heterogeneous system where agents with different values for the relevant parameters coexist and interact.

Concerning innovation frequency, the model results are unambiguous: if we compare two different homogeneous systems, one in which agents innovate frequently and the other where innovation takes place more rarely, the former system gives rise to a much higher number of artefacts. Agents produce more and are more robust in this case. And if we introduce in the same heterogeneous system both frequent innovators and rare innovators, the former ones outperform the latter (they produce and sell more artefacts).

On the other hand, the dependency of the system results on the jump range is different, and homogeneous systems where this parameter takes a high value are more fragile than others (a higher percentage of agents die out in the simulations). Indeed, since a higher range increases the chance of developing an artefact in an empty region of artefact space, this behaviour can be easily understood and might even have been guessed in advance (although it is indeed difficult to guess a priori the outcomes of this model).

There is however an interesting behaviour which is observed in heterogeneous systems, which are inhabited by two kinds of agents which differ for the jump range. In this case the system shows a robust behaviour and those agents which belong to the group with the higher range not only survive, but perform better than the others.

This might be explained in terms of a bunch of agents whose goals are distant from the existing artefacts, which are followed by other agents which perform smaller jumps. These more prudent agents improve the effectiveness of the "long jumpers", which could not even survive if they were all of the same kind. One indeed observes a kind of cooperation between daring innovators and more prudent ones, which follow the former in the new regions of artefact space.

This behaviour seems to resemble some cases which have actually been observed in industrial districts and market systems. However, the Lane and Maxfield theory has been developed with a focus on the interaction among few agents. The use of a model with several agents shows that this behaviour emerges out of their interactions, and thus bridges a gap between the phenomena already described by the theory and system-level interesting properties.

Let us now focus on the important issue of the relationships among agents in order to build new artefacts. The model allows us to compare different strategies for the choice of the partners in such an endeavour. It is for example possible to compare a purely random choice (where the partner is chosen with uniform probability among all the agents which are known) and a choice directed by some criterion. In particular, we have tested a choice driven by the vote, which is a function of the past interactions between two agents and which can be considered a model analogue of the notion of "mutual directedness" (cfr. the previous Section).

It has been observed that this latter way of choosing a partner performs consistently better than the random choice, thus confirming the theoretical expectations. This point requires perhaps clarification: the theory makes claims concerning relationships between two, or among a few agents, while the model deals with tens or hundreds of them. So the model provides support to the claim that a large set of agents seeking their innovation partners on the basis of certain types of relationship can give rise to a higher rate of innovation than random search.

Moreover, collaborations tend to last long. These results may suggest that stable collaborations can be useful for improving the system performances, a point which should stimulate further theoretical developments.

On the other hand, it is also interesting to observe that different criteria for partner selection (based either on the heterogeneity between the agents, or on the closeness of their goals) lead to performances which are quantitatively similar to those observed using the vote, although the average number of lasting collaborations per agent, and their average duration, may be remarkably different. Although the differences in performances are small, there are some preliminary indications that the alignment in artefact space, i.e. the fact that one chooses a partner with a goal close to its own, may be the most effective criterion among those which have been tested. This may also be a starting point for theoretical developments, since the theory has not yet provided any particular claim about the relative weight of the different factors which contribute to the generativity of a relationship.

Another interesting example of the possible dialogue between model and theory concerns the role of goal-driven behaviour. What would happen if the agents had no goal, but produced their innovations by simply applying one of their recipes to inputs chosen at random?

Note that this problem touches a very deep issue, i.e. the relationship between biological and social evolution. One of the facets of this problem which is particularly intriguing is the comparison between the different mechanisms that drive the introduction of novelties: random changes (the rule in biology) versus goal-oriented i.e. intentional changes in human systems. But what happens in this latter case in an unpredictable world, where both the system's own dynamics and the results of our intentional behaviour cannot be forecasted? Is there any difference between the two cases?

Note that the mechanism of the jump, which implies multiplication of the initial target times a random number, can mimick the choice of a goal in a situation which is essentially unpredictable. So, it is interesting to consider what are the main differences between a system endowed with this kind of goal and another one which operates in a biological-like fashion.

This latter condition can be simulated by supposing that, when an agent innovates, it simply takes the ordered list of operators of one of its existing recipes and chooses new inputs - without defining a goal and trying to reach it. We will call these no-goal (NG for short) systems.

First of all, it turns out that a crucial parameter for the fate of the goal-oriented system is the persistence of the goal, i.e. the probability that it maintains the same goal even if one or more attempts at reaching it have failed. It has been observed that systems with high persistence are really fragile, and many agents die out quickly; therefore it is better to compare low-persistence goal-oriented (LPG) systems with those without a goal.

The main results are summarized in the following remarks. LPG systems are much faster than NG systems in exploring new regions of artefact space: both the artefact diversity (i.e. the number of different types) and the diameter of artefact space (i.e. the difference between the highest and the lowest number) are much higher than those of the NG system in the first few thousand steps of the simulation. However, LPG systems seem to become saturated, and the number of recipes per agent reaches a plateau. NG systems, on the other hand, are much slower in adventuring in new regions, but they show a continuous albeit slow growth of the different performance measures (diversity, diameter, number of recipes per agent). The exploration of the space of artefacts is slower, but the system is more robust than the one with explicit goals.

Using particular initial conditions, LPG systems show a high agent mortality while NG systems do not, thus confirming the hypothesis that the former are more fragile.

These results have been achieved in a particular model, which reflects some aspects of a particular theory. But how general is this finding? Of course we are considering a rather extreme case of an unpredictable world, and goal-oriented behaviour might be more rewarding when the future can be better forecasted (for example, by endowing the agents with greater intelligence). On the other hand, the use of zero-intelligence economic agents has recently been proposed [5]. The observation of these behaviours might lead to a re-thinking of some of the theoretical concepts or, otherwise, to devising more sophisticated mechanisms to locate the goal, e.g. those based on a wider knowledge of other agents and on reasoning methods , and to verify whether they can lead to a markedly different behaviour.

Interestingly enough, studies on the comparison between NG and LPG systems may stimulate theoretical developments concerning not only innovation processes, but also the relationships between biological and social change.

The important aspects to be stressed in all the above examples (and many others, see [19]) is the creation of a loop between model and theory that can strengthen both.

7   Conclusions

In this paper, we have stressed that systems of differential equations can describe interactions among agents, and that they should be preferred whenever applicable. The alternative provided by agent-based models may seem simpler and easier to understand to the non-mathematically oriented researcher, but the strong nonlinearities are only hidden by the ABM language, and certainly not eliminated. We have also shown that systems of equations can describe many of the characteristics that agents must have.

On the other hand, there are cases when the ABM formalism is required: in particular they are unavoidable if the agents must have sophisticated information processing capabilities and an internal structure.

In these case we have argued that one should try to keep the model as simple as possible, but it often turns out that "as simple as possible" means rather complicated. A rigorous approach requires exploring the space of parameters and searching for robust model properties.

Another (complementary) route to rigour can be provided by the relationship with a theory of the social phenomenon. The modelling activity may provide a way to make the statements rigorous, and to unfold the system-level consequences of the theoretical assumptions. What is most important, the modelling activity may engage in a fruitful dialogue with the theoretical aspects which may improve both, as it has been suggested in the brief discussion of a model and a theory of innovation processes.

Acknowledgments. This work, and in particular the development of the model described and discussed in section 5, has been supported in part by the EU-FET project Iscom (The Information Society as a Complex System). We are deeply indebted to David Lane, who has shared with us his deep understanding of social phenomena and his no less deep understanding of complex systems modelling, who also played a crucial role in the development of the I_2M model. David has strongly advocated the need for a dialogue involving models, theory and case studies to understand the dynamics of social systems. We also thank Andrea Ginzburg and other colleagues in Reggio, Sander van der Leeuw, Denise Pumain, Dwight Read and many Iscom partners for helpful comments and suggestions. Last but not least, we acknowledge the contribution of our students Luca Ansaloni (who also contributed to the development of the software simulator of I_2M), Giovanni De Cicco, Silvia Covili and Simone Michelin.

References

1. Arthur, B.W.: Out-of-Equilibrium Economics and Agent-Based Modeling. In Judd, K.Ll, Tesfatsion, L. (editors): Handbook of Computational Economics, Vol.2: Agent-Based C omputational Economics. Amsterdam, North-Holland, 2005

2. Arthur, B.W., Holland, J.J., LeBaron, B., Palmer, R., Taylor, P. : Asset pricing under endogenous expectations in an artificial stock market. In Arthur W.B., Durlauf S., Lane D. (eds): Economy as an evolving complex system II (chap. 4). Redwood city (CA), Addison Wesley, 1997

3. Auyang, S.A.: Foundations of complex systems theories. Cambridge University Press, 1998

4. Di Gregorio, S., Serra, R., Villani, M.: Applying cellular automata to complex environmental problems: the simulation of the bioremediation of contaminated soils. Theoretical computer science 217 (1999):131-156

5. Farmer, J. D., Patelli, P., Zovko, I. I.: The Predicitive Power of Zero Intelligence in Financial Markets, PNAS USA 102 (2005): 2254-2259

6. Fontana, W., Buss, L.W.: The arrival of the fittest: towards a theory of biological organization. Bull. Mathem. Biol. 56 (1994): 1-64

7. Görnerup, O., Crutchfield, J.P.: Objects That Make Objects: The Population Dynamics of Structural Complexity. Santa Fe Institute Working Paper 04-06-020, arxiv.org e-print adap-org/0406058

8. Haken, H.: Synergetics: an introduction. Berlin, Springer, 1983

9. Holland, J.H., Holyoak, K.J., Nisbett, R.E., Thagard, P.R.: Induction. MIT Press, 1986

10. Jain, S. and Krishna, S.: Autocatalytic sets and the growth of complexity in a simple model, Phys. Rev. Lett. 81 (1998): 5684-5687

11. Jain, S., Krishna, S.: Emergence and growth of complex networks in adaptive systems, Computer Phys. Comm. 121-122 (1999): 116-121

12. Lane, D., Maxfield, R. : Ontological uncertainty and innovation. Journal of Evolutionary Economics 15 (2005): 3-50
13. Lane, D., Serra, R., Villani, M., Ansaloni, L.: A theory based dynamical model of innovation processes. In P. Bourgine, F. Kepes & Schoenauer, M. (eds): Proceedings of the European Conference on Complex Systems ECCS05; CD Rom: long paper #88, 2005
14. Lane, D., Serra, R., Villani, M., Ansaloni, L.: A theory based dynamical model of innovation processes. ComplexUs, in press, 2006
15. Maxfield R., Lane D.: Foresight, complexity and strategy. In Arthur W.B., Durlauf S., Lane D. (eds): Economy as an evolving complex system II (chap. 4). Redwood city (CA), Addison Wesley, 1997
16. Nilsson, N.J. : Artificial Intelligence: a new synthesis. Morgan Kaufman, 1998
17. Palmer, R.G, Arthur, B.W., Holland, J.J., LeBaron, B., Taylor, P.: Artificial economic life: a simple model of a stockmarket. Physica D 75 (1994) 264-274
18. Serra, R., Villani, M.: Exploring an agent-based model of innovation, in Lane, D., Pumain, D., van der Leeuw, S., West, G. (eds): Complexity perspectives on innovation and social change, Springer Methodos Series, in preparation, 2006
19. Serra, R., Villani, M., Colacci, A.: Differential equations and cellular automata models of the growth of cell cultures and transformation foci. Complex Systems 13 (2001) : 347-380
20. Siegelmann, H.: Neural Networks and Analog Computation: Beyond the Turing Limit. Birkhäuser Boston, Basel, Berlin, 1999
21. Strogatz, S. H.: Nonlinear dynamics and chaos: With applications to physics, biology, chemistry, and engineering. Perseus Books, 1994

Coordination of Actions in an Autonomous Robotic System

Erik Sandewall

Department of Computer and Information Science
Linköping University
Linköping, Sweden
erisa@ida.liu.se

Abstract. Robots are autonomous agents whose actions are performed in the real world during a period of time. There are a number of general constraints on such actions, for example that the same action can not have two separate instances during overlapping time intervals, or restrictions that are due to which state variables affect the action or are affected by it. Each process in the robot's cognitive system that is to request the initiation of an action must respect those restrictions. In this article we describe the design and formal characterization of a separate process, called an action coordinator, that manages these restrictions.

1 Topic

The familiar three-level architecture for robotic systems with high-level autonomy is defined in terms of a lower 'process' layer, a middle 'reactive' layer, and an upper 'deliberative' layer. Such an architecture may be natural if the activities of the robotic system are defined in terms of 'actions' with extended duration in time. The deliberative layer will then be in charge of prediction and planning in terms of such actions. It will also be capable of invoking actions through a request to the reactive layer. The current state of the reactive layer at each point in time will specify what are the currently ongoing actions and the current state within the action, at least on a qualitative level. This current state, in turn, defines the operational mode for the control algorithms that are used in the process layer.

In such an architecture there are a number of restrictions on how actions can be performed. The exact nature of these restrictions is an aspect of the semantics for the actions, as represented by a particular logic of actions and change. In the Cognitive Robotics Logic [4], the execution of an action a during and interval of time from s to t is expressed by the formula $\mathsf{D}([s,t],a)$, and one of the restrictions is that if $\mathsf{D}([s,t],a)$ and $\mathsf{D}([s',t'],a)$ hold, then the intervals $[s,t]$ and $[s',t']$ can not overlap in more than one single timepoint, unless they are identical. In other words, we allow $s' = t$ but not $s < s' < t$. There are also a number of other restrictions, for example those having to do with the limitations on when a particular action is executable in itself.

O. Stock and M. Schaerf (Eds.): Aiello Festschrift, LNAI 4155, pp. 177–191, 2006.

The task of enforcing such restrictions is assigned to the deliberative layer in the classical three-level architecture. For example, this layer must be designed in such a way that it will not invoke an action that is already going on. This means that the deliberative layer must be one coherent entity. The generic three-layer model is therefore not easily compatible with a distributed-AI approach where the deliberative function is organized as a collection of independent agents each of which has the capability of invoking actions. For example, it is not sufficient to just queue the action requests of individual agents, since an agent may wish to retract or add requests if an action can not be executed at the time or in the way that this agent has requested.

In order to accomodate a collection of independent 'agents' that together constitute the deliberative layer of the robotic system, it is natural to separate the enforcement of restrictions on actions into an architectural unit of its own. We shall refer to it as the *action coordinator*. The architecture will now consist of four layers rather than three: the process layer, the reactive layer, the action coordinator, and the swarm of deliberative agents. In procedural terms, every such agent is able to issue action requests to the action coordinator, and the action coordinator will send action initiation commands to the reactive layer if and when it determines that this is appropriate. It will also inform the deliberative agents about whether their requests have been accepted or not, and about the termination of the actions they requested.

The function of the action coordinator is reminiscent of the concept of electronic institutions that has emerged in the field of distributed AI, particularly in the context of auctions and negotiation in a community of agents[1]. One may think of the action coordinator as a kind of electronic institution that is specialized for robotic applications which are characterized by layered architectures and actions with duration.

The procedures surrounding the action coordinator can be understood in terms of message-passing for the purpose of invocation and information. However, in a logicist framework where both the deliberative agents and the actions themselves are characterized in logic, it is appropriate to also use logic for characterizing the action coordinator and its interactions with the agents. This has the advantage that additional, complex behavior can be 'programmed' in a clear and transparent way into the action coordinator. In the present article we shall describe the action coordinator using Cognitive Robotics Logic, CRL [4].

2   Cognitive Robotics Logic

The present section contains the basic definitions for CRL, using the same presentation as in [3] except that the use of composite actions has been removed. We retain the constructs for success and failure of actions for continuity, and in order to lay the ground for future extension of the results presented here, although those constructs will not be used in the present article.

2.1   Standard Constructs in Logics with Explicit Time

The following is the basic notation that is generally used, with minor variations, when time and action is represented using an explicit time domain. Three primary predicates are used. The predicates H for *Holds* and D for *Do* are defined as follows. $H(t, p)$ says that the "propositional fluent" ([1]) p holds at time t. In other words, p is reified and $H(t, p)$ is the same as $p(t)$ in the case where p is atomic. $D([s, t], a)$ says that the action a is performed over exactly the closed temporal interval $[s, t]$. Open and semiopen intervals are denoted (s, t), $[s, t)$, and $(s, t]$ as usual.

Non-propositional fluents are also admitted, using the notation $H(t, f : v)$ where $f : v$ is the propositional fluent saying that the fluent f has the value v. Fluent-valued functions are allowed, and one of their uses is to define fluents for properties of objects. For example, $ageof(p)$ may be the fluent for the age of the person p, used as in $H(1998, ageof(john) : 36)$.

In the full notation there is also a third predicate, X, that is pronounced *occludes* and is used for characterizing exceptions from the assumption of continuity of the value of fluents. Continuity includes persistence as a special case, for discrete-valued fluents. $X(s, f)$ expresses that at time s, the value of the fluent f is not required to be continuous or to persist.

In all cases, s and t are timepoints (usually s for starting time and t for termination time) and a is an action. We assume that time is discrete and linear, and let θt represent the timepoint that precedes t. $s < t$ is defined as $s = \theta^n t$ for some $n > 0$.

Logics with explicit time are usually used in such a way that each model of the axioms characterizes one possible history in the world, not a tree of possible histories. Alternative histories are represented by different models([2]). Therefore, a timepoint t is sufficient for identifying the state of the world at time t in the present model.

2.2   Ontology for Invocation and Success

For each kind of action, in many applications, there are certain conditions that must be satisfied in order to be able to say that the action can be initiated at all. Once initiated, it can either succeed or fail. The distinctions between inapplicability, failure, and success depend both on the application as such, and

---

[1] We have previously tried to maintain a terminological distinction between *fluent* as a *function* from timepoints to corresponding values, and a *feature* as a formal object that designates a fluent. With that terminology, the p and f that occur in the second argument of H are features, not fluents. Similarly, the functions *inv*, *app*, and *fail* that will be introduced later in this section, are functions from actions to features. However, since it is so common to use the word 'fluent' both for the function and its designator, we follow that practice here.

[2] However, it was shown in [2] that it is straightforward to generalize the time domain so that it also accounts for the case of branching time.

on how one chooses to model it. In this subsection we specify the ontology $(^3)$ for these concepts which will be used in the formalization in the next section.

An *invocation* of an action can cause it to begin its *execution*, which ends with either *success* or *failure*. The matter is complicated by the requirement to represent that it is sometimes impossible to execute an action. In our approach, invocation of an action is possible at any time, but the invocation does not necessarily lead to the execution of the action. In particular, it does not if the action is inapplicable by definition (for example, turning on the light in a room where there is no light) or if the action is already executing.

When an action begins to execute, it is said to *initiate*. Once an action has (been) initiated, it must ultimately either *succeed* or *fail*. The distinction between success and failure is done on the following pragmatic grounds: planning goal achievement is done using the assumption that actions succeed, and using knowledge about their results when they do succeed. The case where an action fails is dealt with on a case-by-case basis once the failure has occurred.

For the same reasons, we assume that applicability is defined in such a way that it can be determined at planning time. Those conditions that prevent an action from having its effect and that can in general not be detected until execution time, must be modelled as failure and not as inapplicability$(^4)$.

Each action has a temporal duration, which must be an interval that is greater than a single point except for some specific cases defined below. Note, in particular, that when an action is not applicable, it is considered not to execute, it is not considered to fail instantly.

2.3   Syntax for Invocation and Success

Two representations will be used for the expression of success, failure, and applicability of actions. In one, we use specially constructed fluents, in the other, variants of the D predicate that distinguish between action success and action failure. The former representation is considered as the basic one, and the latter is introduced as abbreviations or 'macros' that facilitate the writing of effect rules for actions.

The following are three functions from actions to propositional fluents:

inv, where $H(s, inv(a))$ says that the action a is invoked at time s. At all other times, $H(s, inv(a))$ is false.

app, where $H(s, app(a))$ says that the action a is applicable at time s.

fail, where $H(t, fail(a))$ says that the action a terminated with failure at time t. $H(t, fail(a))$ is false at all times when the action is not executing, or when it is executing but not terminating, or when it is terminating successfully.

In addition, we mention one function from propositional fluents to actions:

---

^{3} We mean ontology in the classical sense of the word, not merely a taxonomical structure.

^{4} To be precise, in a well-formed plan, each action must be applicable in any state of the world that may result, according to the effect laws, if the preceding actions are successful. If some of them fail then later actions may be inapplicable in a way that can only be detected at plan execution time.

test, where $test(p)$ or $test(f : v)$ is an action that is always applicable, whose duration is always instantaneous (expressed by $D([s, s], test(p)))$, and that satisfies

$$H(s, fail(test(p))) \leftrightarrow \neg H(s, p)$$

In other words, $test(p)$ succeeds at time s iff p is true at s.

Notice that this function does not represent an *action*, and questions about the executability of the tests, their possible side-effects, and the precision of the results are not relevant for it. It is simply a kind of conditional operator in the logic([5]).

The following abbreviation is introduced:

$D_v(s, a)$ for $H(s, inv(a)) \wedge (\neg H(s, app(a)) \vee \exists s' \exists t[D([s', t], a) \wedge s' < s < t])$: the action a is invoked at time t but it is either not applicable, or already executing at that time. (This is the case where invocation of the action does not initiate an execution).

The priority of the propositional connectives is defined so that $a \rightarrow b \wedge c$ means $a \rightarrow (b \wedge c)$.

2.4   Axiomatic Characterization

The following set of axioms characterizes the obvious properties of these relations.

S1. If an action is being executed, then it must have been invoked and be applicable and non-executing at invocation time:

$$D([s, t], a) \rightarrow H(s, inv(a)) \wedge \neg D_v(s, a)$$

This implies:

$$D([s, t], a) \rightarrow H(s, inv(a)) \wedge H(s, app(a)) \wedge$$
$$\neg \exists s' \exists t[D([s', t], a) \wedge s' < s < t]$$

S2. If an action is invoked, then it is executed from that time on, unless it is inapplicable or already executing:

$$H(s, inv(a)) \rightarrow \exists t[s \leq t \wedge D([s, t], a)] \vee D_v(s, a)$$

The full version of this axiom is slightly larger in order to also allow for composite actions.

S3. An action can not take place during overlapping intervals:

$$D([s, t], a) \wedge D([s', t'], a) \wedge s \leq s' < t \rightarrow s = s' \wedge t = t'$$

---

[5] Actually, the *test* operator is mostly motivated for its use in composite action expressions where it makes it possible to define conditional actions. The use of composite actions is excluded in the present article, but we retain the definitions for the *test* function anyway since it is integrated with the basic axioms.

S4,S5. Actions of the form $test(p)$ are always applicable, and instantaneous:

$$H(s, app(test(p)))$$

$$D([s,t], test(p)) \rightarrow s = t$$

S6. All other actions execute over extended periods of time: never immediately, except for actions of the form $test(p)$:

$$D([s,t], a) \rightarrow s < t \vee \exists p[a = test(p)]$$

S7. Actions only fail at the end of their execution:

$$H(t, fail(a)) \rightarrow \exists s[D([s,t], a)]$$

S8. Definition of success for actions of the form $test(a)$:

$$D([s,s], test(p)) \rightarrow (H(s, fail(test(p))) \leftrightarrow \neg H(s,p))$$

Several of these axioms capture desirable properties directly. For others, all the consequences are not immediately obvious. One useful consequence is the following theorem, previously reported in [3]:

Theorem 1. *In any model for the axiom S3, let* $\{[s_i, t_i]\}_i$ *be the set of all intervals such that* $D([s_i, t_i], a)$ *for a specific action* a. *Then there is some ordering of these intervals such that* $s_i < s_{i+1}$ *and* $t_i \leq s_{i+1}$ *for all* i.

Proof. Suppose the proposition does not hold, and choose an order of the pairs such that $s_i \leq s_{i+1}$, and where each pair only occurs once. Also, choose j so that either $s_j = s_{j+1}$, or $s_j < s_{j+1} < t_j$. If no such j is to be found, then the ordering already satisfies the condition in the proposition.

However, the case $s_j = s_{j+1}, t_j \neq t_{j+1}$ contradicts axiom (S3). The case $s_j < s_{j+1} < t_j$ also contradicts axiom (S3). This concludes the proof. QED.

The value of this observation is that through it, it makes sense to use the fluent $fail(a)$ for characterizing the success or failure of an action with extended duration. If theorem 1 were not to hold, then it would not be clear from $H(t, fail(a))$ which invocation the failure referred to. This consideration is also the reason for the choice manifested in axiom S1: if an action a is invoked while it is already in the midst of executing, then it is not represented as "failing", since this would confuse matters with respect to the already executing instance. Instead, we use the convention that it is invoked, possibly applicable, but it does not get to execute from that starting time.

We also obtain at once:

Theorem 2. *In any model for the axioms S1–S8, if* $D([s,t], a)$ *and* $H(u, fail(a))$ *for some* u *in* $(s,t]$, *then* $t = u$.

Informally, we can think of each model in dynamical terms as a possible history in the world being described, and what this theorem says is that if an action is invoked and begins to execute, then if $H(u, fail(a))$ becomes true at some timepoint u during the execution, the action halts and ends with failure, and if it is able to proceed until its normal ending without $H(u, fail(a))$ becoming true at any time, then it ends with success.

Any use of this logic will naturally be concerned with the effects of actions. In Cognitive Robotics Logic and its background, the Features and Fluents approach, as well as its successors, this is specified using action laws, which in particular make use of the occlusion predicate, and in combination with assumptions of persistence. In [6] we showed how this logic can be used for specifying an architecture for a logic-based cognitive robotic system where rules specifying failure conditions for an action can be written as implications where the consequent has the form $H(u, fail(a))$.

2.5   Examples

The following additional abbreviations are introduced. They are generally useful for writing effect rules and applicability restrictions rules for actions.

$G(s, a)$ for $H(s, inv(a))$: the action a is invoked ("go") at time s

$A(s, a)$ for $H(s, app(a))$: the action a is applicable at time s

$D_s([s, t], a)$ for $D([s, t], a) \wedge \neg H(t, fail(a))$: the action a is executed successfully over the time interval $[s, t]$, it starts at time s and terminates with success at time t.

$D_f([s, t], a)$ for $D([s, t], a) \wedge H(t, fail(a))$: the action a is executed but fails over the time interval $[s, t]$, it starts at time s and terminates with failure at time t.

$D_c([s, t], a)$ for $\exists u[D([s, u], a) \wedge t \leq u]$: the action a is being executed, the execution started at time s and has not been terminated before time t. (It may terminate at t or later).

For both D_s and D_f, s is the time when the action was invoked, and t is the exact time when it concludes with success or failure.

As an example of the use of this notation, the following formula states that a condition φ guarantees that an action always succeeds:

$$H(s, \varphi) \wedge G(s, a) \rightarrow \exists t[D_s([s, t], a)]$$

Ordinary action laws specify the action's effects when it succeeds. They are therefore written as usual and with D_s on the antecedent side: if preconditions apply and the action is performed successfully, then the postconditions result.

As a second example, consider the case of actions that are described in terms of a precondition, a prevail condition, and a postcondition, where the postcondition is at the same time the termination condition for the action [7]. The prevail condition must be satisfied throughout the execution of the action; if it is violated then the action fails. Simple pre/ post/ prevail action definitions can be expressed as follows, if φ_a is the precondition of the action a, ω_a is the postcondition, and ψ_a is the prevail condition:

$$A(s, a) \leftrightarrow H(s, \varphi_a)$$
$$D_s([s, t], a) \rightarrow H(t, \psi_a) \wedge H(t, \omega_a)$$

$$A(s, a) \wedge D_c([s, t], a) \wedge u \in [s, t) \rightarrow H(u, \psi_a) \wedge \neg H(u, \omega_a)$$
$$D_c([s, t], a) \wedge \neg H(t, \psi_a) \rightarrow D_f([s, t], a)$$

The traditional case of only pre- and postconditions is easily obtained by selecting ψ_a as tautology.

3   The Action Coordinator

We now proceed to using the Cognitive Robotics Logic for specifying the action coordinator as described in section 1. We limit the problem to a relatively simple action coordinator that does not take the success and failure of actions into account.

3.1   CRL Formulation

Starting from the CRL formalism that was introduced in the previous section, we define the action coordinator by extending the logic with an additional object domain for agents, and with two new fluent-valued functions, $inva$ and $asta$. The fluent $inva(g, a)$ will express an invocation of the action a by the agent g (i.e., a request for its initiation). The proposition $H(s, inv(g, a))$ is true iff the agent g invokes a at time s.

Notice the difference between $inv(a)$ that was introduced above, and $inva(g, a)$ that is introduced here. The relation between them will be specified and proved below.

The action coordinator's response to an invocation is represented using the funtion $asta$, for 'action state'. At each point in time, the fluent $asta(g, a)$ has a value representing the current response to an invocation of a that has previously been issued by the agent g. The invocation is represented as a momentary condition, but the response is represented as something that applies over an interval of time. This frees the agents from 'remembering'([6]) what responses they have received for their invocations.

The value of $asta(g, a)$ shall be one of the following discrete values. If the action is executed during the interval $[s, t]$ on behalf of the agent g, then the value is stex (for 'start executing') at initiation time s, ex (for 'executing') in the interior of the interval, and nil at time t. Before any such execution has taken place, the value is also nil. If an invocation $inva(g, a)$ occurs when the value of $asta(g, a)$ is nil, then it switches to pend, for 'pending'. It may retain this value for some time, but it can also switch to either of stex meaning that the invocation was honored by initiating the action, or to ref meaning that the invocation was refused. During and after execution the value is ex and nil as already explained, and after execution the action is available for initiation again. If the action has been refused, on the other hand, a renewed invocation $inva(g, a)$ will change the value of $asta(g, a)$ from ref to pend.

---

[6] I.e., from having to retain that information in its local state.

The intended structure of possible transitions is illustrated in figure 1. Notice that the stex state can only be visited during one single timestep at a time, whereas all the other states can remain for several timesteps.

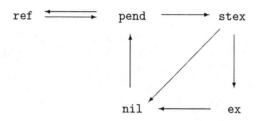

Fig. 1. State transitions for asta(g,a)

The transition from one value to another depends on the following factors. The transitions from pend to stex or ref represent the decisions of the coordinator. The transitions from stex to ex and from ex to nil, or directly from stex to nil reflect the execution of the actions, and normally they are obtained from the process layer where each action is executed. The transitions from nil or ref to pend represent how the action coordinator receives and administrates invocations of actions by the agents.

This transition structure is intended to represent an upper bound on admissible transitions. Specific policies in the action coordinator can be represented by restricting the transitions from pend to stex or ref, but not by relaxing any of the transitions described here.

However, one extension that may be of interest is to have a way for agents to discontinue an ongoing action. In this case the transition from ex to nil is caused by a message to the action coordinator from a deliberative agent, and not from the process layer. This requires an extension to the formalism for expressing how the agent sends that message, and it is not considered in the present article.

3.2   Axioms for the CRL Formulation

The system of transition rules can be expressed using the following axioms. Recall that the function θ represents the predecessor of a given timepoint, and $s < t$ represents that $s = \theta^n t$ for some $n > 0$. We introduce the auxiliary predicate $Blocked(s, a)$ that characterizes those conditions where an action can not initiate even if some agent invokes it. It is formally defined as follows:

NS1. $Blocked(s, a) \leftrightarrow \neg H(s, app(a)) \lor \exists g[H(s, acta(g, a) : ex)]$

The following axioms characterize the generic action coordinator.

K0. Fluents of the form $acta(g, a)$ take (at most) one value at each point in time, chosen among the five values mentioned above.

$H(s, acta(g, a) : r) \wedge H(s, acta(g, a) : r') \rightarrow$
$r = r' \wedge (r = \mathsf{ex} \vee r = \mathsf{pend} \vee r = \mathsf{nil} \vee r = \mathsf{ref} \vee r = \mathsf{stex})$
We suppose that $acta(g, a)$ also *has* a value for each combination of s, g, and a, but an axiom to this effect does not appear to be needed for the proofs being made below.

K1. If an action is being executed, then it must have been initiated, and from the point of view of each initiating agent it goes through the states stex, ex, and nil:
$D([s, t], a) \rightarrow s < t \wedge \exists g[H(s, acta(g, a) : \mathsf{stex})] \wedge$
$\forall g[H(s, acta(g, a) : \mathsf{stex}) \rightarrow H(t, acta(g, a) : \mathsf{nil}) \wedge$
$\forall u[s < u < t \rightarrow H(u, acta(g, a) : \mathsf{ex})]]$

K2. If an action initiates, then it is executed from that time on and in a finite interval of time, so that it has an ending time. It can only initiate if it is applicable at that point in time:
$H(s, asta(g, a) : \mathsf{stex}) \rightarrow H(s, app(a)) \wedge \exists t[D([s, t], a)]$

K3. If an action is executing and not initiating from the point of view of an agent, then it must have been in that state or initiating in the preceding timepoint with respect to the same agent:
$H(s, asta(g, a) : \mathsf{ex}) \rightarrow H(\theta s, asta(g, a) : \mathsf{ex}) \vee H(\theta s, asta(g, a) : \mathsf{stex})$

K4. An action can only initiate for an agent if it was pending for that agent at the preceding timepoint:
$H(s, acta(g, a) : \mathsf{stex}) \rightarrow H(\theta s, asta(g, a) : \mathsf{pend})$

K5. If an action is executing for one agent, or if it is inapplicable, then it can not be initiated for any agent:
$Blocked(s, a) \rightarrow \forall g[\neg H(s, asta(g, a) : \mathsf{stex})]$

K6. Consider an action a that is inert (nil) or refused for a particular agent at a particular timepoint. If it is invoked by the agent then it must be pending at the next timepoint, otherwise it must retain the same value.
$H(\theta s, asta(g, a) : r) \wedge (r = \mathsf{nil} \vee r = \mathsf{ref}) \rightarrow$
$(H(s, inva(g, a)) \rightarrow H(s, asta(g, a) : \mathsf{pend})) \wedge$
$(\neg H(s, inva(g, a)) \rightarrow H(s, asta(g, a) : r))$

K7. An action can only switch from another state to being pending as the result of an invocation from the agent in question:
$(H(s, asta(g, a) : \mathsf{pend}) \wedge H(\theta s, asta(g, a) : r) \wedge r \neq \mathsf{pend} \rightarrow H(s, inva(g, a))$

K8. If an action is pending from the point of view of an agent, then in the next timestep it must be initiated, refused, or still pending:
$H(\theta s, asta(g, a) : \mathsf{pend}) \rightarrow$
$H(s, asta(g, a) : \mathsf{stex}) \vee H(s, asta(g, a) : \mathsf{ref}) \vee H(s, asta(g, a) : \mathsf{pend})$

K9. There exists a timepoint s_0 such that $asta(g,a)$ has the value nil for all times $\leq s_0$:

$$\exists s_0 \forall s, g, a[s \leq s_0 \rightarrow H(s, asta(g,a) : \mathsf{nil})]$$

We also introduce the following policy rule.

P1. If an action is pending and applicable, then initiate it:

$$H(\theta s, asta(g,a) : \mathsf{pend}) \wedge \neg Blocked(s,a) \rightarrow H(s, acta(g,a) : \mathsf{stex})$$

This policy rule is the first example of a rule that restricts the transitions for an action from being pending, to being initiated or refused. This particular rule *forces* a pending action to initiate as soon as it is not blocked by not being applicable, or by another instance of the same action being executed. One can of course think of alternative rules that instead require the initiation to be delayed or refused in specific circumstances. It is intended that rules K0 through K9 shall remain fixed, whereas policy rules can be exchanged.

3.3   Properties of the CRL Formulation

The logical structure that was defined in the previous subsection allows for a number of interesting cases. In particular, consider a situation where two separate actions are pending and become unblocked at the same time, but where it is not possible to execute them concurrently. With the axioms shown above, including the policy rule P1, both will be initiated. We take the view that in this case, one should set things up so that both actions do execute, but at least one of them will fail, possibly after only one timestep. Although this convention may seem peculiar at first, please notice that the conflict between two concurrent actions may also arise later on during their execution, and it may be due to external events that could hardly have been predicted when the actions started. Since we anyway have to accomodate actions that fail for such reasons in the course of their execution, we can as well represent the starting-time conflict in the same way.

The representation shown above allows one to express that each occurrence of an action is done on the request of one or more agents. There must be at least one agent for which it is being performed, according to axiom K1. If an action a is pending for more than one agent g and then becomes unblocked at a particular timepoint s, then the policy axiom P1 requires that the action initiates for all those agents at the same time. However, if P1 is not used then it is possible to initiate the action for some of the invoking agents but not for all of them, or to not initiate it at all.

The rules K0 through K9 characterize the action coordinator in a number of ways. Rules K1, K2, and K3 specify how the execution of an action, as expressed using the $D()$ predicate, is controlled by the action states as represented by $acta(g,a)$. Rules K6 and K7 specify how those action states interact with the messages from the cognitive agents, as expressed using the $inva(g,a)$ fluents. Axioms K0, K4, K5, K8 and K9 specify the permitted values and permitted transitions for the action states, although the other axioms also imply some restrictions on those transitions.

We shall show below that the 'K' series axiomatization in axioms K0 through K9 restricts the fluent $acta(g, a)$ to the finite-state automaton that was informally described in a previous subsection. However, we first prove some other results since along the way they provide a needed lemma. Notice that the axioms do not *only* represent the automaton; they also characterize the use of multiple invoking agents and the relationship between their invocations.

3.4   Relation to Previous Formulation

We shall now demonstrate that the axioms S1, S2, S3, and S6 that were defined above follow from the proposed axioms for the action coordinator, including the policy axiom P1. In doing so we achieve two objectives. First, the action-based behavior that was described by the 'S' series axioms is replaced by a more finegrained machinery that is arguably a more precise description of how the deliberative layer works in an intelligent autonomous agent. Secondly, we have verified that the proposed specification for the new, four-layer architecture with distributed cognitive capabilities is consistent with the logical architecture that had been introduced before.

In order to properly relate the old and the new axiomatization, we shall need an axiom that relates fluents of the form $inv(a)$ to the constructs used in the new axioms. Since in the 'S' series axioms, an action initiates if and only if $inv(a)$ holds at a timepoint where the action is applicable and not already executing, we adopt the following axiom:

NS2. $H(s, inv(a)) \leftrightarrow \exists g[H(\theta s, acta(g, a) : \mathsf{pend})]$

Using this definition as the bridge, we shall show that the axioms S1, S2, S3 and S6 in the old set of axioms can be obtained as consequences of the new set of axioms, and in particular axioms K0 through K5 plus K9 and the policy axiom P1. Notice that P1 is necessary here, since the 'S' series axiomatization prescribed that an invoked action shall start executing as soon as it is not blocked. Axioms K6 through K8 will not be needed for these proofs, which is not surprising since they represent the decision machinery for the agent coordinator.

The functions *test* and *fail* are outside this consideration, so that axioms S4, S5, S7, and S8 are not to be treated. Also, axiom S6 is modified by removing the reference to the *test* function, becoming

S6'. All actions execute over extended periods of time:
$D([s, t], a) \rightarrow s < t$

We notice at once that S6' is subsumed by the new axiom K1, and proceed with the others.

The following is the definition for the abbreviation $D_v(s, a)$ that was introduced above, for reference:

NS3. $D_v(s, a) \leftrightarrow$
$H(s, inv(a)) \wedge (\neg H(s, app(a)) \vee \exists s' \exists t'[D([s', t'], a) \wedge s' < s < t'])$

We begin with a few lemmas.

Lemma 1. $D([s, t], a) \rightarrow H(s, inv(a))$

Proof. Assume $D([s, t], a)$. By K1, $\exists g[H(s, acta(g, a) : \text{stex})]$. By K4, $\exists g[H(\theta s, acta(g, a) : \text{pend})]$. By NS2, $H(s, inv(a))$. QED.

Lemma 2. $D([s, t], a) \rightarrow \neg Blocked(s, a)$

Proof. Assume $D([s, t], a) \wedge Blocked(s, a)$. By K1 from $D([s, t], a)$, there is some g such that $H(s, acta(g, a) : \text{stex})$. According to K5 this contradicts $Blocked(s, a)$. QED.

Lemma 3. $\exists g[H(s, acta(g, a) : \text{ex})] \leftrightarrow \exists s', t'[D([s', t'], a) \wedge s' < s < t']$

Proof. The right to left direction of the implication follows directly from K1. For the left to right direction, assume $H(s, acta(g, a) : \text{ex})$. It follows from K3 that there are preceding timepoints from s and back where the value of $acta(g, a)$ is ex, until it arrives to one s' where the value is stex, so that $D([s', t'], a)$ for some t', according to K2. Such a timepoint s' must exist according to axiom K9. According to K1 it must be the case that $s' < t'$ and $s < t'$. QED.

Using Lemma 3 the definition of $D_v()$ can be rewritten as
$\quad D_v(s, a) \leftrightarrow H(s, inv(a)) \wedge Blocked(s, a)$.

We proceed now to the proofs of propositions S1, S2, and S3 which had the status of axioms in the earlier articles.

Proposition S1. $D([s, t], a) \rightarrow H(s, inv(a)) \wedge \neg D_v(s, a)$

Proof. Lemmas 1 and 2 give $D([s, t], a) \rightarrow H(s, inv(a)) \wedge \neg Blocked(s, a)$. By tautology, $D([s, t], a) \rightarrow H(s, inv(a)) \wedge (\neg H(s, inv(a)) \vee \neg Blocked(s, a))$ which is equivalent to proposition S1 using the definition of $D_v()$ as rewritten above. QED.

Proposition S2. $H(s, inv(a)) \rightarrow \exists t[s \leq t \wedge D([s, t], a)] \vee D_v(s, a)$

Proof. Assume $H(s, inv(a))$. By the definition of inv, $\exists g[H(\theta s, acta(g, a) : \text{pend})]$. Policy axiom P1 gives $H(s, acta(g, a) : \text{stex}) \vee Blocked(s, a)$, and K2 gives $\exists t[D([s, t], a)] \vee Blocked(s, a)$. Axiom K1 then gives $\exists t[D([s, t], a) \wedge s \leq t] \vee Blocked(s, a)$. The assumption gives $\exists t[D([s, t], a) \wedge s \leq t] \vee (H(s, inv(a)) \wedge Blocked(s, a))$ and the rewritten definition of $D_v()$ concludes the proof. QED.

Proposition S3. $D([s,t],a) \wedge D([s',t'],a) \wedge s \leq s' < t \rightarrow s = s' \wedge t = t'.$

Proof. Assume $D([s,t],a) \wedge D([s',t'],a) \wedge s \leq s' < t$. Furthermore, for the purpose of proof by contradiction, assume $s < s'$.
K1 obtains $\exists g[H(s', acta(g,a) : \mathsf{ex})]$. ¿From NS1 it follows $Blocked(s',a)$.
However K1 also implies $\exists g'[H(s', acta(g',a) : \mathsf{stex})]$, which according to K5 is a contradiction. Therefore $s = s'$.

Next, assume $t < t'$. ¿From $D([s,t],a)$ and using K1, $H(t, acta(g,a) : \mathsf{nil})$ follows. ¿From $D([s,t'],a)$ and using K1, it follows $H(t, acta(g,a) : \mathsf{ex})$. According to K0 this is a contradiction. If $t' < t$ then the same contradiction is obtained due to symmetry. It follows that $t = t'$. QED.

3.5   Finite-State Characterization of Action State

We return now to the finite-state characterization of action state.

Theorem 3. In any model for the axioms specified above, the sequence of values that are assigned by $H()$ to $asta(g,a)$ for given g and a and for successive s, must be restricted to the state transitions that are shown in figure 1. It can stay in the same state for several steps in time, except for state stex where it can only stay for one step in time.

Proof. Axiom K9 specifies that for initial timepoints the value must be nil. In a given model satisfying the axioms, and for given g and a there, consider the set of all intervals $[s_i, t_i]$ such that $D([s_i, t_i], a)$ holds and $H(s_i, asta(g,a) : \mathsf{stex})$. We have already proved that these intervals must be disjoint (proposition S3). Within each of these intervals the transitions in figure 1 are satisfied according to axiom K1. Furthermore, at the endpoint t_i in each of those intervals the value is nil.

Consider now the intervals from the ending time t_i of one interval in this set, to the starting time s_{i+1} of the next interval. The state must be nil at t_i and stex at s_{i+1}. According to axiom K4 it must be pend at θs_{i+1}. Consider now the possible state sequences from nil to pend. Within that interval it can not take the value stex because in that case axiom K2 contradicts the construction. It can also not be ex for the same reason, using lemma 3. It is therefore restricted to nil, ref, and pend, according to axiom K0. A transition from pend to nil would violate axiom K8. Also, transitions between ref and nil in either direction would violate axiom K6. The remaining transitions between these three values are allowed by the diagram.

The timepoints before the first action interval and after the last one satisfy the same restrictions. This concludes the proof. QED.

4   Additional Facilities in the Action Coordinator

The specification of the action coordinator in the previous section provides a simple-minded one that just invokes actions as they are requested while respecting a minimal set of restrictions. It can be programmed by the proper choice of policy rules and by axioms specifying the applicability fluent $app(a)$. The definitions for $app(a)$ are of course domain specific. They were not an issue in the proofs in the previous section since applicability is used in the same way in the 'S' series and the 'K' series of axioms.

The general notion of an action coordinator strongly suggests that a number of other facilities can also be included in it, in particular:

- The use of composite actions, where the action coordinator is in charge of invoking successive sub-actions in a composite action. Plans can be seen as composite actions.
- The definition of goal-directed behavior, where the action coordinator is able to represent the relation between a goal and a sequence of actions that is supposed to lead to that goal. If one of the actions in the sequence fails then the action coordinator should identify a revised plan and start executing it instead.
- A more detailed description of the execution of individual actions in the robot's physical world, for example by relating the logical description of actions to their quantitative description using differential equations.

We have addressed each of these topics in some earlier articles [3,5,6]. It now seems possible that the concept of an action coordinator can provide a unifying framework within which these and other aspects of deliberative behavior can be addressed in a coherent way.

Acknowledgements. The initial ideas for this work occurred during a two-week visit to the IIIA in Barcelona at the invitation of Ramón Lopez de Mantaras. I take this opportunity to thank him and his colleagues there for their hospitality and for many stimulating discussions.

References

1. P. Noriega and C. Sierra. Towards layered dialogical agents. In *Proc. ECAI Workshop on Agents, Theories, Architectures and Languages (ATAL'96)*, pages 69–81. Springer Verlag, 1996.
2. E. Sandewall. *Features and Fluents*. Oxford University Press, 1994.
3. E. Sandewall. Relating high-level and low-level action descriptions in a logic of actions and change. In Oded Maler, editor, *Proc. of International Workshop on Hybrid and Real-Time Systems*, pages 3–17. Springer Verlag, 1997.
4. E. Sandewall. Cognitive robotics logic and its metatheory: Features and fluents revisited. *Electronic Transactions on Artificial Intelligence*, 2:307–329, 1998.
5. E. Sandewall. Logic-based modelling of goal-directed behavior. In Anthony G. Cohn, Lenhart Schubert, and Stuart C. Shapiro, editors, *Proc. of Conference on Principles of Knowledge Representation and Reasoning*, pages 304–315, 1998.
6. E. Sandewall. Use of cognitive robotics logic in a double helix architecture for autonomous systems. In Michael Beetz, Joachim Hertzberg, Malik Ghallab, and Martha Pollack, editors, *Advances in Plan-Based Control of Robotic Agents*, pages 226–248. Springer Verlag, 2001.
7. E. Sandewall and R. Rönnquist. A representation of action structures. In *Proc. of [U.S.] National Conference on Artificial Intelligence*, pages 89–97, 1986.

Artificial Intelligence in RoboCup

Daniele Nardi and Luca Iocchi

Dipartimento di Informatica e Sistemistica
Università "La Sapienza", Roma, Italy
{nardi, iocchi}@dis.uniroma1.it

Abstract. "RoboCup is an international joint project to promote AI, robotics, and related field. It is an attempt to foster AI and intelligent robotics research by providing a standard problem where wide range of technologies can be integrated and examined. RoboCup chose to use soccer game as a central topic of research, aiming at innovations to be applied for socially significant problems and industries."[1] The aim of the paper is to provide an AI research perspective on RoboCup, based on the experience gained partecipating in the competition, within our research group at "La Sapienza".

1 Introduction

RoboCup was launched about ten years ago as a landmark project which identifies "playing soccer" as a standard problem to be faced by research in the fields of AI and Intelligent Robotics and has the rather challenging long-term goal of building robots that play soccer as humans.

Since then, RoboCup has been very successful in organizing yearly World Championships (the RoboCup results have been regularly reported in *AI Magazine*), with associated scientific Symposia (whose proceedings are published in the *Springer Verlag, Lecture Notes*) and several local events, including promotional and educational activities.

The scientific scope of RoboCup spans over the fields of Robotics and Artificial Intelligence. In this paper, we admittedly regard RoboCup mainly from the standpoint of Artificial Intelligence.

There are several motivations for getting involved in the design and development of a RoboCup team, the most significant ones being the following.

First of all, the RoboCup competition poses interesting scientific problems and a significant body of research has been accomplished. As an example, a recent collection of the papers originating from the four legged league played with the Sony AIBO robotic platforms, shows not only outstanding progress, but also a significant number of outcomes that are far beyond the scope of the competition itself.

Secondly, RoboCup is very attractive for students (and very stressful for teachers!), giving them with motivation to work in AI and Intelligent Robotics and providing them with a significant experience of competitive project work.

---

[1] From the RoboCup Web site www.robocup.org

O. Stock and M. Schaerf (Eds.): Aiello Festschrift, LNAI 4155, pp. 193–211, 2006.

In this paper, we present a research perspective on RoboCup based on the experience gained in our laboratory, first building the Azzurra Robot Team, the Italian national team of robots which participated in the Middle-Size league of RoboCup-98 and RoboCup-99, then participating in the cited Four Legged league, starting in RoboCup-00 and finally, in the Rescue Real Robot and Simulated leagues.

This paper is organized as follows: we first provide some information on RoboCup, by describing the league structure, and sketching its overall research goals, activities and results; next, we focus on our own experience by addressing some research issues that we have been pursuing in our RoboCup activity. In particular, we focus on Cognitive Architectures, Vision and Perception, and Multi-Agent/Robot Systems.

2   RoboCup

RoboCup started its activity by focusing on soccer (football for Europeans). However, the scope of competitions has been broadening over the years. The new leagues have introduced new scientific challenges and have promoted the deployment of results in practical applications (see for example the RoboCup@Home league). A special league that is not in the scope of the present paper is the Junior League, which looks at future generations of researchers in Artificial Intelligence and Robotics.

In this section we first sketch the current RoboCup competitions and then address the main research challenges that have been pursued within RoboCup, trying to put them in perspective with the state of the art of research.

2.1   Leagues and Competitions

In the following, we focus on the soccer and rescue RoboCup leagues.

Soccer Leagues. The *Simulation* soccer league is played by computer programs. Each team is formed by 11 players, each of which is controlled by a separate program. The simulation is run by a Soccer Server. Each player has limited resources both in terms of sensing and in terms of motion capabilities. Communication among players is allowed and provides the basis for the development of cooperative play strategies. The simulation league is now evolving towards 3D simulation of the environment and including a coach competition to address the strategic aspect of the game.

There are two leagues based on wheeled robots, which pose different constraints on the robots. The *Small-Size* league is played on a ping-pong table by 5 vs. 5 robots, whose size is 15 cm^3, approximately. The sensing capabilities rely mainly on a global vision system, which allows for tracking the robots in the field in real time, and for implementing both centralized and off-board computation. The *Middle-Size* league is played within a 5x9 meter field by 4 players per team and the body of the robot must be within a cylinder of 50 cm diameter and 80 cm height. All sensing devices must be onboard the robots, in particular global vision and other sensing devices not on board the robots are not allowed.

The *Four Legged Robot* league is played by 4 four-legged SONY AIBO [63]. The 6x4 meter field is equipped with additional color markers. AIBO robots have on board a color camera and their mechanical structure provides 18 degrees of freedom. The availability of a standard platform has significantly contributed to the scientific evaluation of the solutions proposed. Recently, a *Humanoid Robot* league started to approach the ultimate goal of RoboCup to build a humanoid team to play with humans [39]. However, rsearch is still focusing on mechanics and locomotion.

Rescue Leagues. RoboCup Rescue [41,44] aims at the design of systems for search and rescue operations after large scale disasters. The goal of the league is to provide a socially and industrially relevant target to the research carried out in RoboCup. Moreover, this kind of application brings in scientific challenges related to uncertainty about the environment that are not present in the soccer leagues.

The Rescue Simulation league is concerned with the simulation of complex urban environments, in a post-earthquake scenario. The simulator provides a novel view with respect to state-of-the-art tools, since it deals with several types of events that impact on a disaster scenario: fire propagation, building collapse, road obstructions, traffic congestion, lack of communication, casualties and victims. The aim of the competition is to measure the performance of a rescue operation by several rescue parties programmed by the teams in terms of saved lives. The agent model to be implemented has the structure of the classical AI agent. Recently, a new simulation environment modeling robot exploration inside buildings has been introduced.

The Rescue Robot league aims at the design of robots that search victims in an unknown environment representing a disaster scenario. The experimental set-up, called *arena*, is being developed in close cooperation with USAR[2]. The arenas have already been used in various experiments (including RoboCup and AAAI rescue competitions) and nowadays represent a reference for experimental evaluation of rescue robots. The current aim of the competition is twofold: mobility and autonomy. As for the former, the research is focussed on the mechanical design in order to overcome the obstacles present in the environment; the latter is concerned with the design of robots that can autonomously explore the environment, possibly working in a team, build a map, find the victims and locate them in the map.

2.2   Research Challenges

The scientific goals of RoboCup have been described in several papers, discussing general objectives of the RoboCup initiative [40,2,1,52,10], specific research goals for the different leagues [42,39,41,43,51] and annual reports of the competitions [53,4,15,59,47,64,3,54,48].

---

[2] Performance Metrics and Test Arenas for Autonomous Mobile Robots. `www.isd.mel.nist.gov/projects/USAR/`

Below, we briefly look at the research challenges from a different perspective, namely by trying to look at the research topics that have been pursued in RoboCup spanning over the league structure. Obviously, there is no attempt to provide complete coverage, but rather to present a very personal view. For a more comprehensive overview we refer to the RoboCup website.

Control Architectures. As outlined above, robots embody different kinds of sensing and acting devices. The flow of data from the sensors to the actuators is processed by several different modules and the description of the interaction among these modules is usually referred to as the architecture.

The design of the architecture is relevant to all the RoboCup leagues. From the architecture viewpoint the RoboCup teams can be regarded as reactive and deliberative. We recall that the term "reactive" denotes that the robot reacts directly to the stimuli coming from the external environment [9], often without embodying a model of the surrounding world, which, conversely, characterizes deliberative robots.

While the RoboCup settings require the development of systems that exhibit both reactive and deliberative capabilities, a reactive behavior can have a very critical role, and certainly the effectiveness of the hardware significantly impacts on the robots' performance. However, some of the techniques for implementing behavior-based control may be difficult to apply, because of the dynamics of the environment. In this respect, both machine learning techniques (e.g., [61,7]) as well as novel approaches to behavior engineering have been pursued (e.g., [11,8]).

On the other hand, the complexity of the task to perform requires a high level representation of the information acquired from the environment and of the behaviors of the robots. Therefore, although most of the architectures used in RoboCup originates from behavior-based robotics, a few approaches have been proposed that originate from the research in Cognitive Robotics[3] (e.g., [16,65,13,6,19]).

Vision and Perception. Vision and Perception are obviously central to any robot design and therefore robot perception is a main research issue in RoboCup. Soccer games require the robots to perceive the positions of the ball, the other robots, and the field elements that are used for self-localization, while rescue robots need to build a map of the unknown environment and locate victims.

In the Small-Size league the centralized vision system can provide fast and reliable information on the game. On board sensors are used, but their effectiveness is limited due to the size of the robots. In the other soccer leagues, vision is the main source for acquiring information about the objects in the field. In the standardized platforms the emphasis is on processing speed for: colour segmentation, feature extraction, object recognition and tracking. On the self-constructed robots several settings have been investigated. In particular, in the Four-Legged league, AIBO robots use a single color camera on a pan-tilt

---

[3] The term "Cognitive Robotics" was first introduced by the research group at the University of Toronto led by Ray Reiter [46].

"neck" that allows for quick change of view points during the games; in the Middle-size, the common choice is the use of omni-directional cameras using mirrors of different shapes that allow for a 360 degree field of view [50] and thus for higher reactivity in the game. The real time constraints of image processing can be successfully met both with specialized hardware and with processing on conventional machines. Anyways, the amount of information that is extracted from the images can be very different and context dependent special processing is sometimes performed.

Moreover, several kinds of sensors have been used to increase the performance in object recognition, and other tasks related to perception. In particular, the use of laser range finders combined with inter-robot communication has been shown extremely effective for tracking opponents and teammates [27,28].

The main research issues that have been addressed in the soccer leagues are mostly related to real-time object recognition and tracking (e.g., [45]), self-localization (e.g., [29,35,56,49]), and vision-based navigation (see Section 4 for details). Interesting approaches to multi-robot cooperative perception have also been considered [18].

In contrast to soccer, in the rescue scenario different sensors are used on the robots: cameras are still used to examine the scene, but due to the difficulty in interpreting general scenes in this environment, images are also transmitted remotely to a human operator who examines them.

To create the map of the environment, laser range finders are a common choice. In some cases these sensors are mounted on a moving device enabling 3D scanning and, consequently, producing 3D maps of the environment [30]. Finally, thermic sensors are used to find heat sources that can be generated by the victims.

In the rescue scenario, object recognition becomes much more difficult due to the uncertainty and lack of structure of the environment. Specifically, the task of victim detection should address a number of issues that arise in this scenario, such as differences in size, shape, posture, color of skin, degrees of freedom of the human body, occlusions [24].

Moreover, the use of different kind of sensors (sonars, bumpers, infrared, thermic and vision sensors) either for object recognition or for other purposes (i.e., navigation) raises the problem of sensor fusion.

One specific aspect of perception is related to *localization*: we recall that localization amounts to knowing the robot's pose (position and orientation) in the environment [60]. This is a crucial feature for autonomous robots performing complex tasks over long periods of time and it is thus a main requirement for mobile robots involved in the soccer leagues. Due to the structure of the environment, both vision and laser range finders can be used, the former being more accurate, but requiring additional on board equipment and processing and several techniques have been experimented and compared [25,26,49].

Localization is combined with map construction in the autonomous robots designed for the rescue league, thus leading to the Simultaneous Localization and Mapping (SLAM) problem [60]. The features of the rescue environment make it challenging for state-of-the-art approaches, both in terms of real time performance and robustness. Moreover, the extentions to 3D mapping have recently been addressed to improve the quality of the rescue mission result.

Multi-agent systems. Coordination of robotic agents with distributed control is considered one of the central research issues in RoboCup competitions. In a highly dynamic and uncertain environment, such as the one provided by RoboCup games, the centralized coordination of activities does not seem to be adequate. In particular, the possible communication failures as well as the difficulty of constructing a global reliable view of the environment require full autonomy on each robot.

Coordination in multi-agent systems and team work in the context of RoboCup have a central role in simulation leagues, because of the high number of players. In particular, the rescue simulation league provides an ideal testbed for the work on large scale team coordination. Clearly, the presence of the opponents introduces substantial differences in the approaches to coordination.

In the soccer leagues (except for the Small-Size), all the sensors are on board the robots, thus distributed approaches to coordination are preferred since they offer increased robustness with respect to centralized ones. Moreover, coordination among the players is a critical issue for the performance of the team [66,36], because the dynamics of the game make it necessary to avoid interferences among players' actions and because of the difficulty of reconstructing global information about the environment.

Teams of robots can also be deployed in the rescue robot league. In this setting, the search can be greatly improved by using multiple robots. However, deploying a team of robots requires a robust autonomous behavior. In RoboCup rescue, all the basic aspects of cooperation within a team of robots are relevant and this makes it quite challenging from a research perspective.

RoboCup really has a large impact on the research in multi-agent, and, in particular, on multi-robot systems. In fact, the competitions provide very challenging scenarios for the experimentation and all the main technical issues arising in designing within cooperative teams, namely: cooperative perception, cooperative localization, task assignment, action synchronization and distributed planning, have been addressed by RoboCup related research.

Learning. Learning approaches are being applied to many problems arising in all the leagues of RoboCup using several techniques: genetic programming, reinforcement learning, and neural networks.

In the robot soccer leagues, learning process must face the challenges of the experiments with real robots, and is thus more suited to the learning of basic skills (e.g., [23,55,57,67]), but also of cooperative behaviors (e.g., [61]). In the simulation leagues experiments on learning strategies, behaviors and adaptation to the opponents' model have been performed (e.g., [7]). Moreover, complex forms of learning can also be combined in a layered learning approach [58], according to a task decomposition structure.

It turns out that learning and adaptation of basic skills, such as walking (e.g., [55]), and vision calibration (e.g., [57]), have shown to be much more effective than parameter tuning by hand.

Moreover, rather interesting experiments have been performed in learning collaborative behaviors, in order to achieve coordination without communication that is also viewed as implicit communication. See [62] for a survey of learning in multi robot systems and RoboCup, specifically.

Fig. 1. Soccer and Rescue Robots in action. In the upper left, robots of the Azzurra Robot Team in the Middle-Size league; in the lower left, AIBO soccer robots of the SPQR team in the Four-Legged League; on the right, our robot used in RoboCup Rescue competitions.

In the next sections we look more closely at the issues that we have directly addressed in our experience in RoboCup, within the research group at "La Sapienza" (Figure 1 shows our teams of robots during RoboCup competitions). In particular, here we shall address our contributions according to the structure above, except for learning, which is, to us, left for future research.

3   Cognitive Architectures

In this section we address our proposals to apply a cognitive robotics approach to the design of soccer players and rescue robots.

Cognitive Robotics aims at designing and realizing actual agents (in particular mobile robots) that are able to accomplish complex tasks in real, and hence dynamic, unpredictable and incompletely known environments without human assistance. For this purpose, they can be controlled at a high level by providing them with a description of the world and expressing the tasks to be performed in the form of goals to be achieved. However, in RoboCup, due to the difficulty of acquiring information from the environment, instead of using complex strategical reasoning, the cognitive approach has been useful as a metaphor for clean design and quick development.

The implementation of a Cognitive Robotics approach requires addressing a formal framework for representation and reasoning about actions and the system organization. This enables the execution of complex actions while retaining the ability to quickly react to the changes of the environment, namely the architecture.

Let us first briefly address the architecture of our robots, a heterogeneous layered hybrid architecture [34], with two levels: the *deliberative* (cognitive) level, in which a high-level state of the agent is maintained and decisions on which actions are to be performed are taken, and the *operative* (reactive) level, in which low-level conditions on the world are verified and actions are actually executed. To provide an effective integration of reasoning and reactivity, we adopt an asynchronous execution model. The actions in the operative level are expressed as control programs. Such programs are generated by the planner (possibly off-line) and turned into sequences of actions, or, more generally, into execution structures (including behaviors and low level control actions).

Our formal framework [17] originates from Propositional Dynamic Logics and exploits their formal correspondence with Description Logics. In [37] an extension of such a framework is presented including both concurrency on primitive actions and autoepistemic operators for explicitly representing the robot's epistemic state. The resulting formal setting allows for the representation of actions with context-dependent effects, sensing actions, and concurrent actions, and properly addresses both the presence of exogenous events and the characterization of the notion of an executable plan in such a complex setting.

The proposed framework has been implemented in a system which is capable of generating plans that are actually been used to describe the knowledge of robotic soccer players. In the implementation, the output of the planner is used to generate control programs, that can be directly executed on the mobile base. Specifically, we have been able to formalize at the logical level several situations arising in the RoboCup scenario and to generate, through the planner, a significant portion of the control programs that were executed on our soccer players.

Recently, the approach has been extended to deal with non-determistic actions and actions with probabilistic effects [32]. A probabilistic representation raises, at the symbolic level, some of the uncertainty that must be faced in the RoboCup environments. Consequently, quantitative information to choose the possible courses of action becomes available at planning time, allowing an improvement in the agents' performance.

Although the above mentioned planning formalism is quite expressive, programming high level behaviors for complex mobile robots performing complex tasks requires a more powerful description language. For example, in programming soccer behaviors for AIBO robots, additional features need to be considered: action interruption due to changes in the environment, parallel execution of complex behaviors (sub-plans).

To address these new requirements we have devised a planning language based on Petri Nets [68] that allows for specifying complex constructs of actions, such as ordinary actions, sensing actions, conditional structures, loops, interrupts, and concurrent execution of plans.

For this formalism we have implemented a plan executor that controls the evolution of the plan according to the internal state of the robot, as well as a set of graphical tools for easily drawing, verifying and debugging plans.

4   Robot Perception

In this section we sketch our contributions in robot perception applied in RoboCup competitions. In particular, we describe methods for color segmentation and object recognition that are in use in our Four-Legged League, methods for localization and mapping that are in use in both the soccer and the rescue teams, and human body detection through stereo vision that has been implemented on our rescue robot.

4.1   Color Segmentation and Object Recognition

RoboCup soccer is a color-coded environment, where colors are used to define principal objects needed by the robots to perform their tasks. Consequently, color segmentation is typically the first step of the vision system of a robot playing RoboCup soccer. Since good color segmentation allows for easy implementation of object recognition and localization, most of the robot vision systems are based on fast and accurate implementation of such process.

Many approaches for on-line color segmentation have been proposed in the RoboCup community. However, some of these approaches can be implemented in real-time only on robots with adequate computational resources (e.g., Middle-size robots). In the Four-Legged League, AIBO robots have very limited computational power (576 MHz CPU) and a low quality low resolution image acquisition device; moreover, real-time implementation of the vision process is extremely important for a correct execution of behaviors. An effective implementation of on-line color segmentation on AIBO robots has been reported in [38].

The method developed within our team [31] is an on-line segmentation method that dynamically computes a transformation of the color distribution of the image. The approach has been actually implemented in such a way to obtain real-time performance on AIBOs, by performing dynamic computation only periodically. Moreover, it can be used for off-line generation of color tables that can be manually refined, speeding up the manual calibration process.

 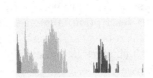 

Fig. 2. Dynamic Color Segmentation

Figure 2 shows an example of segmentation. The method is mainly based on fast analysis of a mono dimensional color space: the H component of the HSV color space. By analyzing color distribution of the current image, we are able to compute a transformation function that transforms such distribution in another distribution. The resulting distribution provides for efficient color segmentation based on static thresholds, but being computed according to the current image is also very robust to illumination changes.

During RoboCup 2005, we have tested the method not only during the variable lighting challenge, but also in some games (test matches). We set parameters the first day and never changed them during the competition. The method has been evaluated also on a large data set and with a novel evaluation mechanism [31]. Such evaluation shows that the method achieves performance that are comparable with those obtained by manual calibration, being at the same time robust to illumination changes.

4.2   Localization and Mapping

In all the RoboCup leagues (except for the Small-Size League), localization is a critical problem, since global positioning sensors are not allowed. This is especially true for those approaches that attempt to build an explicit model of the state of the robot (i.e., excluding the purely reactive ones). The Middle-size and Four-Legged RoboCup environments assume the following characteristics that must be considered for the choice of localization methods: (i) the geometry of the lines drawn on the field are known, (ii) the environment is highly dynamic (there are many robots and the ball i moving in the field); (iii) the task must be performed continuously for a "long" time (the length of each period is 10 minutes); (iv) the environment cannot be modified; (v) crashes among robots are possible. In addition, for the Four-Legged league it is necessary to consider that the only sensor available is a low quality camera, that odometry is very poor, and that robots have limited computational power. All these factors determine a difficult scenario for localization methods.

Self-localization is a well known and well studied problem in robotics. Probabilistic approaches (see [60]) have been successfully used in many application fields (and also in RoboCup) and comparisons among different methods [25,26,49] have been performed in order to evaluate the characteristics and the performance of the different methods.

For robots in the Middle-Size League we proposed a localization method [33], that is based on matching a geometric reference map with a representation of range information acquired by the robot's sensors. The method exploits the properties of the Hough Transform for recognizing lines from a set of points, as well as for calculating the displacement between the estimated and the actual pose of the robot. The Hough Transform enables for a representation of lines that makes the matching process computationally fast and robust to noise.

In the Four Legged League, we have experimented a different approach, based on particle filters [49], that results in better performance in a setting where robots have very poor perception abilities.

Fig. 3. Map

An extension of localization based on the Hough Transform has been used for scan matching and mapping [14], without the assumption of polygonal environments. This method has been experimented also on our rescue robot [5] for the implementation of a scan matching process that is used within the mapping module. Figure 3 shows the output of a rescue mission of our robot at RoboCup 2004, the map, the victims found and the robot trajectory are the results of partial autonomous exploration and victim detection [12].

4.3 Human Body Detection

Human body detection is a central problem in the rescue league, where the objective is to find victims and place their position on a map. The characteristics of the rescue scenario make this task even more difficult than the general one of detecting the presence of people in an environment. In fact, besides the general difficulties of this process, such as the high degree of freedom of a human body, self occlusions, appearance variation due to clothing, shape variations due to the different gender, race and age, in search and rescue missions it is necessary also to consider additional problems due to lighting variations and to the fact that it is not always possible to achieve the position of the best viewpoint.

Our method for human body detection [24] makes use of a stereo vision camera and exploits 3D range information in order to compute 3D measures for human body limbs and match them against a human body model. The method works reasonably well even in situations of severe occlusions (that are very common in a rescue scenario); however, it requires the subject to be quite close to the camera.

Given a stereo pair of images, the left image is processed with an edge detection module and then with contour extraction. These steps provide a segmentation of the image based on closed contours. The stereo pair is also processed by a stereo algorithm to compute the disparity and 3D information about the segments identified before. The body part classifier takes these segments and uses a similarity measure to detect body limbs, that are finally used for matching a human body model.

The method uses an iterative algorithm that performs split and merge operations between the detected segments and uses human body models at different levels in order to find the best association [24]. A goodness function is used to match found segments and human body parts. If such a goodness is above a predefined threshold, then the hypothesis is accepted and the presence of a human figure in the image is declared, otherwise a new iteration is attempted merging/splitting detected segments. The process terminates with the result that a human figure is not present in the image when, after a limited number of iterations, none of the segments has a sufficient score for a human body limb, so the final score of the goodness function is low.

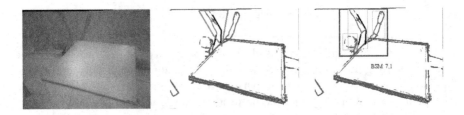

Fig. 4. Example of victim detection in a rescue scenario

Figure 4 shows an example of such a process including the left image of the stereo pair, the segmented image with closed contours and the similarity measures.

5   Multi-robot Coordination

Multi-Robot Coordination is a fundamental aspect for an effective realization of teams of intelligent systems, therefore it is a central research issue in RoboCup.

This section describes two approaches to multi-robot coordination: the first related to the soccer task, the second concerning large scale multi agent systems and in general situations where limited communication bandwidth is available.

5.1   Coordination of Heterogeneous Soccer Robots

We have already pointed out that the idea of the soccer leagues is that robots are autonomous, since each robot acquires the information about the game only through on-board devices. While a centralized approach is possible, in most cases robot control is fully distributed. In fact, a distributed approach provides for increased robustness to unavoidable measurement errors, sensor noise, network failures, and other possible malfunctioning of one element in the system. Moreover, due to the difficulties of reconstructing precise and reliable information about the environment (with the exception of [27]), coordination needs to be achieved without making strong assumptions on the knowledge of the single players, but typically relying on explicit communication to exchange information among the players.

In addition, team members may differ both in the hardware and in the software. Consequently, coordination requires not only a distributed coordination protocol, but also a very flexible one that allows to accommodate the various configurations that can arise by forming teams with different basic features.

Besides the above constraints, coordination should deal both with roles (defender, attacker etc.) and with strategy (defensive, offensive). While the strategic level is currently demanded to an external selection (the human coach), roles are dynamically assigned (see [58]) to the various team elements during the game, depending on the configuration present on the field.

In this scenario we have developed a distributed coordination method that has been successfully employed by all the members of the ART team during the 1999 RoboCup competition [36] and then on the SPQR Legged team. The ART team was formed by players that have been developed by 6 research groups, operating in various Italian universities. Therefore, it ART can be considered a multiagent robotic system that is heterogeneous both from the hardware and from the software viewpoint, thus making coordination rather challenging. The effectiveness of the proposed method has been proved by the fact that we were always ready to substitute any robot with another one, without endangering the coordinate behavior of the overall team.

From a technical viewpoint, the proposed protocol is based on the explicit exchange of data about the status of the environment and is based on simple forms of negotiations. Simplicity in the protocol stems from the need to make rather weak assumptions on each robot's capabilities. An increase of such capabilities would lead to more complex protocols. However, we believe that a major issue in coordination is to find a suitable balance between the robot's individual capabilities and the form of cooperation realized.

5.2   A Token Passing Approach for Dynamic Task Assignment

Coordination protocols using broadcast messages among the agents to share utility values at every cycle (as the one discussed in the previous section) is effective

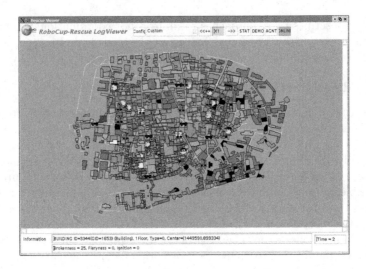

Fig. 5. Rescue simulation agents

in very dynamic domains, but does not scale well with the size of the team, since it requires a large network bandwidth. In many applications (and also in the RoboCup rescue simulation and real robots), communication bandwidth is limited and coordination approaches relying too much on communication may easily decrease their performance due to network latency of temporary failures. Moreover, rescue simulation involves a higher number of agents to be controlled (than soccer robot leagues) and tasks have additional constraints on the execution that are usually not present in the soccer environment (e.g., deploying too many firemen for extinguishing a fire in a building is not effective).

To address these problems, we have devised a coordination method that requires very low communication bandwidth, is able to guarantee a dynamic conflict-free task allocation, and to deal with constraints on task execution, while maintaining performance that is comparable to approaches using broadcast communication at every cycle. In particular, we have considered a scenario in which: i) tasks are not known beforehand, they have to be discovered during mission execution; ii) each task may require exactly n agents in order to be performed; iii) once agents are allocated to a particular task they must synchronize their actions.

The basic idea of our approach [20,21] is based on a Token Passing technique in which tokens are used to represent tasks that must be executed by the agents, and each team member creates, executes and propagates these tokens based on its knowledge of the environment. Tokens are not statically predefined, but generated on-line during mission execution as a result of agent perceptions. An asynchronous distributed algorithm is used to detect and solve conflicts due to simultaneous or erroneous task perception by several agents. Our approach guarantees a conflict free allocation of exactly n agents for each task. Moreover, this new coordination method significantly reduces network usage for coordination thus scaling better with the number of agents in the team.

Experiments have been performed with simulated agents in the RoboCup Rescue simulation environment [20]. Figure 5 shows agents in a map representing the city of Foligno, that has been developed within our project and used in the RoboCup 2003 competition. In particular, in our experiments we have evaluated different strategies for the firemen engaged with fire extinguishing tasks, showing that the assignment of exactly n agents to each task (i.e., to each firing building) effectively improves the performance of the team. Moreover, the use of token passing guarantees a low communication bandwidth. This makes it thus suitable to work in the RoboCup Rescue simulator, which embodies many limitations on communication between agents, simulating communication problems that are often present in rescue scenarios.

The token passing approach has also been used with AIBO robots in a cooperative foraging task [22]. The robots have to collect several objects scattered in the environment. The collection of each object requires that exactly two robots help each other to grab it (a helper robot and a collector robot). After the grabbing phase, only one robot is needed to transport the object. Object number and position in the environment is not known, and objects are identical in color and shape, therefore they can only be distinguished by their position in the environment.

6   Conclusion

In this paper we have addressed research in Artificial Intelligence carried out within the RoboCup initiative, focussing in particular on the work of the group at "La Sapienza". We have discussed some of the challenges and addressed some of the research results. Overall, we believe that the paper shows that the RoboCup framework is well suited for developing interesting research work in Artificial Intelligence and it will be even more so in the future.

We conclude the paper with a few additional considerations on the RoboCup initiative that are not directly specific to the research achievements, but are related to a broader view of the research development.

Over the years, the RoboCup activity has involved a large number of students. They have been a major resource of the project and the main investments have been dedicated to them, through schools, through several preparation meetings where the technical solutions developed were presented and carefully evaluated, and, finally, through the participation in the competitions held in a scientifically qualified international environment. In addition to conventional lectures, such activities allowed the students to interact with colleagues and teachers of other academic institutions and discuss their own ideas in a very stimulating and competitive framework. The overall training experience for the students has no counterpart in the Italian university curricula. While RoboCup may not be viewed as a self-contained research framework, its value in the recruitment of new young and enthusiastic researchers is extremely valuable.

The costs of developing and maintaining RoboCup teams are not negligible as compared to the funding that Universities and other agencies can provide for this kind of education and research activities. This drove us to enter the rescue competion to be able to combine the competition with application projects. Here is a tradeoff between application oriented and basic research. While basic research plays a fundamental role, it is more difficult to obtain funding. However, we hope that appreciation of the RoboCup initiative, which combines, very attractively education with the development of basic research in Artificial Intelligence, can bring new resources and new funding to the field.

After about ten years of RoboCup, it is worth considering how the initiative will proceed in the remaining 44 years to the ultimate goal of humanoid soccer players in 2050. This year is also the sixthieth birthday of Artificial Intelligence, and we can realize how difficult it was to foresee the achievements of today's AI 40 years ago. We simply conclude by wishing that in the next 40 years the research stream of AI that has been developed through RoboCup, steadily continues to advance our knowledge and achievements in the field.

Acknowledgments. We would like to thank all the members of the ART team and of the SPQR teams for their work and their contribution to the research view presented in the paper. But, foremost, we would like to thank Gigina Carlucci Aiello for encouraging, supporting and appreciating our efforts. Grazie, Gigina.

References

1. M. Asada, H. Kitano, I. Noda, and M. Veloso. Robocup: Today and tomorrow – what we have learned. *Artificial Intelligence*, 110:193–214, 1999.
2. M. Asada, H. Kitano, P. Stone, A. Drougoul, D. Duhaut, M. Veloso, H. Asama, and S. Suzuki. The robocup physical agent challenge: goals and protocols for phase 1. In H. Kitano, editor, *RoboCup 2001: Robot Soccer World Cup V*, volume 1365 of *Lecture Notes in Computer Science*, pages 41–61. Springer-Verlag, 1998.
3. M. Asada, O. Obst, D. Polani, B. Browning, A. Bonarini, M. Fujita, T. Christaller, T. Takahashi, S. Tadokoro, E. Sklar, and G. A. Kaminka. An overview of robocup-2002 fukuoka/busan. *AI Magazine*, 24(2):21–40, 2003.
4. M. Asada, M. M. Veloso, M. Tambe, I. Noda, H. Kitano, and G. K. Kraetzschmar. Overview of robocup-98. *AI Magazine*, 21(1):9–19, 2000.
5. S. Bahadori, D. Calisi, A. Censi, A. Farinelli, G. Grisetti, L. Iocchi, and D. Nardi. Autonomous systems for search and rescue. In A Birk, S. Carpin, D. Nardi, Jacoff A., and S. Tadokoro, editors, *Rescue Robotics*. Springer-Verlag, 2005.
6. A. Bonarini, M. Matteucci, and M. Restelli. Filling the gap among coordination, planning, and reaction using a fuzzy cognitive model. In *RoboCup 2003: Robot Soccer World Cup VII*, pages 662–669, 2003.
7. M. Bowling and M. Veloso. Simultaneous adversarial multi-robot learning. In *Proceedings of IJCAI'03*, 2003.
8. A. Brendenfeld, T. Christaller, W. Ghring, H. Gnther, H. Jaeger, H.U. Kobialka, P.G. Plger, P. Schll, A Siegberg, A Striet, C. Verbeek, and J. Wilberg. Behavior engineering with "dual dynamics" models and design tolls. In M. Veloso, editor, *RoboCup 2003: Robot Soccer World Cup VII*, pages 57–62, Stockholm, Sweden, Aug. 1999. IJCAI Press.
9. R. A. Brooks. A robust layered control system for a mobile robot. *IEEE Journal of Robotics and Automation*, RA-2(1), 1986.
10. H.-D. Burkhard, D. Duhaut, M. Fujita, P. Lima, R. Murphy, and R. Rojas. The road to robocup 2050. *IEEE Robotics and Automation Magazine*, 9(2):31 – 38, June 2002.

11. H.D. Burkhard, M. Hannebauer, J. Wendler, H. Myritz, G. Sander, and J. Meinert. Bdi design priciples and cooperative implementation - a report on robocup agents. In M. Veloso, editor, *RoboCup 2003: Robot Soccer World Cup VII*, pages 68–73, Stockholm, Sweden, Aug. 1999. IJCAI Press.

12. D. Calisi, A. Farinelli, L. Iocchi, and D. Nardi. Autonomous navigation and exploration in a rescue environment. In *Proc. of the 2nd European Conference on Mobile Robotics (ECMR)*, pages 110–115, September 2005.

13. C. Castelpietra, A. Guidotti, L. Iocchi, D. Nardi, and R. Rosati. Design and implementation of cognitive soccer robots. In *RoboCup 2001: Robot Soccer World Cup V*, pages 312–318, 2002.

14. A. Censi, L. Iocchi, and G. Grisetti. Scan matching in the hough domain. In *Proceedings of the IEEE International Conference on Robotics and Automation (ICRA'05)*, 2005.

15. S. Coradeschi, L. Karlsson, P. Stone, T. Balch, G. Kraetzschmar, and M. Asada. Overview of RoboCup-99. *A.I. Magazine*, 1999.

16. A. C. P. L. da Costa and G. Bittencourt. UFSC-Team: A cognitive multi-agent approach to the RoboCup'98 simulator league. In *RoboCup-98: Robot Soccer World Cup II*, pages 371–376, 1998.

17. G. De Giacomo, L. Iocchi, D. Nardi, and R. Rosati. A theory and implementation of cognitive mobile robots. *Journal of Logic and Computation*, 5(9):759–785, 1999.

18. M. Dietl, J.-S. Gutmann, and B. Nebel. Cooperative sensing in dynamic environments. In *IROS01*, Maui, Hawaii, 2001.

19. F. Dylla, A. Ferrein, G. Lakemeyer, J. Murray, O. Obst, T. Rofer, F. Stolzenburg, U. Visser, and T. Wagner. Towards a league-independent qualitative soccer theory for robocup. In *RoboCup 2004: Robot Soccer World Cup VIII*, 2005.

20. A. Farinelli, G. Grisetti, L. Iocchi, S. Lo Cascio, and D. Nardi. Design and evaluation of multi agent systems for rescue operations. In *Proc. of IEEE/RSJ International Conference on Intelligent Robots and Systems (IROS'03)*, pages 3138–3143, 2003.

21. A. Farinelli, L. Iocchi, D. Nardi, and V. A. Ziparo. Task assignment with dynamic perception and constrained tasks in a multi-robot system. In *Proc. of the IEEE Int. Conf. on Robotics and Automation (ICRA)*, pages 1535–1540, 2005.

22. A. Farinelli, L. Iocchi, D. Nardi, and V. A. Ziparo. Assignment of dynamically perceived tasks by token passing in multi-robot systems. *Proceedings of the IEEE, Special issue on Multi-Robot Systems*, 2006.

23. P. Fidelman and P. Stone. Learning ball acquisition on a physical robot. In *Proc. of Intern. Symposium on Robotics and Automation (ISRA)*, 2004.

24. S. Bahadori Ghouchani. *Human Body Detection in Search and Rescue Missions*. PhD thesis, University of Rome 'La Sapienza', Dipartimento Di Informatica e Sistemistica, 2006.

25. J. S. Gutmann, W. Burgard, D. Fox, and K. Konolige. An experimental comparison of localization methods. In *In Proc. of the IEEE/RSJ International Conference on Intelligent Robots and Systems (IROS)*, 1998.

26. J. S. Gutmann and D. Fox. An experimental comparison of localization methods continued. In *In Proc. of the IEEE/RSJ International Conference on Intelligent Robots and Systems (IROS)*, 2002.

27. J.-S. Gutmann, W. Hatzack, I. Herrmann, B. Nebel, F. Rittinger, A. Topor, T. Weigel, and B. Welsch. The cs freiburg robotic soccer team: Reliable self-localization, multirobot sensor integration, and basic soccer skills. In M. Asada, editor, *RoboCup 2002: Robot Soccer World Cup VI*, Berlin, Heidelberg, New York, 1998. Springer-Verlag.

28. J.-S. Gutmann, T. Weigel, and B. Nebel. Fast, accurate, and robust self-localization in the robocup environment. In *RoboCup 2003: Robot Soccer World Cup VII*, 1999.
29. J.-S. Gutmann, T. Weigel, and B. Nebel. Fast, accurate, and robust self-localization in the robocup environment. In *RoboCup-99: Robot Soccer World Cup III*, pages 304–317, 1999.
30. S Hartmut, R. Woerst, M. Henning, K. Lingemann, A. Nuechter, K. Pervolez, R.T. Kiran, T. Christaller, and J. Hertzberg. Robocuprescue - robot league team kurt3d. In *RoboCup 2004: Robot Soccer World Cup VIII*, 2004.
31. L. Iocchi. Robust color segmentation through adaptive color distribution transformation. In *Proc. of RoboCup Symposium*, 2006.
32. L. Iocchi, T. Lukasiewicz, D. Nardi, and R. Rosati. Reasoning about actions with sensing under qualitative and probabilistic uncertainty. In *Proc. of 16th European Conference on Artificial Intelligence (ECAI'04)*, pages 818–822, Spain, 2004.
33. L. Iocchi and D. Nardi. Hough localization for mobile robots in polygonal environments. *Robotics and Autonomous Systems*, 40:43–58, 2002.
34. L. Iocchi. *Design and Development of Cognitive Robots*. PhD thesis, Univ. "La Sapienza", Roma, Italy, On-line `ftp.dis.uniroma1.it/pub/iocchi/`, 1999.
35. L. Iocchi and D. Nardi. Self-localization in the RoboCup environment. In *RoboCup-99: Robot Soccer World Cup III*, pages 318–330. Springer-Verlag, 1999.
36. L. Iocchi and D. Nardi. SPQR-Legged Team 2003. In *RoboCup 2003: Robot Soccer World Cup VII*. Springer-Verlag, 2003.
37. L. Iocchi, D. Nardi, and R. Rosati. Planning with sensing, concurrency, and exogenous events: logical framework and implementation. In *Proceedings of the Seventh International Conference on Principles of Knowledge Representation and Reasoning (KR'2000)*, pages 678–689, 2000.
38. M. Jüngel. Using layered color precision for a self-calibrating vision system. In *RoboCup 2004: Robot Soccer World Cup VIII*, 2004.
39. H. Kitano and M. Asada. Robocup humanoid challenge: That's one small step for a robot, one giant leap for mankind. In *Proc. of IEEE/RSJ International Conference on Intelligent Robots and Systems 1998 (IROS '98)*, pages 419–424, 1998.
40. H. Kitano, M. Asada, Y. Kuniyoshi, I. Noda, E. Osawa, and H. Matsubara. Robocup: A challenge problem of ai. *AI Magazine*, 18:73–85, 1997.
41. H. Kitano and et al. Robocup-rescue: Search and rescue for large scale disasters as a domain for multi-agent research. In *Proceedings of IEEE Conference on Man, Systems, and Cybernetics(SMC-99)*, 1999.
42. H. Kitano, M. Tambe, P. Stone, M. Veloso, S. Coradeschi, E. Osawa, H. Matsubara, I. Noda, and M. Asada. The robocup synthetic agent challenge 97. In *Proceedings of IJCAI-97*, Nagoya, Japan, August 1997.
43. H. Kitano, S. Suzuki, and J. Akita. Robocup jr.: Robocup for edutainment. In *Proc. of Int. Conf. on Robotics and Automation (ICRA)*, pages 807–812, 2000.
44. H. Kitano and S. Tadokoro. RoboCup-Rescue: A grand challenge for multi-agent and intelligent systems. *AI Magazine*, 22(1):39–52, 2001.
45. C. Kwok and D. Fox. Map-based multiple model tracking of a moving object. In *RoboCup 2003: Robot Soccer World Cup VII*, 2003.
46. Y. Lesperance, H.J. Levesque, F. Lin, D. Marcu, R. Reiter, and R.B. Scherl. A logical approach to high-level robot programming. In *AAAI FAll Symposium on Control of the Physical World by Intelligent Systems*, 1994.
47. P. Lima, T. Balch, M. Fujita, R. Rojas, M. Veloso, and H. Yanco. Robocup 2001. *IEEE Robotics and Automation Magazine*, 9(2):20 – 30, June 2002.

48. P. Lima, L. Custdio, L. Akin, A. Jacoff, G. Kraezschmar, N. B. Kiat, O. Obst, T. Röfer, Y. Takahashi, and C. Zhou. RoboCup 2004 competitions and symposium: A small kick for robots, a giant score for science. *AI Magazine*, 26(2):36–61, 2005.

49. L. Marchetti, G. Grisetti, and L. Iocchi. A comparative analysis of particle filter based localization methods. In *Proc. of RoboCup Symposium*, 2006.

50. E. Menegatti, F. Nori, E. Pagello, C. Pellizzari, and D. Spagnoli. Designing an omnidirectional vision system for a goalkeeper robot. In *RoboCup 2001: Robot Soccer World Cup V*, 2001.

51. R. Murphy, J. G. Blitch, and J. L. Casper. RoboCup/AAAI urban search and rescue events: Reality and competition. *AI Magazine*, 1(23):37–42, 2002.

52. D. Nardi. Artificial Intelligence in RoboCup. In *Proc. of ECAI*, 2000.

53. I. Noda, S. Suzuki, H. Matsubara, M. Asada, and H. Kitano. Overview of robocup-97. In *RoboCup-97: Robot Soccer World Cup I*, pages 20–41, 1998.

54. E. Pagello, E. Menegatti, D. Polani, A. Bredenfel, P. Costa, T. Christaller, A. Jacoff, M. Riedmiller, A. Saffiotti, E. Sklar, and T. Tomoichi. RoboCup-2003: New scientific and technical advances. *AI Magazine*, 25(2):81–98, 2004.

55. T. Röfer. Evolutionary gait-optimization using a fitness function based on proprioception. In *RoboCup 2004: Robot Soccer World Cup VIII*, pages 310–322, 2005.

56. T. Röfer and M. Jüngel. Fast and robust edge-based localization in the sony four-legged robot league. In *Proc. 7th International Workshop on RoboCup 2003*, 2004.

57. M. Sridharan and P. Stone. Autonomous color learning on a mobile robot. In *Proc. of AAAI*, 2005.

58. P. Stone and M. Veloso. Task decomposition, dynamic role assignment, and low-bandwidth communication for real-time strategic teamwork. *Artificial Intelligence*, 110(2):241–273, 1999.

59. P. Stone et al. Overview of RoboCup-2000. In *RoboCup 2000: Robot Soccer World Cup IV*, pages 1–28, 2001.

60. S. Thrun, W. Burgard, and D. Fox. *Probablilistic Robotics*. MIT press, 2005.

61. E. Uchibe, M. Asada, and K. Hosoda. Cooperative behaviour acquisition in multi mobile robots environment by reinforcement learning based on state vector estimation. In *Proc. of IEEE Int. Conf. on Robotics and Automation*, pages 1558–1563, 1998.

62. M. Veloso and P. Stone. *A Survey of Multiagent and Multirobot Systems*, chapter in Robot Teams: From Diversity to Polymorphism, T. Balch and L. E. Parker, eds. AK Peters, 2002.

63. M. Veloso, W. Uther, M. Fujita, M. Asada, and H. Kitano. Playing soccer with legged robots. In *Proceedings of IROS-98, Intelligent Robots and Systems Conference*, Victoria, Canada, October 1998.

64. M. Veloso et al. RoboCup-2001: The fifth robotic soccer world championships. *AI Magazine*, 23(1):55–68, 2002.

65. T. Weigel at al. CS Freiburg: Doing the right thing in a group. In *RoboCup 2000: Robot Soccer World Cup IV*, 2001.

66. K. Yokota, K. Ozaki, N. Watanabe, A. Matsumoto, D. Koyama, T. Ishikawa, K. Kawabata, H. Kaetsu, and H. Asama. Cooperative team play based on communication. In *RoboCup-98: Robot Soccer World Cup II*, pages 491–496, 1998.

67. J. C. Zagal and J. Ruiz-del Solar. Learning to kick the ball using back to reality. In *RoboCup 2004: Robot Soccer World Cup VIII*, 2005.

68. V. A. Ziparo and L. Iocchi. Petri net plans, 2006. Technical Report.

Planning Under Uncertainty and Its Applications

Paolo Traverso

ITC-irst - Via Sommarive 18 - 38050 Povo - Trento, Italy
traverso@itc.it

Abstract. Several application domains require planning techniques that model uncertainty in the results of both actions and observations. Actions may have different effects that cannot be predicted at planning time. Observations may result into uncertainty about the current state of the world. In this paper, we first discuss the problem of planning with uncertainty in action execution and observations. We then discuss how this problem can be relevant to different application domains that represent rather different characteristics, like planning for controlling a robot that has to perform a surveillance task, as well as planning for the automated composition of web services for e-commerce.

1 Introduction

Planning is concerned with choosing and organizing actions for changing the state of a dynamic system. A *planning domain* is a conceptual and abstract model for describing the system that can evolve by executing actions. For instance, a planning domain can model the fact that a robot can transport objects from one room to another one, thus changing the objects' positions. A railway interlocking system can issue commands that change the state of switches, rail-road crossings and signals to trains. A web service for on-line payments can activate a money transaction thus changing the state of the bank account.

Two key elements of a planning domain are *actions* and *observations*. Actions make the system change its state. Observations acquire information about the current state of the system. Two common sources of uncertainty are the results of actions and observations:

- **Uncertainty about Actions:** An action may have different effects, and it is impossible for the planner to predict at planning time which effect will actually take place at execution time. The uncertainty is about which is the actual state after an action has been executed.
- **Uncertainty about Observations:** In some situations the state of the world cannot be completely observed, and thus cannot be uniquely determined. An observation at execution time can result is a set of states. The uncertainty is about the current state of the world.

In this paper, we first discuss the state of the art in planning under uncertainty (Section 2). We then describe a general teoretical framework for planning with

O. Stock and M. Schaerf (Eds.): Aiello Festschrift, LNAI 4155, pp. 213–228, 2006.

uncertainty in actions and observations (Section 3). We consider two examples of problems with rather different characteristics: planning for a robot's surveillance tasks (Section 4), and planning for the automated composition of a purchase and deliver web service (Section 5). In these two sections, we discuss how planning under action and observation uncertainty is relevant in both cases.

2   State of the Art and Related Work

Within the automated planning community, it is widely recognized that several application domains need modeling and reasoning tools different from classical planning. Classical planning relies indeed on several restrictive assumptions. Among them, three main assumptions are not realistic for several application domains. First, the hypothesis that actions have deterministic effects, i.e., that each action applicable in a state leads to a single new state. Second, the assumption of full observability, i.e., the fact that plan execution has complete knowledge about the current state of the system and observations determine the state univocally. Finally, the restriction that goals are sets of states, i.e., the objective is to build a plan that leads to one of the goal states.

One of the first approaches to automated planning that has relaxed these assumption is planning based on Markov Decision Processes (MDP) (see [12] for an extensive survey). MDP is essentially planning over stochastic domains, i.e., non deterministic domains that allow for actions with multiple effects and where probabilities can be assigned to transitions. The MDP framework is rather expressive. It allows for representing and dealing with information about costs and probabilities of action transitions and with rewards associated to states. In MDP, algorithms that optimize a utility function are defined on the basis of costs/rewards.

In this paper we propose a framework that does not deal with probabilities, costs, rewards, and utility functions. We are convinced that, while in some applications statistical information about probabilities can be collected with experiments and historical data, there is a rather large set of domains where *uncertainty has nothing to do with probabilities*. This is the case of the two examples and application domains that are discussed in the following sections, and, from our experience, of several applications in different sectors, e.g., safety critical systems, industrial controllers, e-government, e-banking and e-business.

Indeed, the MDP approach provides the ability to find solutions that have detailed requirements on costs and rewards. However, in some applications, it is rather important that the plan guarantees some requirements, rather than optimizes a given utility. Moreover, from the practical point of view, probabilities are rather difficult to be managed in the case of large state spaces. Probabilities do not allow to represent uncertainty as a set of states, and to deal with them in an effective and efficient way.

The problem of planning in nondeterministic domains and under partial observability has also been addressed with approaches different from MDP and probabilistic planning. One of the early attempts is described in [19], which presents an off-line planning algorithm based on a breadth-first search on an and-or graph. Other initial attempts have extended classical planning to the problem of planning in nondeterministic domains (see e.g., [28,31]). More recent approaches extend planners based of planning graphs and satisfiability . Most of them are limited either to full observability, or to conformant planning. Among those that can deal with partial observability, the first significant result was SGP [35], an extension of GRAPHPLAN, which provides significant improvements in performances compared with previous extensions to classical planners. SGP produces acyclic conditional plans, but it is unable to deal with nondeterministic action effects, i.e. uncertainty is limited to the initial condition, which is a set of states rather than a single state.

Among the planners based on satisfiability, QBFPLAN [32] can deal with uncertainty in action execution and partial observability. The planning problem is reduced to a QBF satisfiability problem, that is then given in input to an efficient solver [33]. As all planners based on satisfiability, QBFPLAN is limited to bounded-length planning, i.e. it looks for a solution of specified length l. When this does not exist, it iteratively increases l until a solution is found or a specified limit is reached. QBFPLAN is thus unable to detect when the problem is unsolvable. QBFPLAN exploits its symbolic approach to avoid exponential blow up caused by the enumeration of states, but it can hardly scale up to large problems.

A different approach to the problem of planning under partial observability is the idea of "Planning at the Knowledge Level", implemented in the PKS planner [29], This approach is based on a representation of incomplete knowledge and sensing at a higher level of abstraction than that provided by ground modeling. While this can effectively reduce the complexity of the problem, it makes it impossible to model a variety of problems that can only be dealt with world-level reasoning.

Planning via symbolic model checking has been devised to deal with the special case of full observability (both with reachability goals [13] and with extended goals [30,27]) and with the special case of conformant planning. Planning for reachability goals under partial observability via symbolic model checking has been addressed in [7]. The more general problem of dealing with partial observability and temporal goals is addressed in a preliminary extension [5,10], which however does not exploit any symbolic machinery.

Other planners that are based on symbolic model checking techniques restrict to the case of full observability, see, e.g., UMOP [23,24], or to classical planning, see, e.g., MIPS [16]. Other approaches are based on different model checking techniques, e.g., on explicit-state representations, and most of them are also limited to the case of full observability. This is the case of SIMPLAN [25]. [15] presents an automata based approach to formalize planning in deterministic domains. The work in [20,22,21] presents a method where model checking with timed automata is used to verify that generated plans meet timing constraints.

In our approach, we deal with uncertainty at planning time. Methods that interleave planning and execution (see, e.g., [26,9,3]) can be considered alternative (and orthogonal) approaches to the problem of planning off-line with large state spaces. On one side, they open up the possibility to deal with larger state spaces. On the other side, these methods cannot guarantee to find a solution, unless assumptions are made about the domain.

3   Planning Under Action and Observation Uncertainty

We model a dynamic system as a planning domain $\mathcal{D}$ that is defined in terms of a set states $\mathcal{S}$ of the system that is modeled, some of which are the initial states $\mathcal{I} \subseteq \mathcal{S}$ of the system, a set of actions $\mathcal{A}$ that can change the state of the system, and a set of observations $\mathcal{O}$ that can result by observing the system. Given these basics elements of a planning domain, we model uncertainty in actions and observations by means of a *transition function* that describes how (the execution of) an action leads from one state to possibly many different states, and by means of an *observation function* that describes which observations are associated to the actual state of the domain:

- **Transition Function:** $\mathcal{R} : \mathcal{S} \times \mathcal{A} \rightarrow 2^{\mathcal{S}}$. The transition function $\mathcal{R}$ associates to each current state $s \in \mathcal{S}$ and to each action $a \in \mathcal{A}$ the set $\mathcal{R}(s, a) \subseteq \mathcal{S}$ of next states. We say that action a is *executable* in state s if $\mathcal{R}(s, a) \neq \emptyset$. We require that in each state $s \in \mathcal{S}$ there is some executable action, that is some $a \in \mathcal{A}$ such that $\mathcal{R}(s, a) \neq \emptyset$. The transition function allows us to model uncertainty in action execution since an action may lead to different states.
- **Observation Function:** $Obs : \mathcal{S} \rightarrow 2^{\mathcal{O}}$. The observation function Obs associates to each state s the set of possible observations $Obs(s) \subseteq \mathcal{O}$. We require that some observation is associated to each state $s \in \mathcal{S}$, that is, $Obs(s) \neq \emptyset$. The observation associated to a given state is not unique. This allows us to model noisy sensing and lack of information, since different states may be indistinguishable.

In this paper we focus on finite state planning domains; we say that $\mathcal{D}$ is *finite state* if sets $\mathcal{S}$, $\mathcal{A}$, and $\mathcal{O}$ are finite.

Given a planning domain and a goal, the result of the planning activity is a plan that organizes actions in order to achieve the goal. Actions can be organized in different ways. For instance, moving a robot to a given room and then to the corridor is an example of a sequential plan; executing an instruction depending on the result of the execution of a previous one is an example of a conditional plan; requesting a reservation to a web service for booking hotels until some requirements on room quality and price are satisfied, is an example of an iterative plan.

As a consequence of modeling uncertainty in actions and observations, we have to model plans that go beyond the simple sequential plans proposed by classical planning models, see, e.g., [18,36,38,37]. Even the case of plans that determine the next action depending on the current state of the system are not enough. Indeed, we may need to execute different actions in the same state, e.g., depending on the states that have been previously visited, or the actions that have been previously executed. For instance, a robot performing a surveillance task, may take different directions in the same room depending on which other rooms have been previously visited.

In general, a plan should identify the action to be performed depending on the previous states of the system and the previous actions. Moreover, since we have uncertainty in observations, the action to be executed should depend on the past observations, and not on the past states, since some states may be indistinguishable.

A plan can be therefore described as something that starts with an initial observation, say o_0, which provides the information to determine the action to be executed, say a_1, which leads to a new observation o_1. Given o_0 and o_1, the plan decides for action a_2 to be executed, and so on. A plan can be therefore defined as a partial function that, given a sequence of observations, returns an action.

In order to define when a plan satisfies a goal, we need to give "semantics" to execution of plans in planning domains. Uncertainty in action execution as well as uncertainty in observations lead to plan executions that cannot be limited to sequences of states (as it is usually done in planning without uncertainty). Informally, uncertainty leaves to multiple possibilities, i.e., to branching to different states. As a consequence, the execution of a plan can be described as a tree, rather than a sequence. The branching in the tree corresponds to the different states resulting from the execution of an action as well as to the different states resulting from an observation.

Formally, following [14], executions can be defined as *labeled execution trees*, i.e., trees with nodes labeled with states and observations. A *planning problem* can thus be formalized as a planning domain that induces a labeled execution tree and a *goal* that is a set of labeled execution trees, thus identifying the desired behavior. A *solution* to a planning problem is a plan such that its labeled execution tree is one of the (or *satisfies*) the labeled execution trees represented by the goal.

For an extensive discussion of how this framework can be instantiated to planning for reachability gaols under full observability, to planning for temporally extended goals, and to conformant planning, i.e., to planning under null observability, see [14].

4   Planning for Robot's Surveillance Tasks

In this section, we consider a simple example where a robot has to accomplish a surveillance task by keeping visiting different rooms of a building (see figure 1). The building of five rooms, namely a store, a department dep, a laboratory lab, an office, and a corridor corr. The robot can move between the rooms. The laboratory is a dangerous room it is not possible to exit from. Between rooms office and dep, there is a door that the robot cannot control. Therefore, an *east* action from room office successfully leads to room dep only if the door is open.

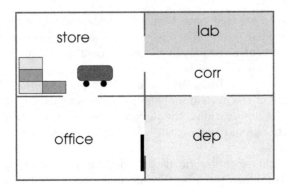

Fig. 1. A simple example of a surveillance task

This is a first source of uncertainty about action execution. Another source of uncertainty occurs when the robot tries to move *east* from the store: in this case, the robot may end either in room corr or in room lab. The planning domain is depicted in figure 2.

For this example, for simplicity, we assume that we do not have uncertainty in observation. Under this assumption, plans can use states rather than observations, and a plan can be defined as a function from sequences of states to actions to be executed, i.e., a plan is a partial function $\pi : \mathcal{S}^+ \rightharpoonup \mathcal{A}$. Notice that these kinds of plans can be represented finitely, in terms of a function that given a state and a "context", specify the action to be executed, and in terms of a function that, depending on the action outcome, specifies the next context [30].

An example of this kind of plan is shown in Figure 3. The plan leads the robot from room store to room dep going through the office, and then back to the store, again going through the office. Two contexts are used, namely c_0 when the robot is going to the dep and c_1 when the robot is going back to the store. This allows the plan to execute different actions in state office and in state store.

The execution of a plan can be described as a labeled tree as shown in the previous section. In the case of no uncertainty in observation the plan execution can be described as a $\mathcal{S}$-labeled tree $\mathcal{T}$ such that:

- $s_0 \in \mathcal{T}$, where $s_0 \in \mathcal{I}$;
- if $s_0 s_1 \cdots s_n \in \mathcal{T}$, $\pi(s_0 s_1 \cdots s_n) = a_n$, $s_{n+1} \in \mathcal{R}(s_n, a_n)$, then $s_0 s_1 \cdots s_n s_{n+1} \in \mathcal{T}$.

We can describe goals for a surveillance task by means of formulas in a temporal logic. In this setting, we use Computation Tree Logic (CTL) [17] that enables us to characterize the corresponding set of trees. A CTL goal is defined by the following grammar, where s is a state of the domain $\mathcal{D}$:

$$g ::= p \mid g \wedge g \mid g \vee g \mid \mathrm{AX}\, g \mid \mathrm{EX}\, g$$
$$\mathrm{A}(g\, \mathrm{U}\, g) \mid \mathrm{E}(g\, \mathrm{U}\, g) \mid \mathrm{A}(g\, \mathrm{W}\, g) \mid \mathrm{E}(g\, \mathrm{W}\, g)$$
$$p ::= s \mid \neg p \mid p \wedge p$$

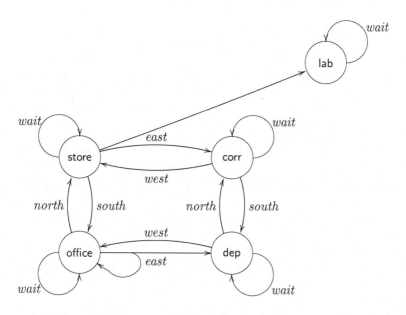

Fig. 2. The planning domain for the surveillance task

STATE	CONTEXT	NEXT ACTION	STATE	CONTEXT	NEXT STATE	NEXT CONTEXT
store	c_0	south	store	c_0	office	c_0
office	c_0	east	office	c_0	dep)	c_1
			office	c_0	office	c_0
dep	c_1	west	dep	c_1	office	c_1
office	c_1	north	office	c_1	store	c_1
store	c_1	wait	store	c_1	store	c_1

Fig. 3. An example of plan

CTL combines temporal operators and path quantifiers. "X", "U", and "W" are the "next time", "(strong) until", and "weak until" temporal operators, respectively. "A" and "E" are the universal and existential path quantifiers, where a path is an infinite sequence of states. They allow us to specify requirements that take into account uncertainty in action execution. Intuitively, the formula AX g means that g holds in every immediate successor of the current state, while the formula EX g means that g holds in some immediate successor. The formula A(g_1 U g_2) means that for every path there exists an initial prefix of the path such that g_2 holds in the last state of the prefix and g_1 holds in all the other states along the prefix. The formula E(g_1 U g_2) expresses the same condition, but only on some of the paths. The formulas A(g_1 W g_2) and E(g_1 W g_2) are similar to A(g_1 U g_2) and E(g_1 U g_2), but allow for paths where g_1 holds in all the states and g_2 never holds. Formulas AF g and EF g (where the temporal operator "F" stands for "future" or "eventually") are abbreviations of A($\top$ U g) and E($\top$ U g), respectively. AG g and EG g (where "G" stands for "globally" or "always") are abbreviations of A(g W $\perp$) and E(g W $\perp$), respectively.

A remark is in order. Even if negation $\neg$ is allowed only in front of basic propositions, it is easy to define $\neg g$ for a generic CTL formula g, by "pushing down" the negations: for instance $\neg$AX $g \equiv$ EX $\neg g$ and $\neg$A(g_1 W g_2) $\equiv$ E($\neg g_2$ U($\neg g_1 \wedge \neg g_2$)).

We can use the expressivity of CTL to specify both temporal conditions, by using the temporal operators "X", "U", "W", "F", and "G", and conditions on the "strength" of the requirement, by distinguishing, by means of the path quantifiers "A" and "E", whether a temporal requirement should hold for all the possible or just some of the possible executions. For instance, we might require that the system "tries" to reach a certain state, and if it does not manage to do so, it guarantees that some safe state is maintained. As an example, we can require that a mobile robot tries to reach a given location, but guarantees to avoid dangerous rooms all along the path. Or we can require that the robot keeps moving back and forth between location A, and, if possible, location B. We thus specify that the robot must pass through location A at each round, while it should just pass through location B if possible: the "strength" of the requirements for the two locations is different. We can therefore classify goals according to these two dimensions: temporal requirements and "strength" requirements. For instance, goals with different temporal requirements are *reachability goals*, i.e., requirements on which states should be reached, and *mainteinability goals*, i.e., requirements on some property that should be maintained all along the execution path. (Weak) Untill operators allow us to combine conditions on the states to be reached and on the properties to be maintained until a given condition is reached. Reachability and mainteinability goals can be *strong* or *weak*, depending on whether the temporal condition should hold on all possible or just for some execution path. *Strong reachability goals* are goals of the form AX g and AF g. While *Weak reachability goals* are of the form EX g and EF g. We can represent *Strong* and *Weak maintainability goals* with formulas AG g and EG g, respectively. Of course, we can nest strong, weak, reachability, and maintenability requirements to obtain more complex and structured goal, like with the goals AG (AF $g_1 \wedge$ EF g_2), AG EF g, and A(EF g_1 W g_2).

In the domain of Figure 1, let us consider first some examples of reachability goals. The strong goal AF dep requires a plan that guarantees that the robot reaches the department, in spite of nondeterminism. Notice that, intuitively, there is no plan that satisfies such goal starting from the store[1]. The weak goal EF dep requires a plan that has a possibility to reach the department. The plan that moves the robot east, either to the lab or to the corridor, and from the corridor goes south to the department, is a plan that satisfies EF dep. Reasonable reachability requirements that are weaker than AF dep and stronger than EF dep are Strong Cycling reachability goals, like A(EF dep W dep): it allows for those execution loops that have always a possibility of terminating, and when they do, the department is guaranteed to be reached. The goal A($\neg$lab U dep) is not satisfiable since it states that the robot should reach the department by avoiding

---

[1] Such goal can be expressed in CTL as store $\rightarrow$ AF dep.

the laboratory, while the goal A(¬lab W dep) is weaker, since it is also satisfied by any plan that never goes to the department and never goes to the laboratory.

The mainteinability goal AG EF dep intuitively means "maintain the possibility of reaching the department". It is different from A(EF dep W dep), since the latter, after the department is reached, allows for any behaviour, e.g., staying in a room different from the department forever.

The goal AF dep ∧ ¬AG lab states that the robot is guaranteed to reach the department and is guranteed to never be in the lab. It is different from A(¬lab U dep), which does not impose any condition after the department is reached.

An interesting strong mainteinability goal is AG (AF store ∧ EF dep), which requires that the robot keeps moving back and forth between the store, and, if possible, the department. We can add the condition that the laboratory is never reached with the goal AG (AF store∧EF dep)∧¬AG lab. Both these two goals are satisfied by a plan that keeps trying to move back and forward the robot between the store and the department, i.e., it tries to move the robot to the department and then, independently of whether the robot manages to go through the door, it moves back the robot to the store, and so on.

We can also compose reachability and maintainability goals in arbitrary ways. For instance, lab → AF AG ¬lab states that a robot that is in the lab, should eventually move out of it, and never come back. The weaker goal lab → EF AG ¬lab states that after moving out of the lab, the robot should have always a possibility not to come back.

We can now define valid plans, i.e., plans that satisfy CTL goals, i.e., we define $\mathcal{T} \models g$, where $\mathcal{T}$ is the execution tree of a plan π for domain $\mathcal{D}$, and g is a CTL goal. The definition of predicate $\models$ is based on the standard semantics of CTL [17]. A planning algorithm can thus search the state space by progressing CTL goals, see, e.g., [30]. A conceptually similar, even if technically different, approach has been applied to the problem of planning under uncertainty in observations.

The framework can be extended to deal with uncertainty in observation by considering that some states are indistinguishable, and observations and observation functions must be considered in place of states and state transitions, according to the lines presented in [8,6,11].

5 Planning for Composing Purchase and Delivery Web Services

In this section we consider a rather different planning domain, first introduced in [34]. The problem is to automatically compose a furniture purchase & delivery service, say the P&S service, which satisfies some user request. We do so by combining two separate, independent, and available services: a furniture producer Producer, and a delivery service Shipper. The idea is that of combining these two services so that the user may directly ask the composed service P&S to purchase and deliver a given item at a given place. To do so, we exploit a description of the expected interaction between the P&S service and the other actors. In the case of the Producer and of the Shipper the interactions are defined in terms of the service requests that are accepted by the two actors. In the case of the User, we describe the interactions in terms of the requests that the user can send to the P&S. As a consequence, the P&S service should interact with three available services: Producer, Shipper, and User (see Figure 4). These are the three available services W_1, W_2, and W_3, which can be described in standard languages for web services, like BPEL4WS [4].

The problem is to automatically generate the implementation of the P&S service, say W. Producer accepts requests for providing information on a given product and, if the product is available, it provides information about its size. The Producer also accepts requests for buying a given product, in which case it returns an offer with a cost and production time. This offer can be accepted or refused by the external service that has invoked the Producer. The Shipper service receives requests for transporting a product of a given size to a given location. If delivery is possible, Shipper provides a shipping offer with a cost and delivery time, which can be accepted or refused by the external service that has invoked Shipper. The User sends requests to get a given item at a given location, and expects either a negative answer if this is not possible, or an offer indicating the price and the time required for the service. The user may either accept or refuse the offer. Thus, a typical interaction between the user, the combined purchase & delivery service P&S, the producer, and the shipper would go as follows:

1. the user asks P&S for an item i, that he wants to be transported at location l;
2. P&S asks the producer for some data about the item, namely its size, the cost, and how much time does it take to produce it;
3. P&S asks the delivery service the price and time needed to transport an object of such a size to l;
4. P&S provides the user an offer which takes into account the overall cost (plus an added cost for P&S) and time to achieve its goal;
5. the user sends a confirmation of the order, which is dispatched by P&S to the delivery and producer.

Of course this is only the nominal case, and other interactions should be considered, e.g., for the cases the producer and/or delivery services are not able to satisfy the request, or the user refuses the final offer.

The component web services described in the example above can be modeled as state transition systems: web services can be in one of their possible states (some of which are marked as initial states) and can evolve to new states as a result of performing some actions, which in this case are the actions of sending or receiving messages, as well as internal actions that cannot be observed by external web services. A transition function describes how this exchange of messages or internal evolutions lead from one state to possibly many different states. Web services evolutions can be monitored through observations describing the visible part of a web service. An observation function defines the observation associated to each state of the service.

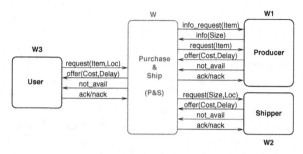

Fig. 4. The Purchase & Ship Example

State transition systems modeling web services are nondeterministic, i.e., one action may result in several different outcomes. This is modeled by the fact that the transition function returns sets of states. Nondeterminism is needed since the system cannot often know *a priori* which outcome will actually take place, e.g., whether it will receive a confirmation or a cacellation from an external service. Moreover, our state transition systems are partially observable, i.e., external services can only observe part of their system state, e.g., its external communications can be observed but other services do not have access to its internal status and variables. Partial observability is modeled by the fact that different states may result in the same observation.

Intuitively, we associate to each available web service a state transition system as follows. The states S are used to codify the different steps of evolution of the service (e.g., what position has been reached inside the composite process of the shipper) and the values of the predicates defined internally to the service. The actions A are used to model the invocation of the external processes. The actions also model the invocations by the external actors of the services that the composed service should provide. The observations O are used to model the outputs of the invoked external processes (and the inputs of the external invocations).

Composition goals express requirements for the service to be automatically generated. They should represent conditions on the temporary evolution of services, and, as shown by the next example, requirements of different strengths and preference conditions.

In our example (see Figure 4), a reasonable composition goal for the P&S service is the following:

Goal 1: The service should try to reach the ideal situation where the user has confirmed his order, and the service has confirmed the associated (sub-)orders to the producer and shipper services. In this situation, the data associated to the orders have to be mutually consistent, e.g., the time for building and delivering a furniture shall be the sum of the time for building it, and that for delivering it.

However, this is an ideal situation that cannot be enforced by the P&S service: the product may not be available, the shipping may not be possible, the user may not accept the total cost or the total time needed for the production and delivery of the item... We would like the P&S service to behave properly also in these cases, and get to a consistent situation, where the P&S confirms none of the two services for production and delivering, otherwise P&S is likely, e.g., to loose money. More precisely, we have also the following goal:

Goal 2: The P&S service should absolutely reach a fall-back situation where every (sub-)order has been canceled. That is, there should be no chance that the service has committed to some (sub-)order if the user can cancel his order.

Some remarks are in order. First of all, there is a difference in the "strength" in which we require Goal 1 and Goal 2 to be satisfied. We know that it may be impossible to satisfy Goal 1: we would like the P&S service to *try* (do whatever is possible) to satisfy the goal, but we do not require that the service guarantees to achieve it in all situations. The case is different for Goal 2: there is always a possibility for the P&S service to cancel the orders to the producer and shipper, and to inform the user. We can require a guarantee of satisfaction of this goal, in spite of any behavior of the other services. Moreover, Goal 1 and Goal 2 are not at the same level of desire. Of course we would not like a P&S service that satisfies always Goal 2 (e.g., by refusing all requests from the user) even when it would be possible to satisfy Goal 1. We need then to express a strong preference for Goal 1 w.r.t. Goal 2. Informally, we can therefore describe the composition goal as follows:

Composition Goal: *Try* to satisfy Goal 1, *upon failure*, *do* satisfy Goal 2.

As the previous example shows, composition goals need the ability to express conditions on the whole behaviour of a service, conditions of different strengths, and preferences among different subgoals. The EAGLE language [27] has been designed with the purpose to satisfy such expressiveness requirements. Let propositional formulas $p \in \mathcal{P}rop$ define conditions on the states of a state transition system. *Composition goals* $g \in \mathcal{G}$ over $\mathcal{P}rop$ are defined as follows:

$$g := p \mid g \textbf{ And } g \mid g \textbf{ Then } g \mid g \textbf{ Fail } g \mid \textbf{Repeat } g \mid$$
$$\textbf{DoReach } p \mid \textbf{TryReach } p \mid \textbf{DoMaint } p \mid \textbf{TryMaint } p$$

Goal **DoReach** p specifies that condition p has to be eventually reached in a strong way, for all possible non-deterministic evolutions of the state transition system. Similarly, goal **DoMaint** q specifies that property q should be maintained true despite non-determinism. Goals **TryReach** p and **TryMaint** q are weaker versions of these goals, where the plan is required to do "everything that is possible" to achieve condition p or maintain condition q, but failure is accepted if unavoidable. Construct g_1 **Fail** g_2 is used to model preferences among goals and recovery from failure. More precisely, goal g_1 is considered first. Only if the achievement or maintenance of this goal fails, then goal g_2 is used as a recovery or second-choice goal. Consider for instance goal **TryReach** c **Fail DoReach** d.

The sub-goal **TryReach** c requires to find a plan that tries to reach condition c. During the execution of the plan, a state may be reached from which it is not possible to reach c. When such a state is reached, goal **TryReach** c fails and the recovery goal **DoReach** d is considered. Goal g_1 **Then** g_2 requires to achieve goal g_1 first, and then to move to goal g_2. Goal **Repeat** g specifies that sub-goal g should be achieved cyclically, until it fails. Finally, goal g_1 **And** g_2 requires the achievement of both subgoals g_1 and g_2. A formal semantics and a planning algorithm for EaGLe goals in fully observable nondeterministic domains can be found in [27].

The automated composition task can be performed by the planner as follows. The planner has two inputs: the composition goal and the planning domain which represents all the ways in which the component services W_1, W_2, and W_3 can evolve. The component services are represented by state transition systems, say, Σ_1, Σ_2, and Σ_3. Formally, this combination is a synchronous product.

$$\Sigma = \Sigma_1 \times \Sigma_2 \times \Sigma_3$$

The automated composition task consists in finding a plan that satisfies the composition goal G over a domain Σ. We are interested in complex plans, that may encode sequential, conditional and iterative behaviors, and are thus expressive enough for representing the flow of interactions of the synthesized composed service with the other services and expressive enough for representing the required observations over the other services. We therefore model a plan as an automaton. The automaton is defined from a set of *plan contexts*, one of which is the initial context, and that represents the goal to be satisfied. Transitions in the automaton defined through functions that determine which action has to be executed (depending on the context) and how the context change. An *action function*, given a context and an observation, returns action. A *context function*, given a context and an observation, returns a context..

The contexts are therefore the internal states of the plan; they permit to take into account, e.g., the knowledge gathered during the previous execution steps. Notice that actions to be executed depend on the observation and on the context, and, once an action is executed, the context may change. Action and context functions are deterministic (we do not consider nondeterministic plans), and can be partial, since a plan may be undefined on the context-observation pairs that are never reached during plan execution. If the plan is obtained from a composition goal, then the contexts correspond to the sub-formulas of the goal. The initial context is the original goal, the subsequent contexts obtained by applying the context function are the subgoals that need still to be solved to find a plan. For instance, in a plan for the composition goal of the previous example one would have some contexts associated to Goal1 and other contexts associated to Goal2.

The execution of a plan over a domain can be described in terms of transitions between configurations that describe the state of the domain and of the plan. Intuitively, a configuration is a snapshot of the domain controlled by the plan. Due to the nondeterminism in the domain, we may have an infinite number of different executions of a plan. We provide a finite presentation of these executions with an *execution structure*, i.e, a Kripke Structure [17] with configurations as states. The execution structure represents the evolutions of the domain $\mathcal{D}$ controlled by the plan π. It is the execution structure $\mathcal{D}_\pi$ that must satisfy the composition goal G. If $\mathcal{D}_\pi \models G$, we say that π *is a valid plan for* G *on* $\mathcal{D}$. A formal definition of $\mathcal{D}_\pi \models G$ can be found in [27].

Given a composition goal and a planning domain, we use algorithms for planning with EAGLE goals, according to the approaches described in [34,27]. We have therefore the algorithms for generating a valid plan π that satisfies the composition goal. Since π is an automaton, it can be easily translated to executable process languages, like BPEL4WS.

6   Conclusion

In this paper, after providing a framework for planning under uncertainty in action execution and observations, we discussed how this problem is relevant to different application domains, such as planning for automating robots' tasks as well as planning for the automated composition of web services. In spite of the different characteristics of the domains, this paper shows that similar techniques can be used to automate the planning and composition tasks. Another application domain in which planning techniques similar to the ones described in this paper have been applied is planning for supporting the development of controllers for space applications, see, e.g., [2,1].

References

1. L. C. Aiello, A. Cesta, E. Giunchiglia, M. Pistore, and P. Traverso. Planning and Verification Techniques for the High Level Programming and Monitoring of Autonomous Robotic Devices. In *Proceedings of the European Space Agency Workshop on On Board Autonoy*, Noordwijk,The Netherlands, October 2001. ESA.

2. L. C. Aiello, A. Cesta, E. Giunchiglia, and P. Traverso. Merging Planning and Verification Techniques for "Safe Planning" in Space Robotics. In *6th International Symposium on Artificial Intelligence, Robotics and Automation in Space: A New Space Odyssey (ISAIRAS01)*, Montreal, Canada, June 2001.

3. A. Albore and P. Bertoli. Generating Safe Assumption-Based Plans for Partially Observable, Nondeterministic Domains. In *Proc. AAAI*, 2004.

4. T. Andrews, F. Curbera, H. Dolakia, J. Goland, J. Klein, F. Leymann, K. Liu, D. Roller, D. Smith, S. Thatte, I. Trickovic, and S. Weeravarana. Business Process Execution Language for Web Services (version 1.1), 2003.

5. P. Bertoli, A. Cimatti, M. Pistore, and M. Pistore. A framework ofr Planning with Extended Goals and Partial Observability. In *Proceedings of ICAPS-03*, 2004.

6. P. Bertoli, A. Cimatti, M. Pistore, and P. Traverso. A Framework for Planning with Extended Goals under Partial Observability. In *Proc. ICAPS'03*, pages 215–224, 2003.

7. P. Bertoli, A. Cimatti, M. Roveri, and P. Traverso. Planning under Partial Observability via Symbolic Model Checking. *Artificial Intelligence*.

8. P. Bertoli, A. Cimatti, M. Roveri, and P. Traverso. Planning in nondeterministic domains under partial observability via symbolic model checking. In B. Nebel, editor, *Proceedings of the Seventeenth International Joint Conference on Artificial Intelligence, IJCAI 2001*, pages 473–478. Morgan Kaufmann Publishers, August 2001.

9. P. Bertoli, A. Cimatti, and P. Traverso. Interleaving Execution and Planning for Nondeterministic, Partially Observable Domains. In *Proceedings of ECAI-04*, 2004.

10. P. Bertoli and M. Pistore. Planning with Extended Goals and Partial Observability. In *Proceedings of ICAPS-04*, 2004.

11. P. Bertoli and M. Pistore. Planning with Extended Goals and Partial Observability. In *Proceedings of ICAPS'04*, 2004. To be published.

12. C. Boutilier, T. Dean, and S. Hanks. Decision-Theoretic Planning: Structural Assumptions and Computational Leverage. *Journal of AI Research (JAIR)*, 11:1–94, 1999.

13. A. Cimatti, M. Pistore, M. Roveri, and P. Traverso. Weak, Strong, and Strong Cyclic Planning via Symbolic Model Checking. *Artificial Intelligence*, 147(1-2):35–84, 2003.

14. A. Cimatti, M. Pistore, and P. Traverso. Automated planning. In F. van Harmelen and V. Lifschitz, editors, *Handbook of Knowledge Representation and Reasoning*, chapter 21. Elsevier, 2006. to appear.

15. G. De Giacomo and M. Y. Vardi. Automata-theoretic approach to planning for temporally extended goals. In S. Biundo, editor, *Proceeding of the Fifth European Conference on Planning*, volume 1809 of *LNAI*, pages 226–238, Durham, United Kingdom, September 1999. Springer-Verlag.

16. S. Edelkamp and M. Helmert. On the implementation of mips. In *AIPS-Workshop on Model-Theoretic Approaches to Planning*, pages 18–25, 2000.

17. E. A. Emerson. Temporal and modal logic. In J. van Leeuwen, editor, *Handbook of Theoretical Computer Science, Volume B: Formal Models and Semantics*, chapter 16, pages 995–1072. Elsevier, 1990.

18. R. E. Fikes and N. J. Nilsson. STRIPS: A new approach to the application of Theorem Proving to Problem Solving. *Artificial Intelligence*, 2(3-4):189–208, 1971.

19. M. Genesereth and I. Nourbakhsh. Time-saving tips for problem solving with incomplete information. In *Proceedings of the National Conference on Artificial Intelligence*, 1993.

20. R. P. Goldman, D. J. Musliner, K. D. Krebsbach, and M. S. Boddy. Dynamic abstraction planning. In *Proceedings of the Fourteenth National Conference on Artificial Intelligence and Ninth Innovative Applications of Artificial Intelligence Conference (AAAI 97), (IAAI 97)*, pages 680–686. AAAI Press, 1997.

21. R. P. Goldman, D. J. Musliner, and M. J. Pelican. Using Model Checking to Plan Hard Real-Time Controllers. In *Proceeding of the AIPS2k Workshop on Model-Theoretic Approaches to Planning*, Breckeridge, Colorado, April 2000.

22. R.P. Goldman, M. Pelican, and D.J. Musliner. Hard Real-time Mode Logic Synthesis for Hybrid Control: A CIRCA-based approach, mar 1999. Working notes of the 1999 AAAI Spring Symposium on Hybrid Control.

23. R. Jensen and M. Veloso. OBDD-based Universal Planning for Synchronized Agents in Non-Deterministic Domains. *Journal of Artificial Intelligence Research*, 13:189–226, 2000.

24. R. M. Jensen, M. M. Veloso, and M. H. Bowling. OBDD-based optimistic and strong cyclic adversarial planning. In *Proceedings of the Sixth European Conference on Planning (ECP'01)*, 2001.

25. F. Kabanza, M. Barbeau, and R. St-Denis. Planning control rules for reactive agents. *Artificial Intelligence*, 95(1):67–113, 1997.
26. S. Koenig and R. Simmons. Solving robot navigation problems with initial pose uncertainty using real-time heuristic search. In *Proceedings of the International Conference on Artificial Intelligence Planning and Scheduling*, 1998.
27. U. Dal Lago, M. Pistore, and P. Traverso. Planning with a language for extended goals. In *In Proceedings of the Eighteenth National Conference on Artificial Intelligence (AAAI-02)*, 2002.
28. M. Peot and D. Smith. Conditional Nonlinear Planning. In James Hendler, editor, *Proceedings of the First International Conference on AI Planning Systems*, pages 189–197, College Park, Maryland, June 15–17 1992. Morgan Kaufmann.
29. R. Petrick and F. Bacchus. A knowledge-based approach to planning with incomplete information and sensing. In *In Proceedings of the Sixth International Conference on AI Planning and Scheduling (AIPS'02)*, 2002.
30. M. Pistore and P. Traverso. Planning as model checking for extended goals in non-deterministic domains. In B. Nebel, editor, *Proceedings of the Seventh International Joint Conference on Artificial Intelligence (IJCAI-01)*, pages 479–486. Morgan Kaufmann Publisher, August 2001.
31. L. Pryor and G. Collins. Planning for Contingency: a Decision Based Approach. *Journal of Artificial Intelligence Research (JAIR)*, 4:81–120, 1996.
32. J. Rintanen. Constructing Conditional Plans by a Theorem-Prover. *Journal of Artificial Intellegence Research*, 10:323–352, 1999.
33. J. Rintanen. Improvements to the Evaluation of Quantified Boolean Formulae. In T. Dean, editor, *16th Iinternational Joint Conference on Artificial Intelligence*, pages 1192–1197. Morgan Kaufmann Publishers, August 1999.
34. P. Traverso and M. Pistore. Automated Composition of Semantic Web Services into Executable Processes. In *Proc. of the International Semantic Web Conference (ISWC)*, 2004.
35. Daniel S. Weld, Corin R. Anderson, and David E. Smith. Extending graphplan to handle uncertainty and sensing actions. In *Proceedings of the 15th National Conference on Artificial Intelligence (AAAI-98) and of the 10th Conference on Innovative Applications of Artificial Intelligence (IAAI-98)*, pages 897–904, Menlo Park, July 26–30 1998. AAAI Press.
36. D. E. Wilkins. Domain-independent Planning: representation and plan generation. *Artificial Intelligence*, 22(3):269–301, 1984.
37. D. E. Wilkins. *Practical Planning: extending the classical AI planning paradigm*. Morgan Kaufmann, San Mateo, 1988.
38. D.E. Wilkins. Recovering from execution errors in SIPE. *Computational Intelligence*, 1:33–45, 1985.

Reasoning About Web Services
in a Temporal Action Logic[*]

Alberto Martelli[1] and Laura Giordano[2]

[1] Dipartimento di Informatica, Università di Torino, Torino
mrt@di.unito.it

[2] Dipartimento di Informatica, Università del Piemonte Orientale, Alessandria
laura@mfn.unipmn.it

Abstract. The paper presents an approach to reasoning about Web services in a temporal action theory. Web services are described by specifying their interaction protocols in an action theory based on a dynamic, linear-time, temporal logic. The proposed framework is based on a social approach to agent communication, where the effects of communicative actions allow changes in the social state, and interaction protocols are defined in terms of the creation and fulfillment of commitments and permissions among the agents. We show how to introduce epistemic operators in the action theory to deal with incomplete information, and we address the problem of verifying properties of Web services, as well as the problem of reasoning about the composition of Web services.

1 Introduction

Autonomous agents can communicate, cooperate and negotiate using commonly agreed communication languages and protocols. One of the central issues in the field concerns the specification of conversation policies (or interaction protocols), which govern the communication between software agents in an agent communication language [4].

To allow for the flexibility needed in agent communication [10,14] new approaches have been proposed, which overcome the limitations of the traditional transition net approach, in which the specification of interaction protocols is done by making use of finite state machines. A particularly promising approach to agent communication, first proposed by Singh [21,22], is the social approach [5,11,14]. In the social approach, communicative actions affect the "social state" of the system, rather than the internal (mental) states of the agents. The social state records social facts, like the permissions and the commitments of the agents.

In this paper, we adopt a social approach in the specification of the interactions among Web services and, in particular, we address the problem of service verification and that of service composition [15,18,23]. In our proposal, Web services are described by specifying their interaction protocols in an action theory

---

[*] This research has been partially supported by the project MIUR PRIN 2005 "Specification and verification of agent interaction protocols".

O. Stock and M. Schaerf (Eds.): Aiello Festschrift, LNAI 4155, pp. 229–246, 2006.

based on a dynamic, linear-time, temporal logic. Such logic has been used in [7,9] to provide the specification of interaction protocols among agents and to allow the verification of protocol properties as well as the verification of the compliance of a set of services with a protocol. The Web service domain is well suited for this kind of formalization. The proposed framework provides a simple formalization of the communicative actions in terms of their effects and preconditions and the specification of an interaction protocol by means of temporal constraints.

To accommodate the needs of the application domain, in which information is inherently incomplete, in the Section 2, we extend the action theory defined in [9] to deal with incomplete information. More precisely, we introduce epistemic modalities in the language to distinguish what is known about the social state from what is unknown. In this context, the communicative actions by means of which the services interact can be regarded as knowledge-producing actions, and are similar to sensing actions in the context of planning. In order to deal with the frame problem, we introduce a completion construction on the epistemic domain description, which defines suitable successor state axioms.

In Section 3, we show how a Web service can be specified by modeling its interaction protocol in a social approach. We consider, as an example, a service for purchasing a good, whose interaction protocol has the following structure: the customer sends a request to the service, the service replies with an offer or by saying that the service is not available, and finally, if the customer receives the offer, he/she may accept or refuse it. Communicative actions, such as *offer* or *accept*, are modeled in terms of their effects on the social state (action laws). The protocol is specified by putting constraints on the executability of actions (precondition laws) and by temporal constraints specifying the fulfillment of commitments.

In Section 4, we show that several kinds of verification (both runtime and static verification) can be done on the services and the related verification problems can be modeled as satisfiability and validity problems in the logic. We make use of an automata-based approach to solve these problems and, in particular, we work on the Büchi automaton which can be extracted from the logical specification of the protocol.

In Section 5, we then consider the problem of composing Web services, by referring to an example consisting of two services for purchasing and for shipping goods. We define the service composition problem as a planning problem, whose solution requires building a conditional plan and allowing it to interact with the two services. The plan can be obtained from the Büchi automaton derived from the logical specification of the protocol. We address the problem of proving the correctness of a given service implementation with respect to the specification of the component services.

2   The Action Theory

In this section, we describe the action theory that is used in the specification of the services. We first introduce the temporal logic on which our action theory is based. Then we introduce epistemic modalities and domain descriptions.

2.1   Dynamic Linear-Time Temporal Logic

We briefly define the syntax and semantics of DLTL as introduced in [12]. In such a linear-time, temporal logic the next state modality is indexed by actions. Moreover, (and this is the extension to LTL) the until operator is indexed by programs in Propositional Dynamic Logic (PDL).

Let Σ be a finite non-empty alphabet. The members of Σ are actions. Let Σ^* and Σ^ω be the set of finite and infinite words on Σ, where $\omega = \{0, 1, 2, \ldots\}$. Let $\Sigma^\infty = \Sigma^* \cup \Sigma^\omega$. We denote by σ, σ' the words over Σ^ω and by τ, τ' the words over Σ^*. Moreover, we denote by $\leq$ the usual prefix ordering over Σ^* and, for $u \in \Sigma^\infty$, we denote by $prf(u)$ the set of finite prefixes of u.

We define the set of programs (regular expressions) $Prg(\Sigma)$ generated by Σ as follows:

$$Prg(\Sigma) ::= a \mid \pi_1 + \pi_2 \mid \pi_1 ; \pi_2 \mid \pi^*$$

where $a \in \Sigma$ and π_1, π_2, π range over $Prg(\Sigma)$. A set of finite words is associated with each program by the usual mapping $[[\ ]] : Prg(\Sigma) \to 2^{\Sigma^*}$.

Let $\mathcal{P} = \{p_1, p_2, \ldots\}$ be a countable set of atomic propositions containing $\top$ and $\perp$. We define:

$$DLTL(\Sigma) ::= p \mid \neg\alpha \mid \alpha \vee \beta \mid \alpha \mathcal{U}^\pi \beta$$

where $p \in \mathcal{P}$ and α, β range over $DLTL(\Sigma)$.

A model of $DLTL(\Sigma)$ is a pair $M = (\sigma, V)$ where $\sigma \in \Sigma^\omega$ and $V : prf(\sigma) \to 2^{\mathcal{P}}$ is a valuation function. Given a model $M = (\sigma, V)$, a finite word $\tau \in prf(\sigma)$ and a formula α, the satisfiability of a formula α at τ in M, written $M, \tau \models \alpha$, is defined as follows (we omit the standard definition for the boolean connectives):

- $M, \tau \models p$ iff $p \in V(\tau)$;
- $M, \tau \models \alpha \mathcal{U}^\pi \beta$ iff there exists $\tau' \in [[\pi]]$ such that $\tau\tau' \in prf(\sigma)$ and $M, \tau\tau' \models \beta$. Moreover, for every τ'' such that $\varepsilon \leq \tau'' < \tau'^{1}$, $M, \tau\tau'' \models \alpha$.

A formula α is satisfiable iff there is a model $M = (\sigma, V)$ and a finite word $\tau \in prf(\sigma)$ such that $M, \tau \models \alpha$.

The formula $\alpha \mathcal{U}^\pi \beta$ is true at τ if "α until β" is true on a finite stretch of behavior which is in the linear-time behavior of the program π. The derived modalities $\langle \pi \rangle$ and $[\pi]$ can be defined as follows: $\langle \pi \rangle \alpha \equiv \top \mathcal{U}^\pi \alpha$ and $[\pi]\alpha \equiv \neg\langle \pi \rangle \neg\alpha$. Furthermore, if we let $\Sigma = \{a_1, \ldots, a_n\}$, the $\mathcal{U}$, $\bigcirc$ (next), $\diamond$ and $\square$ operators of LTL can be defined as follows: $\bigcirc \alpha \equiv \bigvee_{a \in \Sigma} \langle a \rangle \alpha$ (i.e., α holds in the state obtained by executing any action in Σ), $\alpha \mathcal{U} \beta \equiv \alpha \mathcal{U}^{\Sigma^*} \beta$, $\diamond \alpha \equiv \top \mathcal{U} \alpha$, $\square \alpha \equiv \neg \diamond \neg \alpha$, where, in $\mathcal{U}^{\Sigma^*}$, Σ is taken to be a shorthand for the program $a_1 + \ldots + a_n$. Hence both LTL(Σ) and PDL are fragments of DLTL(Σ).

As shown in [12], DLTL(Σ) is strictly more expressive than LTL(Σ) and the satisfiability and validity problems for DLTL are PSPACE complete problems.

---

[1] We define $\tau \leq \tau'$ iff $\exists \tau''$ such that $\tau\tau'' = \tau'$. Moreover, $\tau < \tau'$ iff $\tau \leq \tau'$ and $\tau \neq \tau'$.

2.2 Epistemic Modalities

In the following, we need to describe the effects of communicative actions on the social state of the agents. In particular, we want to represent the fact that each agent can see only part of the social state as it is only aware of some of the communicative actions in the conversation (namely those it is involved in as a sender or as a receiver). For this reason, we introduce knowledge operators to describe the knowledge of each agent as well as the knowledge shared by groups of agents. More precisely, we introduce a modal operator $\mathcal{K}_i$ to represent the knowledge of agent i and the modal operator $\mathcal{K}_A$, where A is a set of agents, to represent the knowledge shared by agents in A. Groups of agents acquire knowledge about social facts when they interact by exchanging communicative actions. The modal operators $\mathcal{K}_i$ and $\mathcal{K}_A$ are both of type KD. They are normal modalities ruled by the axiom schema $\mathcal{K}\varphi \rightarrow \neg\mathcal{K}\neg\varphi$ (seriality). Though the usual modal logic used to represent belief operators is $KD45$, in this formalization we do not add the positive and negative introspection axioms to belief modality $\mathcal{K}$, because, following the solution proposed in [1], we restrict epistemic modalities to be used in front of literals. In particular, epistemic modalities neither can occur nested nor can be applied to a boolean combination of literals.

The relations between the modalities $\mathcal{K}_i$ and $\mathcal{K}_A$ are ruled by the following interaction axiom schema: $\mathcal{K}_A\varphi \rightarrow \mathcal{K}_i\varphi$, where $i \in A$, meaning that what is knowledge of a group of agents is also knowledge of each single agent in the group. As usual, for each modality $\mathcal{K}_i$ (respectively, $\mathcal{K}_A$) we introduce the modality $\mathcal{M}_i$ (resp. $\mathcal{M}_A$), which is defined as the dual of $\mathcal{K}_i$, i.e. $\mathcal{M}_i\varphi$ is $\neg\mathcal{K}_i\neg\varphi$.

2.3 Domain Descriptions

The social state of the protocol, which describes the stage of execution of the protocol from the point of view of the different agents, is described by a set of atomic properties called *fluents*, whose epistemic value in a state may change with the execution of communicative actions.

Let $\mathcal{P}$ be a set of atomic propositions, the *fluent names*. A *fluent literal* l is a fluent name f or its negation $\neg f$. An *epistemic fluent literal* is a modal atom $\mathcal{K}l$ or its negation $\neg\mathcal{K}l$, where l is a fluent literal and $\mathcal{K}$ is an epistemic operator $\mathcal{K}_i$ or $\mathcal{K}_A$. We will denote by Lit the set of all epistemic literals.

An *epistemic state* (or, simply, a state) is defined as a *complete and consistent set of epistemic fluent literals*, and it provides, for each agent i (respectively for each group of agents A) a *three-valued* interpretation in which each literal l is *true* when $\mathcal{K}_i l$ holds, *false* when $\mathcal{K}_i\neg l$ holds, and *undefined* when both $\neg\mathcal{K}_i l$ and $\neg\mathcal{K}_i\neg l$ hold. Observe that, given the property of seriality, consistency guarantees that a state cannot contain both $\mathcal{K}f$ and $\mathcal{K}\neg f$, for some epistemic modality $\mathcal{K}$ and fluent f. In fact, from $\mathcal{K}f$ it follows by seriality that $\neg\mathcal{K}\neg f$, which is inconsistent with $\mathcal{K}\neg f$.

In the following we extend the action theory defined in [9] to accommodate epistemic literals. A *domain description* D is defined as a tuple $(\Pi, \mathcal{C})$, where Π is a set of (epistemic) *action laws* and *causal laws*, and $\mathcal{C}$ is a set of *constraints*.

The *action laws* in Π have the form:

$$\Box(\mathcal{K}l_1 \wedge \ldots \wedge \mathcal{K}l_n \rightarrow [a]\mathcal{K}l) \tag{1}$$

$$\Box(\mathcal{M}l_1 \wedge \ldots \wedge \mathcal{M}l_n \rightarrow [a]\mathcal{M}l) \tag{2}$$

with $a \in \Sigma$, and $\mathcal{K}$ is a knowledge modality. The meaning of (1) is that executing action a in a state where $l_1, \ldots, l_n$ are known (to be true) causes l to become known, i.e. it causes the effect $\mathcal{K}l$ to hold. As an example the law $\Box(\mathcal{K}fragile \rightarrow [drop]\mathcal{K}broken)$ means that, after executing the action of dropping a glass the glass is known to be broken, if the action is executed in a state in which the glass is known to be fragile. (2) is necessary in order to deal with *ignorance* about preconditions of the action a. It means that the execution of a may affect the beliefs about l, when executed in a state in which the preconditions are considered to be possible. When the preconditions of a are unknown, this law allows to conclude that the effects of a are unknown as well. $\Box(\mathcal{M}fragile \rightarrow [drop]\mathcal{M}broken)$ means that, after executing the action of dropping a glass, the glass may be broken, if the action is executed in a state in which the glass may be fragile (i.e. $\mathcal{K}\neg fragile$ does not hold).

The *causal laws* in Π have the form:

$$\Box((\mathcal{K}l_1 \wedge \ldots \wedge \mathcal{K}l_n \wedge \bigcirc(\mathcal{K}l_{n+1} \wedge \ldots \wedge \mathcal{K}l_m) \rightarrow \bigcirc\mathcal{K}l) \tag{3}$$

$$\Box((\mathcal{M}l_1 \wedge \ldots \wedge \mathcal{M}l_n \wedge \bigcirc(\mathcal{M}l_{n+1} \wedge \ldots \wedge \mathcal{M}l_m) \rightarrow \bigcirc\mathcal{M}l) \tag{4}$$

The meaning of (3) is that if $l_1, \ldots, l_n$ are known in a state and $l_{n+1}, \ldots, l_m$ are known in the next state, then l is also known in the next state. Such laws are intended to expresses "causal" dependencies among fluents. The meaning of causal law (4) can be defined accordingly.

The *constraints* in $\mathcal{C}$ are, in general, arbitrary temporal formulas of DLTL. Constraints put restrictions on the possible correct behaviors of a protocol. The kind of constraints we will use in the specification of a protocol include the observations on the value of epistemic fluent literals in the *initial state* and the precondition laws. The initial state $Init$ is a (possibly incomplete) set of epistemic literals, which is made complete by adding $\neg\mathcal{K}l$ to $Init$ when $\mathcal{K}l \notin Init$.

The *precondition laws* have the form:

$$\Box(\alpha \rightarrow [a]\bot),$$

with $a \in \Sigma$ and α an arbitrary non-temporal formula containing a boolean combination of epistemic literals. The meaning is that the execution of an action a is not possible if α holds (i.e. there is no resulting state following the execution of a if α holds). Observe that, when there is no precondition law for an action, the action is executable in all states.

In order to deal with the frame problem, we extend the solution proposed in [9] to the epistemic case. We define a completion construction which, given a domain description, introduces frame axioms for all frame fluents in the style of the successor state axioms introduced by Reiter [20] in the situation calculus. The completion construction is applied only to the action laws and causal laws in Π and not to the constraints. The value of each epistemic fluent persists from a state to the next one unless its change is caused by the execution of an action as an immediate effect (of an action law) or an indirect effect (of the causal laws). We call $Comp(\Pi)$ the completion of a set of laws Π.

Let Π be a set of action laws and causal laws. Π may contain action laws of the form:

$$\Box(\mathcal{K}\alpha_i \to [a]\mathcal{K}f) \qquad \Box(\mathcal{K}\beta_j \to [a]\mathcal{K}\neg f),$$
$$\Box(\mathcal{M}\alpha_i \to [a]\mathcal{M}f) \qquad \Box(\mathcal{M}\beta_j \to [a]\mathcal{M}\neg f),$$

as well as causal laws of the form

$$\Box((\mathcal{K}\alpha \wedge \bigcirc\mathcal{K}\beta) \to \bigcirc\mathcal{K}l),$$
$$\Box((\mathcal{M}\alpha \wedge \bigcirc\mathcal{M}\beta) \to \bigcirc\mathcal{M}l),$$

where $a \in \Sigma$ and, as a shorthand, $\mathcal{K}\alpha, \mathcal{K}\beta, \mathcal{K}\alpha_i, \mathcal{K}\beta_j$ are conjunctions of epistemic fluents of the form $\mathcal{K}l_1 \wedge \ldots \wedge \mathcal{K}l_n$ and $\mathcal{M}\alpha, \mathcal{M}\beta, \mathcal{M}\alpha_i, \mathcal{M}\beta_j$ are conjunctions of epistemic literals of the form $\mathcal{M}l_1 \wedge \ldots \wedge \mathcal{M}l_n$.

Observe that, given the definition of the next operator $\bigcirc$ (namely, $\bigcirc\alpha \equiv \bigvee_{a\in\Sigma}\langle a\rangle\alpha$), the first causal law above can be written as follows:

$$\Box((\mathcal{K}\alpha \wedge \bigvee_{a\in\Sigma}\langle a\rangle\mathcal{K}\beta) \to \bigvee_{a\in\Sigma}\langle a\rangle\mathcal{K}l),$$

Observe also that, when a given action a is executed in a state (i.e. in a world of a model), this is the only action executed in it, since models of DLTL are linear (and each models describes a single run on the protocol). Hence, from the formula above it follows:

$$(*) \quad \Box((\mathcal{K}\alpha \wedge \langle a\rangle\mathcal{K}\beta) \to \langle a\rangle\mathcal{K}l).$$

Moreover, as the axioms $\langle a\rangle\phi \to [a]\phi$ and $\langle a\rangle\top \wedge [a]\phi \to \langle a\rangle\phi$ hold in DLTL (see [12]), from (*) we can get:

$$(**) \quad \Box(\langle a\rangle\top \to ((\mathcal{K}\alpha \wedge [a]\mathcal{K}\beta) \to [a]\mathcal{K}l)).$$

This formula has a structure very similar to action laws. We call these formulas *normalized causal laws*. A similar transformation can be applied to the second causal law, giving: $\Box(\langle a\rangle\top \to ((\mathcal{M}\alpha \wedge [a]\mathcal{M}\beta) \to [a]\mathcal{M}l))$.

The action laws and causal laws for a fluent f in Π can then have the following forms:

$$\Box(\langle a\rangle\top \to (\mathcal{K}\alpha_i \wedge [a]\mathcal{K}\gamma_i \to [a]\mathcal{K}f)) \qquad \Box(\langle a\rangle\top \to (\mathcal{K}\beta_j \wedge [a]\mathcal{K}\delta_j \to [a]\mathcal{K}\neg f))$$
$$\Box(\langle a\rangle\top \to (\mathcal{M}\alpha_i \wedge [a]\mathcal{M}\gamma_i \to [a]\mathcal{M}f)) \quad \Box(\langle a\rangle\top \to (\mathcal{M}\beta_j \wedge [a]\mathcal{M}\delta_j \to [a]\mathcal{M}\neg f))$$

We define the completion of Π as the set of formulas $Comp(\Pi)$ containing, for all actions a and fluents f, the following axioms:

$$\Box(\langle a\rangle\top \to ([a]\mathcal{K}f \leftrightarrow (\bigvee_i(\mathcal{K}\alpha_i \wedge [a]\mathcal{K}\gamma_i)) \vee (\mathcal{K}f \wedge \bigwedge_j(\mathcal{K}\neg\beta_j \vee \neg[a]\mathcal{M}\delta_j))))$$
$$\Box(\langle a\rangle\top \to ([a]\mathcal{K}\neg f \leftrightarrow (\bigvee_j(\mathcal{K}\beta_j \wedge [a]\mathcal{K}\delta_j)) \vee (\mathcal{K}\neg f \wedge \bigwedge_i(\mathcal{K}\neg\alpha_i \vee \neg[a]\mathcal{M}\gamma_i)))).$$

These laws say that a fluent $\mathcal{K}f$ ($\mathcal{K}\neg f$) holds either as (direct or indirect) effect of the execution of some action a, or by persistency, since $\mathcal{K}f$ ($\mathcal{K}\neg f$) held in the state before the occurrence of a and its negation is not a result of a. Observe that the two frame axioms above also determine the values in a state for $[a]\mathcal{M}f$ and for $[a]\mathcal{M}\neg f$.

Observe that, as a difference with [9], in a domain description we do not distinguish between frame and non-frame fluents and in the following we assume that all epistemic fluents are frame, that is, they are fluents to which the law of inertia applies. The kind of non-determinism that we allow here is on the choice of the actions to be executed, which can be represented by the choice construct of regular programs.

3   Web Service Specification

In this section, we describe how the interface of a Web service can be defined by specifying its interaction protocol. In the social approach [22,24] an interaction protocol is specified by describing the effects of communicative actions on the social state, and by specifying the permissions and the commitments that arise as a result of the current conversation state. These effects, including the creation of new commitments, can be expressed by means of *action laws*.

The action theory introduced above will be used for modeling communicative actions and for describing the social behavior of agents in a multi-agent system. In defining protocols, communicative actions will be denoted by *action_name(s,r)*, where s is the sender and r is the receiver. In particular, two special actions are introduced for each protocol Pn

$$begin\_Pn(s,r) \quad \text{and} \quad end\_Pn(s,r),$$

which are supposed to start and to finish each *run* of the protocol. For each protocol, we introduce a special fluent Pn (where Pn is the "protocol name") which has to be true during the whole execution of the protocol: Pn is made true by the action $begin\_Pn(s,r)$ and it is made false by the action $end\_Pn(s,r)$.

The use of social commitments has long been recognized as a "key notion" to allow coordination and communication in multi-agent systems [13]. Among the most significant proposals to use commitments in the specification of protocols (or more generally, in agent communication) are those by Singh [22], Guerin and Pitt [11], Colombetti [5].

In order to handle commitments and their behavior during runs of a protocol Pn, we introduce two special fluents. One represents *base-level commitments* and has the form $C(Pn, i, j, \alpha)$ meaning that in the protocol Pn agent i is committed to agent j to bring about α, where α is an arbitrary non-temporal formula not containing commitment fluents. The second commitment fluent models *conditional commitments* and has the form $CC(Pn, i, j, \beta, \alpha)$ meaning that in the protocol Pn the agent i is committed to agent j to bring about α, if the condition β is brought about.

Commitments are created as effects of the execution of communicative actions in the protocol and they are "discharged" when they have been fulfilled. A commitment $C(Pn, i, j, \alpha)$, created at a given state of a run, is regarded to be fulfilled in a run if there is a later state in the run in which α holds.

We introduce the following *causal laws* for automatically discharging fulfilled commitments[2]:

(i) $\Box(\bigcirc\alpha \rightarrow \bigcirc\mathcal{K}_{i,j}(\neg C(Pn, i, j, \alpha)))$

(ii)$\Box((\mathcal{K}_{i,j}(CC(Pn, i, j, \beta, \alpha)) \wedge \bigcirc\beta) \rightarrow \bigcirc\mathcal{K}_{i,j}(C(Pn, i, j, \alpha)))$

(iii)$\Box((\mathcal{K}_{i,j}(CC(Pn, i, j, \beta, \alpha)) \wedge \bigcirc\beta) \rightarrow \bigcirc\mathcal{K}_{i,j}(\neg CC(Pn, i, j, \beta, \alpha)))$

A commitment to bring about α is considered fulfilled and is discharged (i) as soon as α holds. A conditional commitment $CC(Pn, i, j, \beta, \alpha)$ becomes a base-level commitment $C(Pn, i, j, \alpha)$ when β has been brought about (ii) and the conditional commitment is discharged (iii).

We can express the condition that a commitment $C(Pn, i, j, \alpha)$ has to be fulfilled before the "run" of the protocol is finished by the following *fulfillment constraint*:

$$\Box(\mathcal{K}_{i,j}(C(Pn, i, j, \alpha)) \rightarrow Pn \, \mathcal{U} \, \alpha)$$

We will call Com_i the set of constraints of this kind for all commitments of agent i. Com_i states that agent i will fulfill all the commitments of which it is the debtor.

At each stage of the protocol only some of the messages can be sent by the participants, depending on the social state of the conversation. *Permissions* allow to determine which messages are allowed at a certain stage of the protocol. The permissions to execute communicative actions in each state are determined by social facts. We represent them by precondition laws. Preconditions on the execution of action a can be expressed as: $\Box(\alpha \rightarrow [a]\bot)$ meaning that action a cannot be executed in a state if α holds in that state. We call $Perm_i$ (permissions of agent i) the set of all the precondition laws of the protocol pertaining to the actions of which agent i is the sender.

Let us consider as an example a service for purchasing a good.

Example 1. There are two roles: A customer, denoted by C, and a producer, denoted by P. The communicative action of the protocol are: $request(C, P)$, meaning that the customer sends a request for a product, $offer(P, C)$ and not_avail (P, C), the producer sends an offer or says that the product is not available, $accept(C, P)$ and $refuse(C, P)$, the customer accepts or refuses the offer. Furthermore, as pointed out before, there will be the actions $begin\_Pu(C, P)$ and $end\_Pu(C, P)$ to start and finish the protocol.

As mentioned before, the social state will contain only epistemic fluents. We denote the social knowledge by $\mathcal{K}_{C,P}$, to mean that the knowledge is shared by C and P.

The social state will contain the following fluents, which describe the protocol in an abstract way: *requested*, the product has been requested, *offered*, the product is available and an offer has been sent (we assume that $\neg offered$ means

---

[2] We omit the three similar rules with $\mathcal{K}$ replaced by $\mathcal{M}$.

that the product is not available), *accepted*, the offer has been accepted. The fluent Pu means that the protocol is being executed.

Furthermore, we introduce some base-level commitments (to simplify the notation, in the following we will use $\mathcal{K}^w_{C,P}(f)$ as a shorthand of the formula $\mathcal{K}_{C,P}(f) \vee \mathcal{K}_{C,P}(\neg f)$):

$C(Pu, C, P, \mathcal{K}_{C,P}(requested))$
$C(Pu, P, C, \mathcal{K}^w_{C,P}(offered))$
$C(Pu, C, P, \mathcal{K}^w_{C,P}(accepted))$

We also need the following conditional commitments:

$CC(Pu, P, C, \mathcal{K}_{C,P}(requested), \mathcal{K}^w_{C,P}(offered))$
$CC(Pu, C, P, \mathcal{K}_{C,P}(offered), \mathcal{K}^w_{C,P}(accepted))$

For instance, the first conditional commitment says that the producer is committed to send an offer, or to say that the product is not available, if a request for the product has been made.

We can now give the action rules for the action of the protocol. We assume all fluents to be undefined in the initial state (i.e., for each fluent f, for each epistemic modality $\mathcal{K}$, $\neg \mathcal{K} f$ and $\neg \mathcal{K} \neg f$ hold in the initial state), except for fluent Pu which will be known to be false. The execution of $begin\_Pu(C, P)$ and $end\_Pu(C, P)$ will have the following effects:

$\Box[begin\_Pu(C, P)]\mathcal{K}_{C,P}(Pu) \wedge$
$\quad \mathcal{K}_{C,P}(C(Pu, C, P, \mathcal{K}_{C,P}(requested))) \wedge$
$\quad \mathcal{K}_{C,P}(CC(Pu, P, C, \mathcal{K}_{C,P}(requested), \mathcal{K}^w_{C,P}(offered))) \wedge$
$\quad \mathcal{K}_{C,P}(CC(Pu, C, P, \mathcal{K}_{C,P}(offered), \mathcal{K}^w_{C,P}(accepted)))$
$\Box[end\_Pu(C, P)]\mathcal{K}_{C,P}(\neg Pu)$

After starting the protocol, the customer is committed to make a request, and the conditional commitments are created.

The action laws for the remaining actions are the following:

$\Box[request(C, P)]\mathcal{K}_{C,P}(requested)$
$\Box[offer(P, C)]\mathcal{K}_{C,P}(offered)$ $\Box[accept(C, P)]\mathcal{K}_{C,P}(accepted)$
$\Box[not\_avail(P, C)]\mathcal{K}_{C,P}(\neg offered)$ $\Box[refuse(C, P)]\mathcal{K}_{C,P}(\neg accepted)$

We can now give the preconditions for the actions of the protocol.

$\Box(\neg \mathcal{K}_{C,P}(\neg Pu) \rightarrow [begin\_Pu(C, P)]\bot)$
$\Box((\neg \mathcal{K}_{C,P}(Pu) \vee \mathcal{K}_{C,P}(requested)) \rightarrow [request(C, P)]\bot)$
$\Box((\neg \mathcal{K}_{C,P}(Pu) \vee \neg \mathcal{K}_{C,P}(requested) \vee \mathcal{K}^w_{C,P}(offered)) \rightarrow [offer(P, C)]\bot)$
$\Box((\neg \mathcal{K}_{C,P}(Pu) \vee \neg \mathcal{K}_{C,P}(requested) \vee \mathcal{K}^w_{C,P}(offered)) \rightarrow [not\_avail(P, C)]\bot)$
$\Box((\neg \mathcal{K}_{C,P}(Pu) \vee \neg \mathcal{K}_{C,P}(offered) \vee \mathcal{K}^w_{C,P}(accepted)) \rightarrow [accept(C, P)]\bot)$
$\Box((\neg \mathcal{K}_{C,P}(Pu) \vee \neg \mathcal{K}_{C,P}(offered) \vee \mathcal{K}^w_{C,P}(accepted)) \rightarrow [refuse(C, P)]\bot)$
$\Box(\neg \mathcal{K}_{C,P}(Pu) \rightarrow [end\_Pu(C, P)]\bot)$

For instance, action $request(C, P)$ cannot be executed if it is not known that the protocol has been started or if it is known that the request has already been achieved (to avoid repeating the action).

A protocol is specified by giving a domain description, defined as follows:

Definition 1. *A domain description D is a pair $(\Pi, \mathcal{C})$ where*

- *Π is the set of the action and causal laws containing:*
  - *the laws describing the effects of each communicative actions on the social state;*
  - *the causal laws defining the commitment rules.*
- *$\mathcal{C} = Init \wedge \bigwedge_i (Perm_i \wedge Com_i)$ is the conjunction of the constraints on the initial state of the protocol and the permissions $Perm_i$ and the commitment constraints Com_i of all the agents i.*

Given a domain description D, we denote by $Comp(D)$, the completed domain description, the set of formulas: $(Comp(\Pi) \wedge Init \wedge \bigwedge_i (Perm_i \wedge Com_i))$.

Definition 2. *Given the specification of a protocol by a domain description D, the runs of the system according the protocol are exactly the models of $Comp(D)$.*

Note that protocol "runs" are always finite, while the logic DLTL is characterized by infinite models. To take this into account, we assume that each domain description of a protocol will be suitably extended with an action *noop* which does nothing and which can be executed only after termination of the protocol, so as to allow a computation to go on forever after termination of the protocol.

For instance in our example we have the following runs:

$$begin\_Pu(C, P); request(C, P); offer(P, C); accept(C, P); end\_Pu(C, P)$$
$$begin\_Pu(C, P); request(C, P); offer(P, C); refuse(C, P); end\_Pu(C, P)$$
$$begin\_Pu(C, P); request(C, P); not\_avail(P, C); end\_Pu(C, P)$$

4   Reasoning About Web Services

Once the interface of a service has been defined by specifying its protocol, several kinds of verification can be performed on it as, for instance, the verification of service compliance with the protocol at runtime, the verification of properties of the protocol and the verification that a given implemented service, whose behavior is known, is compliant with the protocol.

The verification that the interaction protocol has the property φ amounts to show that the formula

$$(Comp(\Pi) \wedge Init \wedge \bigwedge_i (Perm_i \wedge Com_i)) \rightarrow \varphi, \qquad (5)$$

is valid, i.e. that all the admitted runs have the property φ.

Verifying that a set of services are compliant with a given interaction protocol at runtime, given the history $\tau = a_1, \ldots, a_n$ describing the interactions of the services (namely, the sequence of communicative messages they have exchanged), amounts to checking if there is a run of the protocol containing that sequence of communications. This can be done by verifying that the formula

$$(Comp(\Pi) \wedge Init \wedge \bigwedge_i (Perm_i \wedge Com_i)) \wedge <a_1; a_2; \ldots; a_n> \top$$

(where i ranges on all the services involved in the protocol) is satisfiable.

In the logic DLTL, a *rigid* protocol like the purchase protocol of Example 1 can be easily represented by means of a regular program, such as the following regular program π_{Pu}:

$$begin\_Pu(C, P); request(C, P);$$
$$((offer(P, C);$$
$$(accept(C, P) + refuse(C, P)) +$$
$$not\_avail(P, C));$$
$$end\_Pu(C, P)$$

The correctness of this formulation of the protocol with respect to the formulation given in Example 1 can be verified by proving that all runs of π_{Pu} satisfy the permissions and commitments of the participants, i.e. that the following formula is valid

$$(Comp(\Pi) \wedge Init \wedge \langle \pi_{Pu} \rangle \top) \rightarrow \bigwedge_i (Perm_i \wedge Com_i) \qquad (6)$$

where $\langle \pi_{Pu} \rangle \top$ constrains each model to begin with an execution of π_{Pu}.

Further examples of property verification will be given in the next section.

Verification and satisfiability problems can be solved by extending the standard approach for verification of linear-time, temporal logic, based on the use of Büchi automata. We recall that a *Büchi automaton* has the same structure as a traditional finite state automaton, with the difference that it accepts infinite words. More precisely a Büchi automaton over an alphabet Σ is a tuple $\mathcal{B} = (Q, \rightarrow, Q_{in}, F)$ where:

- Q is a finite nonempty set of states;
- $\rightarrow \subseteq Q \times \Sigma \times Q$ is a transition relation;
- $Q_{in} \subseteq Q$ is the set of initial states;
- $F \subseteq Q$ is a set of accepting states.

Let $\sigma \in \Sigma^\omega$. Then a run of $\mathcal{B}$ over σ is a map $\rho : prf(\sigma) \rightarrow Q$ such that:

- $\rho(\varepsilon) \in Q_{in}$
- $\rho(\tau) \xrightarrow{a} \rho(\tau a)$ for each $\tau a \in prf(\sigma)$

The run ρ is *accepting* iff $inf(\rho) \cap F \neq \emptyset$, where $inf(\rho) \subseteq Q$ is given by $q \in inf(\rho)$ iff $\rho(\tau) = q$ for infinitely many $\tau \in prf(\sigma)$.

As described in [12], the satisfiability problem for DLTL can be solved in deterministic exponential time, as for LTL, by constructing for each formula $\alpha \in$

$DLTL(\Sigma)$ a Büchi automaton $\mathcal{B}_\alpha$ such that the language of ω-words accepted by $\mathcal{B}_\alpha$ is non-empty if and only if α is satisfiable.

A more efficient approach for constructing a Büchi automaton from a DLTL formula making use of a tableau-based algorithm has been proposed in [6]. Given a formula φ, the algorithm builds a graph $\mathcal{G}(\varphi)$ whose nodes are labelled by sets of formulas. States and transitions of the Büchi automaton correspond to nodes and arcs of the graph. As for LTL, the number of states of the automaton is, in the worst case, exponential in the size of the input formula, but in practice it is much smaller.

Since the nodes of the graph $\mathcal{G}(\varphi)$ are labeled by sets of formulas, what we actually obtain by the construction is a labeled Büchi automaton, which can be defined by adding to the above definition a *labeling function* $\mathcal{L} : S \to 2^{Lit}$, where *Lit* is the set of all epistemic literals[3]. It is easy to obtain from an accepting run of the automaton a set of models of the given formula, by completing the label of each state in all consistent ways.

The validity of a formula α can be verified by constructing the Büchi automaton $\mathcal{B}_{\neg\alpha}$ for $\neg\alpha$: if the language accepted by $\mathcal{B}_{\neg\alpha}$ is empty, then α is valid, whereas any infinite word accepted by $\mathcal{B}_{\neg\alpha}$ provides a counterexample to the validity of α.

For instance, given a completed domain description

$$(Comp(\Pi) \wedge Init \wedge \bigwedge_i (Perm_i \wedge Com_i))$$

specifying a protocol, we can construct the corresponding labeled Büchi automaton, such that all runs accepted by the automaton represent runs of the protocol. In [9], we show how to take advantage of the structure of the problems considered in this paper to optimize the construction of the Büchi automaton.

5   Composing Web Services

Assume now that we have a service Sh for shipping goods, and that the customer wants to reason about the composition of the producer service of the previous section and of this service. For simplicity we assume that the protocol of the shipping service is the same as that of producer service. To distinguish the two protocols we will add the suffix Pu or Sh to their actions and fluents, while the role of the shipper will be denoted by S.

The domain description D_{PS} of the composed service can be obtained by taking the union of the sets of formulas specifying the two protocols: $D_{PS} = D_{Pu} \cup D_{Sh}$. Since we want to reason from the side of the customer, we will replace the epistemic operators $\mathcal{K}_{P,C}$ and $\mathcal{K}_{S,C}$ with $\mathcal{K}_C$, representing the knowledge of the customer. Thus the runs of the composed service PS are given by the interleaving of all runs of the two protocols.

The aim of the customer is to extract from the domain description of PS a *plan* allowing it to interact with the two services. The goal of the plan will be

---

[3] Note that epistemic literals are considered as atomic propositions.

specified by means of a set of constraints *Constr* which will take into account the properties of the composed service. For instance, the customer cannot request an offer to the shipping service if it has not received an offer from the producer. This can be easily expressed by adding a new precondition to the action *request_Sh(C, S)*:

$$\Box(\neg \mathcal{K}_C(\mathit{offered\_Pu}) \rightarrow [\mathit{request\_Sh}(C,S)]\bot)$$

Other constraints cannot be easily expressed by means of preconditions, since they involve more "global" properties of a run. For instance we expect that the customer cannot accept only one of the offers of the two services. This property can be expressed by the following formula

$$\Diamond\langle \mathit{accept\_Pu}(C,P)\rangle \leftrightarrow \Diamond\langle \mathit{accept\_Sh}(C,S)\rangle$$

stating that the customer must accept both offers or none of them.

Then, the specification of the interaction protocol of the composed service is given by $D_{PS} \cup \mathit{Constr}$, from which the customer will extract the plan. To do this, however, we must first discuss an important aspect of the protocol, i.e. *nondeterminism*.

We assume that, if a protocol contains a point of choice among different communicative actions, the sender of these actions can choose freely which one to execute, and, on the other hand, the receiver cannot make any assumption about which of the actions it will receive. Therefore, from the viewpoint of the receiver, that point of choice is a point of nondeterminism to care about. For instance, the customer cannot know whether the service *Pu* will reply with *offer_Pu* or *not_avail_Pu* after receiving the request. Therefore the customer cannot simply reason on a single choice of action, but he will have to consider all possible choices of the two services, thus obtaining alternative runs, corresponding to a *conditional plan*. An example of conditional plan is the following[4]

begin_Pu; request_Pu;
  (offer_Pu; begin_Sh; request_Sh;
    (offer_Sh; accept_Pu; accept_Sh; end_Pu; end_Sh +
    not_avail_Sh; refuse_Pu; end_Pu; end_Sh)) +
  (not_avail_Pu; end_Pu).

This plan is represented as a regular program, where, in particular, "+" is the choice operator.

Since we are using a linear-time, temporal logic, the constraints in *Constr* can only express properties dealing with a single run. For instance, the run *begin_Pu; request_Pu; offer_Pu; accept_Pu; begin_Sh; request_Sh; offer_Sh; accept_Sh; end_Pu; end_Sh* is correct with respect to the above constraints, since both offers are accepted. However, assume that the customer chooses to execute this plan, and, after executing action *request_Sh*, the shipping service replies with *not_avail_Sh*. At this point there is no other way of continuing the execution, since the customer has already accepted the offer by the producer, while it should have refused it.

---

[4] We omit sender and receiver of communicative actions.

The first step for obtaining a *conditional plan* consists in building the Büchi automaton obtained from the domain description D_{PS} and the constraints *Constr.*. During the construction of the automaton, we will mark as AND states those states whose outgoing arcs are labeled with actions whose sender is one of the services, such as *offer_Pu* or *not_avail_Pu*[5]. The plan can be obtained by searching the automaton with a forward-chaining algorithm which considers all AND states as branching points of the plan.

In this example, and in many similar cases, the size of the Büchi automaton obtained from the specification of the protocol is small enough to be directly manageable. In this case we might adopt a different approach to the construction of a conditional plan, consisting of "pruning" once and for all the automaton by removing all arcs which do not lead to an accepting state, and all AND states for which there is some outgoing arc not leading to an accepting state. In this way we are guaranteed that, if there is a run σ_1; *offer_Sh*; σ_2, where σ_1 and σ_2 are sequences of actions, there must also be a run σ_1; *not_avail_Sh*; σ_3, for some sequence of actions σ_3. Therefore the customer can execute the first part σ_1 of the run, being sure that it will be able to continue with run σ_3 if the shipping service replies with *not_avail_Sh*. In other words, the customer will be able to act by first extracting a linear plan, and begin executing it. If, at some step, one of the services executes an action different from the one contained in the plan, the customer can build a new plan originating from the current state, and restart executing it.

In the construction of the conditional plan, we have taken into account only the nondeterministic actions of the two services. However there are some choices regarding the actions of the customer, such as   *accept_Pu* or *refuse_Pu*, that cannot be made at planning time. These nondeterministic choices can also be considered in a conditional plan. In our example we might have the following conditional plan π_{PS}

begin_Pu; request_Pu;
    ((offer_Pu; begin_Sh; request_Sh;
        (offer_Sh;
            (accept_Pu; accept_Sh; end_Pu; end_Sh +
            refuse_Pu; refuse_Sh; end_Pu; end_Sh) +
        not_avail_Sh; refuse_Pu; end_Pu; end_Sh)) +
    (not_avail_Pu; end_Pu))

Note that, in the case of nondeterministic actions of the customer, we are not imposing all choices to be present in the conditional plan, as we did for the actions of the other participants, because some choices might not be possible due to the constraints. For instance, after *accept_Pu* the customer must necessarily execute *accept_Sh*.

A different problem, which can be tackled in our formalism when the conditional plan π_{PS} is given, is that of verifying its correctness with respect to the

---

[5] For simplicity we assume that there is no state whose outgoing arcs are labeled with actions sent and received by the same agent.

protocols of the composed services. This requires to verify that, in every run of the conditional plan all the permissions and commitments of the component services are satisfied, and can be done by proving that the formula

$$\bigwedge_k (Comp(\Pi_k) \wedge Init \wedge \langle \pi_{PS} \rangle \top \rightarrow \bigwedge_{k,j} (Perm_j^k \wedge Com_j^k))$$

is valid, where k ranges over the different services and, for each k, j ranges over all the participants of service k. In a similar way, it can be verified that the plan π_{PS} satisfies the constraints $Constr$ defined above, by showing the validity of the formula:

$$\bigwedge_k (Comp(\Pi_k) \wedge Init \wedge \langle \pi_{PS} \rangle \top \rightarrow Constr).$$

Up to now the kind of reasoning performed on composed protocols has taken into account only the "public" actions, i.e. the communicative actions of the component protocols. However, in general, the customer should be able to use "private" actions to reason about the information received from the services and to decide what action to execute. Since the information sent by the services will be available only at runtime, such an action should be considered as a nondeterministic action at planning time. We might easily extend our approach to this case by extending the specification of the composed services with "private" actions and fluents of the customer.

The approach described in this section can be applied to the more general problem of building a new service that manages all interactions between the customer and the two services, so that the customer interacts only with the new service through a suitable protocol [19]. Given the protocol Cu specifying the interactions between the customer and the new service, the new service can be obtained by putting together the three protocols Cu, Pu and Sh, and by adding suitable constraints similar to the ones given above. For instance we may state that the offers of each of the two services can be accepted if and only if the customer accepts them:

$$(\Diamond \langle accept\_Pu \rangle \leftrightarrow \Diamond \langle accept\_Cu \rangle) \wedge (\Diamond \langle accept\_Sh \rangle \leftrightarrow \Diamond \langle accept\_Cu \rangle)$$

We can then proceed as before by building the Büchi automaton from the composed protocol and extracting from it a conditional plan, as for instance:

$begin\_Cu;$ $request\_Cu;$
  $begin\_Pu;$ $request\_Pu;$
    $((offer\_Pu;$ $begin\_Sh;$ $request\_Sh;$
      $((offer\_Sh;$ $offer\_Cu;$
        $(accept\_Cu;$ $accept\_Pu;$ $accept\_Sh;$ $end\_Pu;$ $end\_Sh;$ end_Cu +
          $refuse\_Cu;$ $refuse\_Pu;$ $refuse\_Sh;$ $end\_Pu;$ $end\_Sh;$ $end\_Cu)) +$
        $not\_avail\_Sh;$ $not\_avail\_Cu;$ $refuse\_Pu;$ $end\_Pu;$ $end\_Sh;$ $end\_Cu)) +$
      $not\_avail\_Pu;$ $not\_avail\_Cu;$ $end\_Pu;$ $end\_Cu)$

This plan can be considered as a specification of the (abstract) behavior of the new service.

6   Conclusions and Related Work

In this paper we have presented an approach for the specification and verification of interaction protocols in a temporal logic (DLTL). Our approach provides a unified framework for describing different aspects of multi-agent systems. Programs can be expressed as regular expressions, (communicative) actions can be specified by means of action and precondition laws, social facts can be specified by means of commitments whose dynamics are ruled by causal laws, and temporal properties can be expressed by means of temporal formulas. To deal with incomplete information, we have introduced epistemic modalities in the language, to distinguish what is known about the social state from what is unknown. In this framework, various verification problems can be formalized as satisfiability and validity problems in DLTL, and they can be solved by developing automata-based techniques.

Our proposal is based on a social approach to agent communication, which allows a high level specification of the protocol and does not require a rigid specification of the correct action sequences. For this reason, the approach appears to be well suited to reason about composition of Web services. In [8] we have addressed the problem of combining two protocols to define a new more specialized protocol. Here we have shown that service composition can be modeled by taking the formulas giving the domain descriptions of the services, by adding suitable temporal constraints to them, and translating the set of formulas into a Büchi automaton from which a (conditional) plan can be obtained.

The proposal of representing states as sets of epistemic fluent literals is based on [1], which presents a modal approach for reasoning about dynamic domains in a logic programming setting. A similar "knowledge-based" approach has been used to define the PKS planner, allowing to plan under conditions of incomplete knowledge and sensing [16]. PKS generalizes the STRIPS approach, by representing a state as a set of databases that model the agent's knowledge.

The problem of the automated composition of Web services by planning in asynchronous domains is addressed in [19], and extended to the "knowledge level" in [18]. Web services are described in standard process modeling and execution languages, like BPEL4WS, and then automatically translated into a planning domain that models the interactions among services at the knowledge level. The planning technique [19] consists of the following steps. The first step constructs a parallel *state transition system* that combines the given services in a planning domain. The next step consists of formalizing the requirements for the composite service as a goal in a specific language which allows to express extended goals [3]. Finally the planner generates a plan that is translated into a state transition system and into a concrete BPEL4WS process. The planning problem is solved by making use of the state-of-the-art planner MBP.

The approach to Web service composition presented in this paper has analogies with the one presented in [18], particularly with respect to the sequence of steps performed to build the plan. However, the approach of [18] is based on a planning technique derived from model checking for branching-time temporal logic CTL [17], while our approach is based on the dynamic, linear-time, temporal logic DLTL, and on the translation of DLTL formulas into Büchi automata.

In [2] the problem of automatic service composition is addressed assuming that a set of available services (whose behavior is represented by finite state transition systems) is given together with a possibly incomplete specification of the sequences of actions that the client would like to realize. The problem of checking the existence of a composition is reduced to the problem of checking the satisfiability of a PDL formula. This provides an EXPTIME complexity upper bound. In contrast to [2], in our approach client requirements are specified by providing a set of conditions that the target service must satisfy. The composition problem considered in [2] is a generalization of the verification problem we have addressed at the end of section 5 for the case when the protocol of the target service is underspecified and the component e-services that will provide the services required by the client are not known. The extension of our approach to deal with underspecified specifications of the target service will be the subject of further investigation.

References

1. M. Baldoni, L. Giordano, A. Martelli, and Viviana Patti. Reasoning about complex actions with incomplete knowledge: a modal approach. In *Proc. ICTCS'01- LNCS 2202*, 405–425, 2001.
2. D. Berardi, G. De Giacomo, M. Lenzerini, M. Mecella and D Calvanese. Syntesis of Underspecified Composite e-Services based on Automated Reasoning. In *Proc. ICSOC'04*, 105–114, 2004.
3. U. Dal Lago, M. Pistore, P. Traverso: Planning with a Language for Extended Goals. *AAAI 2002*, 447-454, 2002.
4. F. Dignum and M. Greaves, "Issues in Agent Communication:An Introduction". In F.Dignum and M.Greaves (Eds.), *Issues in Agent Communication*, LNAI 1916, 1-16, 1999.
5. N. Fornara and M. Colombetti. Defining Interaction Protocols using a Commitment-based Agent Communication Language. *Proc. AAMAS'03*, Melbourne, 520–527, 2003.
6. L. Giordano and A. Martelli. Tableau-based Automata Construction for Dynamic Linear Time Temporal Logic. Annals of Mathematics and Artificial Intelligence, Vol.46, n. 3, pages 289–315, Springer, 2006.
7. L. Giordano, A. Martelli, and C. Schwind. Verifying Communicating Agents by Model Checking in a Temporal Action Logic. *Proc. Logics in Artificial Intelligence, 9th European Conference, JELIA 2004*, Lisbon, Portugal, Springer LNAI 3229, 57-69, 2004.
8. L. Giordano, A. Martelli, and C. Schwind. Specialization of Interaction Protocols in a Temporal Action Logic. *LCMAS05 (3rd Int. Workshop on Logic and Communication in Multi- Agent Systems)*, ENTCS 157, 4, 1-138, 2006.
9. L. Giordano, A. Martelli and C. Schwind. Specifying and Verifying Interaction Protocols in a Temporal Action Logic Journal of Applied Logic (Special issue on Logic Based Agent Verification), Elsevier, to appear 2006.
10. M. Greaves, H. Holmback and J. Bradshaw. What Is a Conversation Policy?. *Issues in Agent Communication*,LNCS 1916 Springer, 118-131, 2000.

11. F. Guerin and J. Pitt. Verification and Compliance Testing. *Communications in Multiagent Systems*, Springer LNAI 2650, 98–112, 2003.
12. J.G. Henriksen and P.S. Thiagarajan. Dynamic Linear Time Temporal Logic. in Annals of Pure and Applied logic, vol.96, n.1-3, 187–207, 1999
13. N.R. Jennings. Commitments and Conventions: the foundation of coordination in multi-agent systems. In *The knowledge engineering review*, 8(3),233–250, 1993.
14. N. Maudet and B. Chaib-draa. Commitment-based and dialogue-game based protocols: new trends in agent communication languages. The Knowledge Engineering Review, 17(2):157-179, June 2002.
15. S. Narayanan and S. McIlraith. Simulation, Verification and Automated Composition of Web Services. In *Proceedings of the Eleventh International World Wide Web Conference (WWW-11)*, 77–88, May 2002.
16. R. Petrick and F. Bacchus. A knowledge-based approach to planning with incomplete information and sensing. In *Proceedings of the International Conference on Artificial Intelligence Planning (AIPS)*, 212-222, 2002.
17. M. Pistore and P. Traverso. Planning as Model Checking for Extended Goals in Non-deterministic Domains. *Proc. IJCAI'01*, Seattle, 479-484, 2001.
18. M. Pistore, P. Traverso and P. Bertoli. Automated Composition of Web Services by Planning in Asynchronous Domains. *ICAPS 2005*. 2–11, 2005.
19. M. Pistore, A. Marconi, P. Bertoli and P. Traverso. Automated Composition of Web Services by Planning at the Knowledge Level. *Proc. International Joint Conference on Artificial Intelligence (IJCAI)*, 1252–1259, 2005.
20. R. Reiter. The frame problem in the situation calculus: a simple solution (sometimes) and a completeness result for goal regression. In *Artificial Intelligence and Mathematical Theory of Computation: Papers in Honor of John McCarthy*, V. Lifschitz, ed., 359–380, Academic Press, 1991.
21. M.P. Singh. Agent communication languages: Rethinking the principles. IEEE Computer, 31(12), 40–47, 1998.
22. M.P. Singh. A social semantics for Agent Communication Languages. In *Issues in Agent Communication*, Springer LNCS 1916, 31–45, 2000.
23. B. Srivastava and J. Koehler. Web Service Composition - Current Solutions and Open Problems, In *ICAPS 2003 Workshop on Planning for Web Services*, 28 - 35, Trento, Italy, June 2003.
24. P. Yolum and M.P. Singh. Flexible Protocol Specification and Execution: Applying Event Calculus Planning using Commitments. In *AAMAS'02*, 527–534, Bologna, Italy, 2002.

Intelligent Search on the Internet

Alessandro Micarelli, Fabio Gasparetti, and Claudio Biancalana

Department of Computer Science and Automation
Artificial Intelligence Laboratory
University of "Roma Tre"
Via della Vasca Navale, 79 00146 Rome, Italy
{micarel, gaspare, claudio.biancalana}@dia.uniroma3.it

Abstract. The Web has grown from a simple hypertext system for research labs to an ubiquitous information system including virtually all
human knowledge, e.g., movies, images, music, documents, etc. The traditional browsing activity seems to be often inadequate to locate information satisfying the user needs. Even search engines, based on the Information Retrieval approach, with their huge indexes show many drawbacks,
which force users to sift through long lists of results or reformulate queries
several times. Recently, an important research activity effort has been focusing on this vast amount of machine-accessible knowledge and on how
it can be exploited in order to match the user needs. The personalization and adaptation of the human-computer interaction in information
seeking by means of machine learning techniques and in AI-based representations of the information help users to address the overload problem.
This chapter illustrates the most important approaches proposed to personalize the access to information, in terms of gathering resources related
to given topics of interest and ranking them as a function of the current
user needs and activities, as well as examples of prototypes and Web
systems.

1   Introduction

The decentralized and unregulated work of millions of authors spread around the
world have allowed the Web to grow constantly since its creation. At the beginning
of 2005, the part of the Web considered as potentially indexable by major engines
was estimated to consist of at least 11.5 billion pages [23]. As the number of Internet
users and the number of accessible Web pages grow, it is becoming increasingly
difficult for users to find documents that are relevant to their current information
needs. Therefore, it is understandable that the research field devoted to applying
AI techniques to the problem of discovering and analyzing Web resources has been
drawing much attention over the last few years [29].

The *browsing* interaction paradigm, where users analyze Web pages one at a
time is a useful approach for reading and comprehending the content of a hypertext, but it is not suitable for locating a specific piece of information. A few years
ago, a second interaction paradigm was introduced: *querying* a search engine. Directly retrieving documents from an index of millions of documents in a fraction of

O. Stock and M. Schaerf (Eds.): Aiello Festschrift, LNAI 4155, pp. 247–264, 2006.

a second is the paramount advantage of this approach, which is based on the classic Information Retrieval (IR) model. Nevertheless, despite having different goals and needs, two users who submit the same query obtain the same result list. Moreover, by analyzing search behavior, it is possible to see that many users are not able to accurately express their needs in exact query terms.

Search engines are the first search approach of users [49]. Personalized search aims at building systems which try to serve up individualized pages to the user; such systems are based on some form of model of the needs and context of the user's activities. For instance, some search engines employ the collaborative or community-based approach in order to suggest pages that other users, who submitted the same query, selected frequently [50].

Because of the vastness of the personalized search domain, in this chapter we focus on some of the most recent algorithms, techniques and approaches strongly related to the AI field. For instance, in order to model the user needs and to assign internal representations to Web documents, approaches based on knowledge representation and machine learning, such as semantic networks and frames [3,33,36], neural networks [2,8,22] and decision trees [27], have been successfully developed obtaining interesting results. Ontology-based search and autonomous crawling based on AI algorithms are two further interesting research activities addressing personalized search.

The rest of this chapter is organized as follows: Section 2 presents some AI-based approaches recently developed in order to model user needs and Web documents for the personalization task. Section 3 discusses some further emerging trends in personalized search. Finally, Section 4 closes this chapter.

2   User and Document Modeling for Personalized Search

The available information on the Web is usually represented through HTML documents, or more complex formats, such as images, audio, video, Flash animation, etc.. These layout-based representations organize and adapt the content to the reading user by means of the Web browser. For this reason, personalized Web information systems need to pre-process those formats in order to extract the real document content, ignoring the layout-related information, e.g., HTML table tags or fonts. Once the original document content is extracted, ad-hoc representations are usually employed to organize the information so that the personalization process is able to retrieve and filter it according to the user needs.

User knowledge and preference acquisition is another important problem to be addressed in order to provide effective personalized assistance. Some approaches exploit *user data*, that is, information about personal characteristics of the user (such as age, sex, education, country where he lives) in order to learn the knowledge needed to provide effective assistance. Other approaches analyze the *usage data* related to the user's behavior while interacting with the system.

User data are usually collected following explicit approaches, where the user constructs the model by describing the information in which he is interested. Nevertheless, because users typically do not understand how the matching process works, the information they provide is likely to miss the best query keywords, i.e., the words that identify documents meeting their information needs. Instead of requiring user needs to be explicitly specified by queries and manually updated by user feedback, an alternative approach to personalize search results is to develop algorithms that infer those needs implicitly.

Basically, *implicit feedback* techniques unobtrusively draw usage data by tracking and monitoring user behavior without an explicit involvement, e.g., by means of server access logs or query and browsing histories [25].

Several user modeling approaches are based on simple sets of keywords extracted from interesting documents or suggested by users. A weight can be assigned to each keyword representing its importance in the user profile. Basically, these approaches are inspired by Information Retrieval approaches , an area of research that enjoys a long tradition in the modeling and treatment of non-structured textual documents [4]. Because of the popularity in the Web domain, Sect. 2.1 briefly discusses this topic.

Further techniques are based on models and methods from AI Knowledge Representation, such as Semantic Networks, Frames, Latent Semantic Indexing and Bayesian classifiers. These latter kinds of user models will be discussed in the following sections.

2.1   IR-Based Modeling

In the Information Retrieval field, a collection's documents are often represented by a set of *keywords*, which can be directly extracted from the document text or be explicitly indicated in an initial summary drawn up by a specialist. Independently of the extraction method, these very keywords provide a logic view of the document [4].

When collections are particularly bulky, though, even modern processors have to reduce the set of representative words. This can be done through *stemming*, reducing several words to their common grammatical root, or through the identification of *word groups* (which removes adjectives, adverbs and verbs).

One of the problems of IR systems is that of predicting which documents are relevant and which are not; this decision usually depends on the ranking algorithm used, which tries to put the retrieved documents in order, by measuring similarity. The documents on top of this list are the ones considered more relevant.

The classic models consider all documents described by a set of representative keywords, also known as index terms. The *Vector Model* assigns w weights to the index terms; these weights are used to calculate the similarity level between every document stored in the system and the submitted query.

The Vector Model appraises the level of similarity between document d_j and query q, as a correlation between vectors $\boldsymbol{d}$ and $\boldsymbol{q}$. This correlation can be quantified, for example, as the cosine of the angle between the two vectors:

$$\text{sim}(d_j, q) = \frac{\boldsymbol{d_j} \cdot \boldsymbol{q}}{|\boldsymbol{d_j}||\boldsymbol{q}|} = \frac{\sum_{i=1}^{t} w_{ij} w_{iq}}{\sqrt{\sum_{i=1}^{t} w_{ij}^2} \sqrt{\sum_{j=1}^{t} w_{jq}^2}}$$

Term weights can be calculated in many different ways [28,47,4]. The basic idea of the most efficient techniques is linked to the *cluster* concept, suggested by Salton in [9]. In literature it is possible to recognize many prototypes for Web search based on weighted or Boolean feature vector, e.g., [30,5,24,40,31].

2.2   Latent Semantic Indexing

The complexity of Information Retrieval is to be clearly seen in two main unpleasant problems:

- **synonymity** all the documents containing term B synonym of term A present in the query are lost, therefore relevant information is not retrieved.
- **polysemy** occurs when a term has several different meanings; it causes irrelevant documents to appear in the result lists.

In order to solve such problems, documents are represented through underlying concepts. The concealed structure is not simply a many-to-many mapping between terms and concepts, but it depends from the body. *Latent Semantic Indexing* (LSI) is a technique that tries to get hold of this hidden semantic structure through the spectral analysis of the terms-documents matrix [15]. It has been applied in different scopes, for example the simulation of human cognitive phenomena, such as modeling human word sorting and category judgments, estimating the quality and quantity of knowledge contained in an essay, or how easily it can be learned by individual students.

The vectors representing the documents are projected into a new subspace obtained by the *Singular Value Decomposition* (SVD) of the terms-document matrix A. This sub-dimension space is generated by the eigenvectors of matrix $A^T A$ corresponding to the highest eigenvalues, thus to the most obvious correlations between terms. A necessary step in implementing LSI in a collection of n documents is the formation of the terms-documents matrix $A_{m \times n}$, where m is the number of distinguished terms present in n documents. Each document is thus represented by an m-dimensional column vector. After having calculated the frequency of each word in each document, it is possible to implement *inter-cluster* and *intra-cluster* weighting functions.

Matrix A is therefore broken down through SVD, which somehow allows to obtain the semantic structure of the document collection by means of orthogonal matrices U_r and V_r (containing the single left and right vectors of A, respectively) and the diagonal matrix Σ of the singular values of A. Once matrices U_r, V_r and Σ are obtained, by extracting the first $k < r$ triple singulars, it is possible to approximate the original terms-document A matrix with matrix A_k of rank k:

$$ A_k = \Sigma_k^{-1} \cdot U_k^T \cdot A $$

By using this formula it is possible to map any two documents with different origins in the m-dimensional space of all different terms in the same vector of the reduced space; the set of basic vectors of k−dimensional space accounts for the set of concepts or for the different meanings the several documents may have, therefore a generic document in k-dimensional space can be represented as a linear combination of concepts or equivalently of the basic vectors of the same space. We thus passed from a representation in a space with m dimensions (the dimension is equivalent to the number of terms found in the analysis of all documents) to a compressed form of the same vectors in $k < m$ dimensional space. It should be pointed out that through A_k one can deliberately reconstruct, even though not perfectly, matrix A, since it contains a certain noise level introduced by the randomness with which terms are chosen to tackle certain discourses or topics; this noise is "filtered" by annulling the singular, less significant values.

Besides recommender systems [48,44], a generalization of the LSI approach has been also applied to the click-through data analysis in order to model the users' information needs by exploiting usage data, such as the submitted query and the visited results [52]. The evaluation shows that thanks to high order associations identified by the algorithm, it achieves better accuracy compared with other approaches although the whole computation is remarkably time-consuming.

2.3 Bayesian Approach

The idea of resorting to Bayesian probabilistic models to represent a document is undoubtedly very interesting. Indeed, the probability theory provides the required tools to better cope with uncertainty, while Bayesian formalism enables to represent the probabilistic relations within a very vast set of relevant variables, which very efficiently encodes the joint distribution of probability, exploiting the conditional independence relations existing between the variables [42]. The Bayesian approach to probability can be extended to the problem of comprehension: this union leads to an extremely powerful theory, which provides a general solution to noise problems, overtraining and good predictions.

In order to implement the Bayesian document-modeling technique, it is necessary to decide what features are mostly useful to represent a document, and thereafter decide how to assess the probability tables associated with the Bayesian network. The problem of document representation has a considerable impact on a learning system's generalization capacity [13,34]. Usually, a document is a string of characters: it must be converted into a representation that is suitable for learning algorithms and classification. Words have been acknowledged to work well as representation units, whereas their positioning seems to be less relevant [19]. It is now a matter of deciding the features to describe a document.

The multinomial model is usually employed to assign a representation to a document based on a set of words, associated with their frequency. The word order is lost, but their frequency is recorded. A simplified model (Bernoulli model) represents a document as a set of binary features which indicate the presence or absence of a certain set of words forming the vocabulary.

In the multinomial model, in order to calculate a document probability, the probabilities of the words appearing in it are multiplied [18]. We illustrate the procedure that leads to the final representation of a document, highlighting the hypotheses required to simplify a domain's description.

As a first approach, we must define a feature for each position of a word in the document, and its value is the word in that very position. In order to reduce the number of parameters, the probability tables associated with the Bayesian network are considered the same for all features. The independence entailed in this context asserts that the probability of a word in a certain position does not depend on the words present in other positions.

In order to further reduce the number of parameters, we can suppose that the probability of finding a word is independent of its position: this means assuming that features are independent, and identically distributed. As far the document model is concerned, the reference is to the multinomial model that represents a document as a set of words with their frequency number associated to them. In this representation, the word order is lost, but their frequency is recorded.

Going back to Naïve Bayes, we can say that this twofold document representation is a *bag of words*: a document is seen as a set of words. That's how a feature-value representation of the text is done: each word differing from the others corresponds to a feature whose value is the number of times that word appears in the document. In order to avoid having sets of words with an excess cardinality, it is possible to follow a feature selection procedure that aims at reducing the number of words appearing in the vocabulary. The choice of a document representation is affected by the domain being analyzed.

Among the various search agents based on the Bayesian approach it is worth mentioning *Syskill & Webert* [41]. It can make suggestions about interesting pages during a browsing session and use the internal user model to query a search engine in order to retrieve additional resources. The default version of Syskill & Webert uses a simple Bayesian classifier to determine the document relevance. Sets of positive/negative examples are used to learn the user profile and keep it up-to-date. A feature selection phase based on the mutual information analysis is employed to determine the words to use as features.

Besides the explicit feedback technique that forces users to explicitly construct the model by describing the information in which he is interested, Syskill & Webert is able to suggest interesting Web pages. Nevertheless, this decision is made by analyzing the content of pages, and therefore it requires pages to be downloaded first.

2.4   Semantic Networks and Frames

Semantic networks consist of concepts linked to other concepts by means of various kinds of relations. It can be represented as a directed graph consisting of vertices which represent concepts and edges, themselves representing the semantic relations between the concepts. They were initially studied as a reasonable view of how semantic information is organized within a human memory [45].

The *SiteIF* project [33] uses semantic networks built from the frequencies of co-occurring keywords. In this approach, keywords are extracted from the pages in the user's browsing history. These keywords are mapped into *synsets* (grouping English words into sets of synonyms) contained in the *WordNet* database [37], a semantic lexicon for the English language, identifying meanings that are used as nodes in the semantic network-based profile.

The SiteIF model of the profile is kept up to date by "looking over the user's shoulder", trying to anticipate which documents in the Web repository the user might find interesting. It builds a representation of the user's interest by taking into account some properties of words in documents browsed by the user, such as their co-occurrence and frequency. This profile is used to provide personalized browsing, the complementary information seeking activity to search.

A further approach based on a semantic network structure is *ifWeb* [3], an intelligent agent capable of supporting the user in the Web surfing, retrieval, and filtering of documents taking into account specific information needs expressed by the user with keywords, free-text descriptions, and Web document examples. The ifWeb system makes use of semantic networks in order to create profiles of users. More specifically, the profile is represented as a weighted semantic network whose nodes correspond to terms found in documents, and textual descriptions given by the user as positive or negative examples, that is, through relevance feedback. The arcs of the network link together terms that co-occurred in some documents. The use of semantic networks and co-occurrence relationships allows ifWeb to overcome the limitations of simple keyword matching, particularly polysemy, because each node represents a concept and not just ambiguous keywords.

The relevance feedback technique helps users to explicitly refine their profile, selecting which of the suggested documents satisfy their needs: ifWeb autonomously extracts the information necessary to update the user profile from the documents on which the user expressed some positive feedback. If some of the concepts in the user profile have not been reinforced by the relevance feedback mechanism over a long period of time, a temporal decay process called *rent* lowers the weights associated with these concepts. This mechanism allows the profile to be maintained so that it always represents the current interests of the user.

A different AI representation, which organizes knowledge into chunks is called *frames* [38]. Within each frame are many kinds of information, e.g., how to use the frame, expectations, etc.. The frame's *slots* are attributes for which fillers (scalar values, references to other frames or procedures) have to be specified and/or computed. Collections of frames are to be organized in frame systems, in which the frames are interconnected. Unlike slots, the frame's *properties* are inherited from superframes to subframes in the frame network according to some inheritance strategy. The processes working on such frame structures are supposed to match a frame to a specific situation, to use default values to fill unspecified aspects, and so on.

The *Wifs* system's goal is to filter Web documents retrieved by a search engine in response to a query input by the user [36]. This system re-rank urls returned by the search engine, taking into account the profile of the user who typed in the query. The user can provide feedback on the viewed documents, and the system accordingly uses that feedback to update the user model.

The user model consists of frame slots that contain terms (topics), each associated with other terms (co-keywords). This information forms a semantic network. Slot terms, that is, the topics, must be selected from those contained in a *Terms Data Base* (TDB), previously created by experts who select the terms deemed mostly relevant for the pertinent domain.

Document modeling is not based on traditional IR techniques, such as the Vector Space Model. The abstract representation of the document may be seen as a set of active terms, or *planets*, $T_1, T_2, ..., T_n$ are the ones contained both in the document and TDB, whereas the *satellite* terms $t_1, t_2, ..., t_m$ are the terms included in the document, but not in the TDB, but which co-occur with T_i's. It is evident that the structure is similar to the user model one, but there are no affinity values between the planets and the satellites. For each of these terms, however, document occurrence is calculated. The occurrence value of a term t appearing in a retrieved document is directly proportional to the frequency with which term t appears in the body, and the frequency with which term t appears in the document title.

In order to evaluate each document, the system builds a vector $\overrightarrow{Rel}$, where the element Rel_i represents a relevant value of term t_i compared to user information needs.

Relevance is calculated taking into account the user model, the query, and the TDB. The relevance value $Rel_{new}(t)$ of term t, which simultaneously belongs to the document and user model, as a slot topic, is calculated by intensifying the old relevance value, $Rel_{old}(t)$. Basically, the new relevance value of term t is obtained from the old value, in this case initially equal to 0, as a proportional contribution to the sum of all semantic network weights of the user model containing term t as topic.

If the term that belongs to the user model and document, also belongs to the q query input by the user, then the term relevance value is further strengthened by means of w_{slot}, the weight associated with topic t. In this way, query q, which represents the user's immediate needs, is used to effectively filter the result set to locate documents of interest.

If term t belongs both to the query q, the document d, and the TDB, but is not included in the user model, then the only contribution to relevance is given by the old relevance plus a given constant. Finally, if term t is a topic for the $slot_j$, all contributions given by co-keywords are considered. This is where the true semantic network contributes: all the co-keywords K connected to topic t give a contribution, even if previously unknown to the system, i.e., not currently belonging to the user model, nor to the TDB, but only to the document.

The system calculates the final relevance score assigned to the document as follows:

$$Score(Doc) = \overrightarrow{Occ} \cdot \overrightarrow{Rel}$$

where $\overrightarrow{Occ}$ is the vector consisting of elements Occ_i, that is, the occurrence value of a term t_i, and $\overrightarrow{Rel}$ is the vector consisting of elements Rel_i.

Relevance feedback dynamically updates the user model upon receipt of the viewed documents provided by the user. The *renting* mechanism is used to decrease the weights of the terms appearing in the model that do not receive positive feedback after a certain period of time.

It should be pointed out how the semantic network knowledge representation, and its possibility to store text allowing semantic processing, is an intriguing feature in user modeling research, but it is not completely clear how much of the original model's formulation has been kept in the developed prototypes. For example, the ifWeb's semantic network differs from that of the knowledge representation domain, since it represents terms and not concepts, and the arcs are not explicit semantic relationships, rather just co-occurrences in some documents. Nevertheless, the evaluation performance obtained by the user modeling systems based on this technique in the Web domain is extremely interesting, in spite of the computational resources for processing the complex structures.

3   Other Approaches for Personalized Search

In the previous section, we discussed some AI-based techniques in order to model the user needs and represent the Web documents in information systems. Further personalization approaches can be devised in order to improve the current human-computer interaction with Web search tools. In this section, we discuss two of these approaches: ontology-based search and adaptive focused crawling.

3.1   Ontology-Based Search

Ontologies and the Semantic Web[1] are two important research fields that are beginning to receive attention in the Web Intelligent Search context. Formal ontologies based on logic languages, such as OWL, are able to publish information in a machine readable format, supporting advanced Web search, software agents and knowledge management. Dolog *et al.* [16] are studying mechanisms based on logical mapping rules and description logics, which allow metadata and ontology concepts to be mapped to concepts stored in user profiles. This logical characterization enables the formalization of personalization techniques in a common language, such as *FOL*, and reasoning over the Semantic Web.

Nevertheless, assigning a unique semantic (meaning) to each piece of information on the Web and stating all the possible relationships among the available data is a difficult task to automate and impossible to perform for a human being. For this reason, some prototypes are based on hierarchical taxonomies, such as YAHOO or OPEN DIRECTORY PROJECT (ODP)[2] directories. In these kinds of ontologies, the objects are simply categories that contain references to Web documents, and their interrelationships allow making inferences and retrieving more relevant information than the IR keyword-based search.

---

[1] ttp://www.w3.org/2001/sw/

[2] http://www.dmoz.org

The goal is to create ontology-based user profiles and use these profiles to personalize the human-computer interaction with the Web, during a browsing session or when the user analyzes the search engine results.

The *misearch* system [51] improves search accuracy re-ranking the results obtained by an traditional search service by giving more importance to the documents related to topics contained in the query histories and/or examined search results. The user profiles are represented as weighted concept hierarchies. The OPEN DIRECTORY PROJECT is used as the reference concept hierarchy for the profiles.

The system collects for each user two different types of data collected by means of a wrapper of the search engine: the submitted queries, and the *snippets* of the results selected by the user. The approach is based on a classifier trained on the OPEN DIRECTORY PROJECT's hierarchy, which chooses the concepts most related to the collected information, assigning higher weights to them.

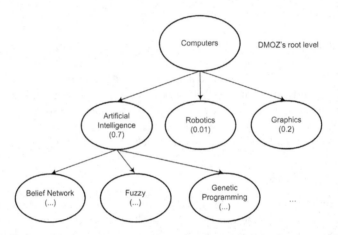

Fig. 1. The **misearch**'s user profile based on the ODP directory, where the categories are weighted according to what the user has seen in the past

A matching function calculates the degree of similarity between each of the concepts associated with result snippet j and the user profile i:

$$sim(user_i, doc_j) = \sum_{k=1}^{N} wp_{i,k} \cdot wd_{j,k} \qquad (1)$$

where $wp_{i,k}$ is the weight of the concept k in the user profile i, $wd_{j,k}$ is the weight of the concept k in the document j, and N is the number of concepts.

Finally, each document is assigned a weight, used for result re-ranking. The results that match the user's interests are placed higher in the list. These new ranks are drawn combining the previous degree of similarity with GOOGLE's original rank.

As for the performance measured in terms of the rank of the user-selected result, it improves by 33%. A user profile built from snippets of 30 user-selected results showed an improvement by 34%. Therefore, even though the text a user provides to the search engine is quite short, it is enough to provide more accurate, personalized results, and the implicit feedback reduces the time users need to spend to learn and keep their profiles up-to-date.

A similar approach is taken by Liu and Yu [32], where user profiles are built by analyzing the search history, both queries and selected result documents, comparing them to the first 3 levels of the ODP category hierarchy.

Because queries are usually very short, they are often ambiguous. For each query and the current context, the system assigns the most appropriate ODP categories. Afterwards, the system performs query expansion based on the top-matching category reducing the ambiguity of the results.

The categories in the user profile are represented by a weighted term vector, where a highly-weighted keyword indicates a high degree of association between that keyword and the category. The system updates the user profile after a query, when the user clicks on a document.

3.2 Adaptive Focused Crawling

Adaptive Focused crawling concerns the development of particular crawlers, that is, a software system that traverses the Web collecting HTML pages or other kinds of resources [43,53], able to find out and collect only Web pages that satisfy some specific topics [12]. This kind of crawler is able to build specialized/focused collections of documents reducing the computational resources needed to store them. Specialized search engines use those collections to retrieve valuable, most reliable and up-to-date documents related to the given topics. Vertical portals, personalized electronic newspapers [6], personal shopping agents [17] or conference monitoring services [26] are examples of implementations of those kinds of search engines in realistic scenarios.

New techniques to represent Web pages, such as the ones discussed in Sect. 2, and match these representations against the user's queries, such as algorithms based on Natural Language Processing (NLP), usually avoided due to the computational resources needed to run them, can be implemented owing to the reduced dimension of the document sets under consideration.

In this section, we discuss two recent approaches proposed in order to build focused crawlers based on sets of autonomous agents that wander the Web environment looking for interesting resources. The first approach is *InfoSpiders* [35] based on genetic algorithms where an evolving population of intelligent agents browses the Web driven by user queries. Each agent is able to draw the relevance of a resource with a given query and reason autonomously about future actions regarding the next pages to download and analyze. The goal is to mimic the intelligent browsing behavior of human users with little or no interaction among agents.

The agent's *genotype* consists of set of chromosomes, which determines its searching behavior. This very genotype is involved in the offspring generation process. The two principal components of the genotype are a set of keywords initialized with the query terms and a vector of real-valued weights, initialized randomly with uniform distribution, corresponding with the information stored in a feed-forward neural network. This network is used to judge what keywords in the first set best discriminate the documents relevant to the user. For each link to visit, the related text that is also included in the genotype set is extracted by the agent. The text is given as input to the neural network. The output of the net is used to draw a probability to choose and visit the given link. The outcome of the agent's behavior, as well as the user feedback, are used to train the neural network's weights. If any error occurs due to the agent's action selection, that is, visiting of irrelevant pages, the network's weights are updated through the backpropagation of error.

Mutations and crossovers among agents provide the second kind of adaptivity to the environment. Offspring are recombined by means of the crossover operation and, along with the mutation operator, they provide the needed variation to create agents that are able to behave better in the environment, retrieving an increased number of relevant resources.

A value corresponding to the agent's energy is assigned at the beginning of the search, and it is updated according to the relevance of the pages visited by that agent. The neural networks and the genetic operators aim at selecting the words that well describe the document that led to the energy increase, modifying the agents' behavior according to prior experience, learning to predict the best links to follow. The energy determines which agents are selected for reproduction and the ones to be killed amidst the population.

Ant foraging behavior research inspired a different approach for building focused crawlers. In [20,21] an adaptive and scalable Web search system is described, based on a multi-agent reactive architecture, derived from a model of social insect collective behavior [7].

Biologists and ethologists created this model to understand how blind animals, such as ants, are able to find out the shortest ways from their nest to the feeding sources and back. This phenomenon can be easily explained, since ants can release pheromone, a hormonal substance to mark the ground. In this way, they can leave a trail along the followed path. This pheromone allows other ants to follow the trails on the covered paths, reinforcing the released substance with their own.

The first ants returning to their nest from the feeding sources are those which chose the shortest paths. The back-and-forth trip (from the nest to the source and back) allows them to release pheromone on the path twice. The following ants leaving the nest are attracted by this chemical substance, therefore are more likely choose the one which has been frequently covered by the previous ants. For this reason, they will direct themselves towards the shortest paths.

In the proposed focused crawling algorithm each agent corresponds to a virtual ant that has the chance to move from the hypertextual resource where it is currently located to another, if there is a link between them. A sequence of links, i.e., pairs of urls, represents a possible agent's route where, at the end of each exploration, the pheromone trail could be released on. The available information for an agent when it is located in a certain resource is: the matching result of the resource with the user needs, e.g., query, and the amount of the pheromone on the paths corresponding to the outgoing links.

The ants employ the pheromone trail to communicate the exploration results one another: the more interesting are the resources an ant is able to find out, the more pheromone trail it leaves on the followed path. As long as a path carries relevant resources, the corresponding trail will be reinforced, thus increasing the number of attracted ants.

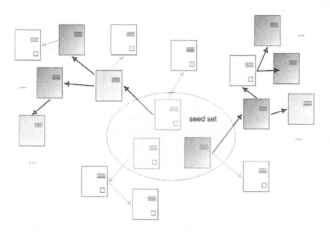

Fig. 2. Starting from a initial URL set, the agents look for interesting pages on the Web. If an ant finds a good page, a trail is left on the path. In this way, during the next cycles other ants will be attracted by the paths that head to good pages.

The crawling process is organized in cycles; in each one of them, the ants make a sequence of moves among the hypertextual resources. The maximum number of allowable moves varies proportionally according to the value of the current cycle. At the end of a cycle, the ants update the pheromone intensity values of the followed path.

The resources from which the exploration starts can be collected from the first results of a search engine, or the user's bookmarks, and correspond to the seed URLs. To select a particular link to follow, a generic ant located on the resource url_i at the cycle t, draws the *transition probability* value $P_{ij}(t)$ for every link contained in url_i that connects url_i to url_j. The $P_{ij}(t)$ is considered by the formula:

$$P_{ij}(t) = \frac{\tau_{ij}(t)}{\sum_{l:(i,l)\in E} \tau_{il}(t)} \qquad (2)$$

where $\tau_{ij}(t)$ corresponds to the pheromone trail between url_i and url_j, and $(i,l) \in E$ indicates the presence of a link from url_i to url_l.

At the end of a cycle, when the limit of moves per cycle is reached, the *trail updating process* is performed. The *updating rule* for the pheromone variation of the k-ant corresponds to the mean of the visited resource scores:

$$\Delta\tau^{(k)} = \frac{\sum_{j=1}^{|p^{(k)}|} score(p^{(k)}[j])}{|p^{(k)}|} \tag{3}$$

where $p^{(k)}$ is the ordered set of pages visited by the k-ant, $p^{(k)}[i]$ is the i-th element of $p^{(k)}$, and $score(p)$ is the function that, for each page p, returns the similarity measure with the current information needs: a $[0, 1]$ value, where 1 is the highest similarity.

Afterworlds, the τ values are updated. The τ_{ij} trail of the generic path from url_i to url_j at the cycle $t+1$ is affected by the ant's pheromone updating process, through the computed $\Delta\tau^{(k)}$ values:

$$\tau_{ij}(t+1) = \rho \cdot \tau_{ij}(t) + \sum_{k=1}^{M} \Delta\tau^{(k)} \tag{4}$$

ρ is the trail evaporation coefficient that avoids unlimited accumulation of substance caused by the repeated positive feedback. The summation widens to a subset of the N ants living in the environment.

Two empirical observations are included in the developed architecture: Web pages on a given topic are more likely to link to those on the same topic [14], and Web pages' contents are often not self-contained [39].

The Ant crawler has two forms of adaptivity: the first concerns the query refinement during the execution when the results are not so satisfactory or, in general, the user does not know how to look for a query able to express what he wants. The second form regards the environment instability due to the updates of Web pages. These two types of adaptivity are possible because the value of the pheromone intensities $\tau_{ij}(t)$ is updated at every cycle, according to the visited resource scores.

Besides Genetic algorithms and the Ant paradigm, further focused crawlers are based on AI techniques and approaches, such as reinforcement learning [46,11] or Bayesian statistical models [1].

4 Conclusions

Through the use of intelligent search, in particular techniques to represent user needs and Web documents, tailoring search engines' results, it is possible to enhance the human-computer interaction that provides useful information to the user.

AI-based knowledge representations prove their effectiveness in this context, outperforming the traditional and widespread IR approach. Moreover, ontologies can be employed to better understand and represent user needs, and intelligent search methods have been included in autonomous crawlers in order to retrieve up-to-date resources regarding particular topics.

Further techniques not mentioned in this chapter; nevertheless, Plan recognition (see for example [10]) and NLP are related to this context. The former attempts to recognize patterns in user behavior, analyzing aspects of their past behavior, in order to predict goals and their forthcoming actions. Language processing aims at understanding the meaning of Web content. How it relates to a user query can fuel the fire of further important research in the personalization domain.

References

1. C. C. Aggarwal, F. Al-Garawi, and P. S. Yu. Intelligent crawling on the world wide web with arbitrary predicates. In *Proceedings of the 10th World Wide Web Conference (WWW10)*, pages 96–105, Hong Kong, 2001.

2. L. Ambrosini, V. Cirillo, and A. Micarelli. A hybrid architecture for user-adapted information filtering on the world wide web. In A. Jameson, C. Paris, and C. Tasso, editors, *Proceedings of the 6th International Conference on User Modeling (UM97)*, pages 59–61. Springer, Berlin, 2–5 1997.

3. F. A. Asnicar and C. Tasso. ifWeb: a prototype of user model-based intelligent agent for document filtering and navigation in the world wide web. In *Proceedings of Workshop Adaptive Systems and User Modeling on the World Wide Web (UM97)*, pages 3–12, Sardinia, Italy, 1997.

4. R. Baeza-Yates and B. Ribeiro-Neto. *Modern Information Retrieval*. Addison-Wesley, 1999.

5. M. Balabanović and Y. Shoham. Fab: content-based, collaborative recommendation. *Communications of the ACM*, 40(3):66–72, 1997.

6. K. Bharat, T. Kamba, and M. Albers. Personalized, interactive news on the web. *Multimedia Syst.*, 6(5):349–358, 1998.

7. E. Bonabeau, M. Dorigo, and G. Theraulaz. Inspiration for optimization from social insect behavior. *Nature*, 406:39–42, 2000.

8. G. Boone. Concept features in re:agent, an intelligent email agent. In *Proceedings of the second international conference on Autonomous agents (AGENTS '98)*, pages 141–148, New York, NY, USA, 1998. ACM Press.

9. C. Buckley, G. Salton, and J. Allan. The effect of adding relevance information in a relevance feedback environment. In *SIGIR '94: Proceedings of the 17th annual international ACM SIGIR conference on Research and development in information retrieval*, pages 292–300, New York, NY, USA, 1994. Springer-Verlag New York, Inc.

10. S. Carberry. Techniques for plan recognition. *User Modeling and User-Adapted Interaction*, 11(1-2):31–48, 2001.

11. S. Chakrabarti, K. Punera, and M. Subramanyam. Accelerated focused crawling through online relevance feedback. In *WWW '02: Proceedings of the 11th international conference on World Wide Web*, pages 148–159, New York, NY, USA, 2002. ACM Press.

12. S. Chakrabarti, M. van den Berg, and B. Dom. Focused crawling: A new approach to topic-specific web resource discovery. In *Proceedings of the 8th World Wide Web Conference (WWW8)*, pages 1623–1640, Toronto, Canada, 1999.

13. W. S. Cooper. A definition of relevance for information retrieval. *Information Storage and Retrieval*, 7:19–37, 1971.

14. B. D. Davison. Topical locality in the web. In *SIGIR '00: Proceedings of the 23rd annual international ACM SIGIR conference on Research and development in information retrieval*, pages 272–279, New York, NY, USA, 2000. ACM Press.

15. S. C. Deerwester, S. T. Dumais, T. K. Landauer, G. W. Furnas, and R. A. Harsh-
man. Indexing by latent semantic analysis. *Journal of the American Society of
Information Science*, 41(6):391–407, 1990.

16. P. Dolog, N. Henze, W. Nejdl, and M. Sintek. Towards the adaptive semantic
web. In François Bry, Nicola Henze, and Jan Maluszynski, editors, *Principles
and Practice of Semantic Web Reasoning, International Workshop, PPSWR 2003,
Mumbai, India, December 8, 2003, Proceedings*, volume 2901 of *Lecture Notes in
Computer Science*, pages 51–68. Springer, 2003.

17. R. B. Doorenbos, O. Etzioni, and D. S. Weld. A scalable comparison-shopping agent
for the world-wide web. In *AGENTS '97: Proceedings of the first international
conference on Autonomous agents*, pages 39–48, New York, NY, USA, 1997. ACM
Press.

18. P. Frasconi, G. Soda, and A. Vullo. Text categorization for multi-page documents :
A hybrid naive bayes HMM approach. In *ACM/IEEE Joint Conference on Digital
Libraries, JCDL 2001*, Roanoke, VA, USA, June 2001. ACM.

19. R. Fung and B. Del Favero. Applying Bayesian networks to information retrieval.
Communications of the ACM, 38(3):42–48, March 1995.

20. F. Gasparetti and A. Micarelli. Adaptive web search based on a colony of cooperative distributed agents. In Matthias Klusch, Sascha Ossowski, Andrea Omicini,
and Heimo Laamanen, editors, *Cooperative Information Agents*, volume 2782, pages
168–183. Springer-Verlag, 2003.

21. F. Gasparetti and A. Micarelli. Swarm intelligence: Agents for adaptive web search.
In *Proceedings of the 16th European Conference on Artificial Intelligence (ECAI
2004)*, 2004.

22. G. Gentili, M. Marinilli, A. Micarelli, and Filippo Sciarrone. Text categorization in
an intelligent agent for filtering information on the web. *IJPRAI*, 15(3):527–549,
2001.

23. A. Gulli and A. Signorini. The indexable web is more than 11.5 billion pages. In
*WWW '05: Special interest tracks and posters of the 14th international conference
on World Wide Web*, pages 902–903, New York, NY, USA, 2005. ACM Press.

24. T. Joachims, D. Freitag, and T. M. Mitchell. Webwatcher: A tour guide for the
world wide web. In *Proceedings of the 15h International Conference on Artificial
Intelligence (IJCAI1997)*, pages 770–777, 1997.

25. D. Kelly and J. Teevan. Implicit feedback for inferring user preference: a bibliog-
raphy. *SIGIR Forum*, 37(2):18–28, 2003.

26. A. Kruger, C. L. Giles, F. M. Coetzee, E. Glover, G. W. Flake, S. Lawrence, and
C. Omlin. Deadliner: building a new niche search engine. In *CIKM '00: Proceedings
of the ninth international conference on Information and knowledge management*,
pages 272–281, New York, NY, USA, 2000. ACM Press.

27. B. Krulwich and C. Burkey. The contactfinder agent: Answering bulletin board
questions with referrals. In *Proceedings of the 13th National Conference on Artificial Intelligence and 8th Innovative Applications of Artificial Intelligence Conference, AAAI96, IAAI96*, pages 10–15, 1996.

28. M. Lan, C.-L. Tan, H.-B. Low, and S.-Y. Sung. A comprehensive comparative study
on term weighting schemes for text categorization with support vector machines. In
*WWW '05: Special interest tracks and posters of the 14th international conference
on World Wide Web*, pages 1032–1033, New York, NY, USA, 2005. ACM Press.

29. A. Y. Levy and D. S. Weld. Intelligent internet systems. *Artificial Intelligence*,
118(1-2):1–14, 2000.

30. H. Lieberman. Letizia: An agent that assists web browsing. In Chris S. Mellish, editor, *Proceedings of the 14th International Joint Conference on Artificial Intelligence (IJCAI1995)*, pages 924–929, Montreal, Quebec, Canada, 1995. Morgan Kaufmann publishers Inc.: San Mateo, CA, USA.

31. H. Lieberman, N. W. Van Dyke, and A. S. Vivacqua. Let's browse: a collaborative web browsing agent. In *Proceedings of the 4th International Conference on Intelligent User Interfaces (IUI'99)*, pages 65–68, Los Angeles, CA, USA, 1998. ACM Press.

32. F. Liu, C. Yu, and W. Meng. Personalized web search for improving retrieval effectiveness. *IEEE Transactions on Knowledge and Data Engineering*, 16(1):28–40, 2004.

33. B. Magnini and . Strapparava. User modelling for news web sites with word sense based techniques. *User Modeling and User-Adapted Interaction*, 14(2):239–257, 2004.

34. Andrew McCallum and K. Nigam. A comparison of event models for naive bayes text classification. In *Proceedings of AAAI-98, Workshop on Learning for Text Categorization*, 1998.

35. F. Menczer and R. K. Belew. Adaptive retrieval agents: Internalizing local context and scaling up to the web. *Machine Learning*, 31(11–16):1653–1665, 2000.

36. A. Micarelli and F. Sciarrone. Anatomy and empirical evaluation of an adaptive web-based information filtering system. *User Modeling and User-Adapted Interaction*, 14(2-3):159–200, 2004.

37. G. A. Miller and C. Fellbaum. Lexical and conceptual semantics. In B. Levin and S. Pinker, editors, *Advances in Neural Information Processing Systems*, pages 197–229. Blackwell, Cambridge and Oxford, England, 1993.

38. M. Minsky. A framework for representing knowledge. Technical report, Massachusetts Institute of Technology, Cambridge, MA, USA, 1974.

39. Y. Mizuuchi and K. Tajima. Finding context paths for web pages. In *Proceedings of the 10th ACM Conference on Hypertext and Hypermedia: Returning to Our Diverse Roots (HYPERTEXT99)*, pages 13–22, Darmstadt, Germany, 1999.

40. A. Moukas and P. Maes. Amalthaea: An evolving multi-agent information filtering and discovery system for the WWW. *Autonomous Agents and Multi-Agent Systems*, 1(1):59–88, 1998.

41. M. J. Pazzani, J. Muramatsu, and D. Billsus. Syskill webert: Identifying interesting web sites. In *Proceedings of the National Conference on Artificial Intelligence (AAAI-96)*, pages 54–61, Portland, OR, USA, August 1996. AAAI Press.

42. J. Pearl. *Probabilistic Reasoning in Intelligent Systems: Networks of Plausible Inference*. Morgan Kaufmann, San Mateo, 1988.

43. B. Pinkerton. Finding what people want: Experiences with the webcrawler. In *Proceedings of the 2nd World Wide Web Conference(WWW2)*, Chicago, USA, 1994.

44. M. H. Pryor. The effects of singular value decomposition on collaborative filtering. Technical report, Hanover, NH, USA, 1998.

45. R. M. Quillian. Semantic memory. In Marvin Minsky, editor, *Semantic information processing*, pages 216–270. The MIT Press, Cambridge, MA, USA, 2–5 1968.

46. J. Rennie and A. McCallum. Using reinforcement learning to spider the web efficiently. In *ICML '99: Proceedings of the Sixteenth International Conference on Machine Learning*, pages 335–343, San Francisco, CA, USA, 1999. Morgan Kaufmann Publishers Inc.

47. S. Robertson and S. Walker. On relevance weights with little relevance information. In *Proceedings of the 20th Annual International ACM SIGIR Conference on Research and Development in Information Retrieval*, Relevance Feedback, pages 16–24, 1997.

48. B. M. Sarwar, G. Karypis, J. A. Konstan, and J. T. Riedl. Application of dimensionality reduction in recommender systems - a case study. In *Web Mining for E-Commerce – Challenges and Opportunities*, Boston, MA, USA, 2000.

49. J. Savoy and J. Picard. Retrieval effectiveness on the web. *Information Processing & Management*, 37(4):543–569, 2001.

50. B. Smyth, E. Balfe, J. Freyne, P. Briggs, M. Coyle, and O. Boydell. Exploiting query repetition and regularity in an adaptive community-based web search engine. *User Modeling and User-Adapted Interaction*, 14(5):383–423, 2005.

51. M. Speretta and S. Gauch. Personalized search based on user search histories. In *Web Intelligence (WI2005)*, France, 2005. IEEE Computer Society.

52. J.-T. Sun, H.-J. Zeng, H. Liu, Y. Lu, and Z. Chen. Cubesvd: a novel approach to personalized web search. In *WWW '05: Proceedings of the 14th international conference on World Wide Web*, pages 382–390, New York, NY, USA, 2005. ACM Press.

53. B. Yuwono, S. L. Y. Lam, J. H. Ying, and D. L. Lee. A World Wide Web resource discovery system. In *Proceedings of the 4th World Wide Web Conference (WWW4)*, pages 145–158, Boston, Massachusetts, USA, 1995.

Cracking Crosswords:
The Computer Challenge

Marco Gori, Marco Ernandes, and Giovanni Angelini

Dipartimento di Ingegneria dell'Informazione
Via Roma 56, 53100 Siena, Italy
marco@dii.unisi.it

Abstract. Crosswords is over 90 years old, yet it is still one of the most popular puzzles around the world. It is in fact a linguistic game which requires a wide knowledge in different domains and the ability to crack enigmatic clues, that are often regarded as inherent human capabilities. Unlike chess, crossword solving does not require strong skills for the actuation of strategic plans, but the linguistic specifications is in itself a source of enormous difficulty for machines.

This paper discusses the problem of automatic crossword solving with special emphasis to the WebCrow project carried out at the University of Siena. After a brief historical description of the evolution of crosswords, the paper gives a formalization of the main problems to be faced and provides a number of relevant architectural issues behind cracking crosswords. In particular, it is claimed that the Web is likely to be the most important source for the development of challenging programs based on clue answering, a sort of question answering mechanism in which the machine is expected to return candidate word solutions.

1   Introduction

Crossword is an extremely popular word game evolved from a long line of word games. The evidence of crossword-like word play dates from the first century A.D. with the word square cryptic game (see Fig. 1), which was found carved in stone on the wall of a building in Pompeii. This particular square, which can be read four ways (left to right, right to left, top to bottom, and bottom to top), is often translated as "Arepo, the sower, watches over his works." The actual significance is not completely understood yet. Another early word square is the Moschion stele, dated around A.D. 300. In the stele, Moschion is honoring Osiris - Egyptian god of the underworld - with this monument, which, again, contains words that one can read in different directions. In the mid-1800's, the clue was introduced but with the definitions had not the modern sense, yet. In 1875, St. Nicholas magazine ran a puzzle with a small grid; for the first time, the across and the down were different, so getting close to nowadays crosswords. However, the truly nature of the popular puzzle only emerged years later when Arthur Wynne, a British journalist based in Liverpool, published the first crosswords on Sunday the 21th of December, 1913 in the New York World. It had a nice diamond-like structure with across and down clues to define the words; one of

O. Stock and M. Schaerf (Eds.): Aiello Festschrift, LNAI 4155, pp. 265–286, 2006.

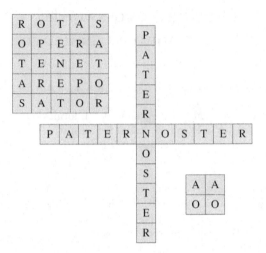

Fig. 1. The enigmatic rotas square. Amongst many interpretation, the "Pater Noster anagram is one of the most popular.

them was filled out to make it clear how to solve the puzzle (see Fig. 2). Not surprisingly, that word was *fun*!

1.1 Motivations and Relevant Literature

Crosswords is probably the most played language puzzle wordwide and provides a very challenging game for human intelligence. *La Settimana Enigmistica*, the main Italian crossword magazine, sells over one million copies weekly. It is estimated that over 50 million Americans solve crosswords[1] with frequency.

Problems like solving crosswords from clues are reputed as AI-complete [7]. This enormous complexity is due to its semantics and the large amount of encyclopaedic knowledge required. The interest in cracking crosswords in the field of artificial intelligence started developing only recently. The first experience reported in the literature is the Proverb system [5] that reached human-like performances on American crosswords using a great number of knowledge-specific expert modules and a crossword database of great dimensions[2].

The recent developments in searching the Web are offering new opportunities to machines to enfold with semantics real-life concepts. That was the motivation for launching the WebCrow project at the University of Siena, where crosswords are attacked (within competition time limits) making use of the Web as its primary source of knowledge. This represents a different approach with respect to Proverb, because WebCrow does not possess any knowledge-specific expert

---

[1] It has been recently observed that this sort of activity helps to prevent from developing mental decline

[2] Before Proverb, AI limited its analysis to the *crossword generation* problem [3]. This makes a closed-world assumption by requiring a predefined dictionary of legal words and results to be an NP-complete task that can be solved in a few seconds.

FUN'S Word-Cross Puzzle.

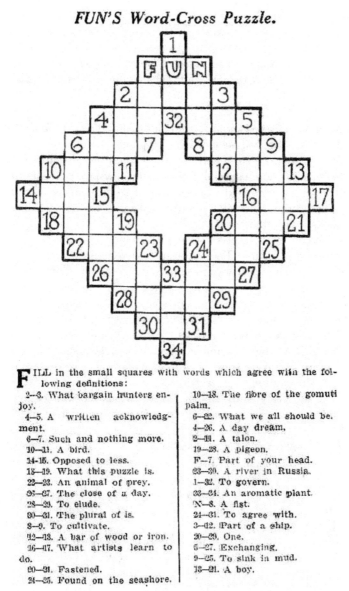

FILL in the small squares with words which agree with the following definitions:

2—3. What bargain hunters enjoy.

4—5. A written acknowledgment.

6—7. Such and nothing more.

10—11. A bird.

14—15. Opposed to less.

18—19. What this puzzle is.

22—23. An animal of prey.

26—27. The close of a day.

28—29. To elude.

30—31. The plural of is.

8—9. To cultivate.

12—13. A bar of wood or iron.

16—17. What artists learn to do.

20—21. Fastened.

24—25. Found on the seashore.

10—18. The fibre of the gomuti palm.

6—22. What we all should be.

4—26. A day dream.

2—11. A talon.

19—28. A pigeon.

F—7. Part of your head.

23—30. A river in Russia.

1—32. To govern.

33—34. An aromatic plant.

N—8. A fist.

24—31. To agree with.

3—12. Part of a ship.

20—29. One.

5—27. Exchanging.

9—25. To sink in mud.

13—21. A boy.

Fig. 2. The first crosswords was created by British journalist Arthur Wynne and appeared in Sunday newspaper New York World, on December 21, 1913

module. Nevertheless, in order to assure the system rubustness to all sorts of clues, WebCrow makes also use of a strict set of other useful modules, which includes a dictionary and a small CrossWord DataBase[3] (CWDB). The web-

---

[3] The database used by Proverb was about one order of magnitude greater than ours.

based clue-answering paradigm aspires to stress the generality of WebCrow's knowledge and its language-independency. We will show in this paper that Web Search, thanks to the fact that in clue-answering we priorly know the exact length of the correct answer, can produce extremely effective results providing the most important source of knowledge for the clue-answering process.

1.2   Problem Setting and Results

Italian crosswords tend to be extremely difficult to handle because they contain a great quantity of word plays, neologisms, compound words, ambiguities and a deep involvement in socio-cultural and political topics, often treated with irony. The latter is a phenomenon, especially present in newspapers, that introduces an additional degree of complexity in crossword solving since it requires the possession of a very broad and fresh knowledge that is also robust to volunteer language vagueness and ambiguity.

We have collected 685 examples of solved Italian crosswords, each one containing an average of 62.7 clues. These examples were mainly obtained from two sources: the main Italian crossword magazine *La Settimana Enigmistica* (due to its popularity and history, this publisher sets a probable standard for Italian crosswords) and an important on-line newspaper's crossword section, *La Repubblica*. Other examples were downloaded from crossword-dedicated web sites. Sixty crosswords (3685 clues, avg. 61.4 each) were randomly extracted from these subsets in order to form the experimental test suite. The remaining crosswords constituted a database (CWDB) of clue-target pairs that was used as an aid for the generation of the candidate-answer lists.

Given this test set WebCrow's challenge was to answer all the clues and to subsequently fill the slots with the highest percentage of correct words. As in many human competitions a 15 minutes time limit was given for each example.

The version of WebCrow that is discussed here is basic but it has already given very promising results. In over two thirds of the clues the correct answer was found by the Web Search Module whithin the downloaded documents and in some cases (nearly 15%) this answer was the most probable (i.e., appearing a the top of the list). The addition of the other modules has raised the coverage to 99% and the probability of having the targeted word in first position to over 35%. Finally, solving the Constraint Satisfaction problem by filling the crossword puzzle, WebCrow averaged on the overall test set around 70% words correct and 80% letters correct. On the examples that experts consider "easy", as the examples from the cover pages of *La Settimana Enigmistica*, WebCrow performed with 80,0% words correct (100% in one case) and 90.1% letters correct. On more difficult examples the percentage of correct words was steadily above 60%: 67,6% with la *La Settimana Enigmistica* (81% letters) and 62.9% with *La Repubblica* (73% letters).

2   The System Architecture

WebCrow is a modular-based system. Therefore, it is also possible to plug in additional *ad hoc* modules in order to increase the system's performances. A sketch of WebCrow's architecture is given in figure 3.

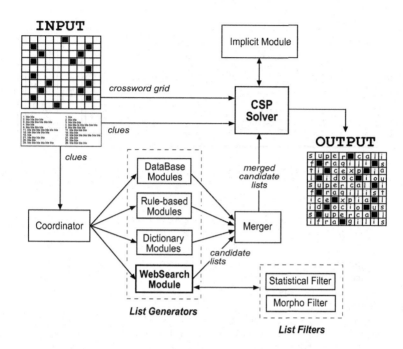

Fig. 3. WebCrow. A general overview of WebCrow's architecture, which was inspired by Proverb's.

The WebCrow solving process can be devided in two phases. During the first one, all the clues of a puzzle are passed by the coordinator to all the "List Generator" modules. Each of them returns for each clue a list of possible solutions. All lists are then merged by the "Merger", using the confidence values of each list and the probabilities associated to each candidate of a list. At the end of this phase, a unique list of candidate-probality pairs is generated for each clue. Finally, WebCrow has to face a constrain-satisfaction problem. From each clue list a candidate has to be chosen and inserted in the crossword-puzzle, trying to satisfy the intrinsic constrains. The aim of this phase is to find an admissible solution which maximize the number of correct words inserted.

3 Using the Web for Clue-Answering: The Web Search Module

The objective of the Web Search Module (WSM) is to find sensible answers to crossword clues, that are expressed in natural language, by exploting the Web and search engines (SE). This task recalls that of a Web-based Question Answering system. However, with crossword clues, the nature of the problem changes sensibly, often becoming more challenging than classic QA[4]. The main differences are:

---

[4] The main reference for standard QA is the TREC competition [16].

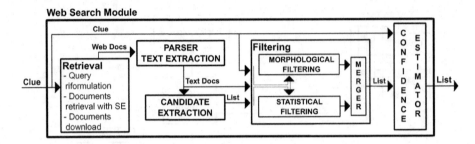

Fig. 4. Web Search Module. A sketch of the internal architecture of the Web Search Module.

- clues are mostly formulated in a non-interrogative form (i.e: ≺*La voce narrante del Nome della Rosa:* **adso**≻[5]) making the task of determining the answer-type more subtle.
- clues can be voluntarily ambiguous and misleading (i.e: ≺*Quello liquido non attacca:* **scotch**≻[6])
- the topic of the questions can be both factoid[7] and non-factoid (i.e: ≺*Ci si va al buio:* **cinema**≻[8])
- there is a unique and precise correct answer: a single word or a compound word (i.e. ≺*Ha cambiato il linguaggio della tv:* **ilgrandefratello**≻[9]), whereas in QA the answers are usually a 50/250-byte-passage in which the target has to be recognizable by humans.
- the list of candidate answers requires also a global confidence score, expressing the probability that the target is within the list.

The only evident advantage in crossword solving is that we priorly know the exact length of the words that we are seeking. We believe that, thanks to this property, web search can be extremely effective and produce a strong clue-answering.

The inner architecture of the WSM is sketched in figure 4. There are four task that have to be accomplished by the WSM: the retrieval of useful web documents, the extraction of the answer candidates from these documents, the scoring/filtering of the candidate lists and, finally, the estimation of the list confidence. In this section all these components will be presented and analyzed.

Despite the fact that the WSM has been implemented only in a basic version, it is clear that this module, among the set of expert modules used by WebCrow, produces the most impressive answering performances, with the best coverage/precision balance. This is evident if we observe tab. 1 (first two columns). In over half of cases the correct answer is found within the first 100 candidates inside a list containing more than 10^5 words.

---

[5] ≺*The background narrator of the Name of the Rose:* **adso**≻
[6] ≺*The liquid one does not stick:* '**scotch**≻, the clue ambiguously refers to the two senses of the target: scotch-whisky and scotch-tape.
[7] As TREC questions, like *Who was the first American in space?*
[8] ≺*We go there in the darkness:* **cinema**≻
[9] ≺*It changed the language used in television:* **thebigbrother**≻

Table 1. Module's coverage (cvr). 1-pos gives the freq. of the target in first position, **5-pos** the freq. within the first five candidates, **100-pos** within the first hundred. **Len** is the avg. list length. The number of documents used the WSM is reported in brackets. Test Set: 60 crosswords (3685 clues).

Module	cvr	1-pos	5-pos	100-pos	Length
WEB (30 docs)	68.1	13.5	23.7	53.2	499
WEB (50 docs)	71.7	13.6	24.0	54.1	735
CWDB-EXACT	19.8	19.6	19.8	19.8	1.1
CWDB-PARTIAL	29.0	10.6	20.1	28.4	45.5
CWDB-DICTIO	71.1	0.4	2.1	21.5	$>10^3$
RULE-BASED	10.1	6.9	8.3	10.1	12.4
DICTIONARY	97.5	0.3	1.6	21.3	$>10^4$
ALL BUT WEB	98.4	34.0	43.6	52.3	$>10^4$
ALL (30 docs)	**99.3**	**36.5**	**50.4**	**72.1**	$>10^4$
ALL (50 docs)	99.4	36.6	52.2	73.0	$>10^4$

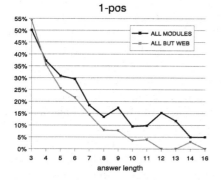

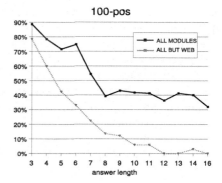

Fig. 5. Target in first position. The frequency of the target in first position in relation to its length with and without the WSM.

Fig. 6. Target in first 100 positions. The frequency of the target in the first 100 positions in relation to its length with and without the WSM.

The contribution of the WSM can be appreciated in the last three rows of tab. 1 where we can observe the loss of performance of the whole system when the WSM is removed. The overall coverage of the system is mainly guaranteed by the dictionary module (sec. 4.4), but the introduction of the WSM is fundamental to increase sensibly the rank of the correct answer. Also interesting is fig. 5 and fig. 6 where we take into consideration the length of the target. It can easily be observed that the WSM guarantees the system to well perform even with long word targets, which are of great importance in the CSP phase.

As it can be seen in table 2 the coverage of the WSM's lists grows sensibly with the first increments in the number of retrieved documents. This growth is imperceptible after 100 docs. We found that optimal balance in the trade off between precision, coverage and time cost is reached using 30 docs. We took this as the standard quantity of sources to be used in the experiments because it allows WebCrow to fulfill the time limit of 15 minutes.

Table 3 gives an insight of the quality of WSM's answers. Several examples of clues are reported along with a small portion of the correspondent candidate list.

3.1   Retrieving Useful Documents

The first goal of the answering process is to retrieve the documents that are better related to the clue. This can be done thanks to the fundamental contribution of search engine's technology (Google^{TM} was used in our testing). In order to increase the informativeness of the search engine the clues go through a reformulation/expansion step. Each clue $C = \{t_1 t_2 ... t_n\}$ generates a maximum of 3 queries: $Q_1 =< t_1 \wedge t_2 \wedge ... t_n >$, $Q_2 =< t_1 \vee t_2 \vee ... t_n >$ and $Q_3 =< (t_1^1 \vee t_1^2 \vee ...) \wedge (t_2^1 \vee t_2^2 \vee ...) \wedge ... (t_n^1 \vee t_n^2 \vee ...) >$ where t_n^i is the i-th derivation (i.e. changing the number, the gender, ...) of term t_n. Q_3 has not been implemented yet. Non informative words are removed from the queries.

A classic QA approach is to make use only of the document snippets in order to stress time efficiency. Unfortunately the properties of the clues make this approach useless (the probability of finding the correct answer to a crossword clue within a snippet has been experimentally observed below 10%) and we decided for a full-document approach.

The interrogation of the search engine and the download the documents represent two tasks that are extremely time consuming, absorbing easily over 90% of time in the entire clue-answering process. Therefore we have implemented it in a highly parallel manner: the WSM simultaneously downloads tens of documents (for one or more clues at a time) adopting a strict time-out for each http request (20 secs.). If a request reaches the time-out then the WSM asks the search engine for a cached copy of the document. If this is unavailable then the document is declared missed and an additional link is requested to the SE.

Table 2. WSM's performance. The performances of the WSM are here reported. The number of documents used is reported in brackets. SF=statistical filter, MF=morphological filter. Time is reported in min:secs. *The growth of the coverage is due the NI submodule.

#docs + filters	Cover	1-pos	5-pos	100-pos	Time
5+SF	46.4	11.1	19.1	41.7	1:25
10+SF	56.2	12.2	21.6	47.5	2:45
20+SF	63.7	12.3	22.1	50.3	5:30
30+SF	67.9	12.3	22.2	52.3	8:10
50+SF	71.6	12.2	22.0	53.4	13:30
100+SF	74.5	11.9	21.5	53.2	26:50
30+SF+MF	**68.1***	**13.5**	**23.7**	**53.2**	**8:45**
50+SF+MF	71.7*	13.6	24.0	54.1	14:15

Table 3. Some examples. On the left: some examples of clues that are "easy" to answer for the WSM. The correct answer is present in the very first candidates. The easiest examples are usually the clues where the topic is directly addressed and where the answer is a factoid. On the write there is a list of "tough" clues. These are tipically very general or ambiguous and the WSM fails to place the correct answer at the head of the list. Nevertheless, all the answers that the system produces tend to be semantically related to the target and to the clue.

"Easy" clues	"Tough" clues
≺*Confina con l'Abruzzo:* **molise**≻	≺*Caratteristica del burlone:* **giocosita**≻
1:*molise* 2:aquila 3:marche 4:umbria	1:simpatica 2:sicurezza 3:compagnia
≺*Il von Klein scrittore:* **heinrich**≻	≺*Documenti per minorenni:* **patentini**≻
1:*heinrich* 2:giovanni 3:kohlhaas	1:necessari 2:richiesti 3:organismi
≺*Atomi elettrizzati:* **ioni**≻	≺*Il verbo di chi ha coraggio:* **lanciarsi**≻
1:*ioni* 2:poli 3:essi 4:sali 5: rame	1:interiore 2:predicato 3:idealismo
≺*Mal d'orecchi:* **otite**≻	≺*Lasciare un segno:* **intaccare**≻
1:*otite* 2:ictus 3:otiti 4:edemi 5:gocce	1:passaggio 2:possibile 3:segnalare
≺*Lo parlano anche in Austria:* **tedesco**≻	≺*Non ha gusto in bocca:* **insapore**≻
1:inglese 2:*tedesco* 3:milione 4:skiroll	1:dialetto 2:prodotto 3:zucchero
≺*Un film di Nanni Moretti:* **carodiario**≻	≺*Larga e comoda:* **ampia**≻
1:palombella 2:portaborse 3:*carodiario*	1:bella 2:sella 3:barca 4:scala 5:valle
≺*Il piú famoso dei Keaton:* **buster**≻	≺*Sembrano ridere:* **iene**≻
1:comico 2:cinema 3:grande 4:*buster*	1:anni 2:loro 3:rane 4:rami 5:voci
≺*Il Giuseppe pittore di Barletta:* **denittis**≻	≺*Una sciagura attraente:* **calamita**≻
1:leontine 2:molfetta 3:ritratto 4:*denittis*	1:passione 2:alcolico 3:fardello

For each example of our test suite we have produced a full retrieval session with a maximum of 200 docs per clue (max. 30 docs with Q_2). 615589 docs were downloaded in 44h 36min (bandwith: 1Mb/s, effective $\approx$100KB/sec, avg. 230 docs/min, 167 docs/clue, 25.6KB/doc). All the test sessions were subsequently made offline exploiting this *web image*.

3.2   Extracting and Ranking the Candidates

The process of generating a list of candidate answers given a collection of relevant documents goes through two important steps. First, the documents are analyzed by a parser which produces as output plain ASCII text[10]. Second, this text is passed to a list generator that extracts the words of the correct length, eliminates doubles and produces an unweighted candidate list. In order to increase the coverage, a list of conpound words (i.e., a sequence of adjacent words fullfilling the lenght requirement) is generated from each document. To avoid noisy information, compound words which occurs only once are ommited.

Both outputs are then passed to two submodules: a statistical filter, based on IR techniques, and a morphological filter, based on machine learning and NLP techniques. Both have been embedded in the WSM.

---

[10] Currenly, the parser handles only HTML scripts. We are planning to implement a PDF parser in the next future.

The candidates are ranked by merging together the information provided by the two list filters. The score-probability associated to each word candidate w is given by

$$p(w, C) = c\,(sf\text{-}score(w, C) \times mf\text{-}score(w, C)) \tag{1}$$

where $sf\text{-}score(w, C)$ is the score attributed to word w by the statistical filter, $mf\text{-}score(w, C)$ is the score provided by the morphological filter, c is the normalizing factor that fulfills the probability requirement $\sum_{i=0}^{n} p(w_n, C) = 1$.

In QA systems it is important to produce very high precision only in the very first (3-5) answer candidates, since a human user will not look further down in the list. For this reason NLP techniques are typically used to remove those answers that are not likely correct. This answer selection policy is not well suited for clue-answering, a more conservative approach is required because the lack of the correct answer makes a greater damage than a low precision. The eq. 1 serves this goal: words that have low scores will appear at the bottom of the list but will not be dropped.

Our future objective is to implement a full battery of filters that can be added to the two already implement: stylistic, morpho-syntactical, lexical and logical. We believe that a robust NLP system could be of great impact in the answering of the clues. In addition to this we are designing a clue classifier that will enable WebCrow's module coordinator to understand when the web search can really be fruitfull and when, conversely, this should not be triggered.

3.3   The Statistical Filtering

This submodule makes use of three types of information: a query (generated by the reformulation of a clue), a collection of ranked documents (parsed and cleaned) provided by the search engine and a list of candidate answers extracted from the documents. We represent this information with the triple (w, Q^n, D_i) where w is a word of the correct length, Q^n (n-th reformulation of clue C) is the query that is given as input to the SE and D_i is the i-th document (containing word w) provided as output by the SE. An additional element is used, $rank(D_i, Q^n)$: the document ranking. To obtain this score we use the position of D_i in the Google's output and then compute $\log(1/pos(D_i))$. It has to be

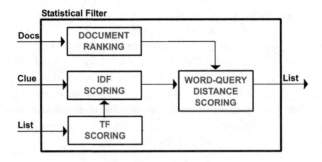

Fig. 7. Statistical Filter. A sketch of the internal architecture of the Statistical Filter.

noted that i does not strictly correspond to $pos(D_i)$ because whenever a document is missed for some reasons (non parsable format, http errors, etc.), the systems looks further down in the list in order to maintain constant the quantity of usable documents.

Finally, we attribute a global score to each word extracted from the documents in the following way:

$$sf\text{-}score(w, Q^n) = \sum_{i=0}^{\#docs} \left(\frac{score(w, Q^n, D_i)}{length(D_i)} rank(D_i, Q^n) \right) \qquad (2)$$

where $length(D_i)$ is the number of words in D_i. The score of a word within a single document is computed in a TF-IDF fashion. TF has been modified in order to take into account the inner-document distance between the word and the query. As shown in eq. 3, each occurrence of a word counts $1/dist(w, Q, D_i)$, whereas in normal TF each occurrence counts equally.

$$score(w, Q, D_i) = idf(w) \sum_{occ(w) \in D_i} \frac{1}{dist(w, Q, D_i)} \qquad (3)$$

$idf(w)$ is the classic inverse document frequency, which provides an immediate interpretation of term specificity. For compound words we take the highest idf value of the word components. $occ(w) \in D_i$ represents all the occurrences of the word w in the document D_i. The distance between word w and query Q is computed as a modified version of the square-root-mean distance between w and each term w_{Q_t} of the query, suggested by [6]. The main bias of the original formula was to weight equally all the words of the query without taking into account that some words are more informative than others. As shown in eq. 4, we decided to overcome this problem by tuning the exponential factor of the square-root-mean distance using the idf value of w_{Q_t} (normalized between 1 and 3). This increases the relevance of those answer candidates that are close to the more informative terms in the query. This novel contribution has resulted experimentally more effective for our goals.

$$dist(w, Q, D_i) = \frac{\sqrt{\sum_{t=0}^{\#terms \in Q} (dist(w, w_{Q_t}, D_i))^{idf(w_{Q_t})}}}{\#terms \in Q} \qquad (4)$$

$dist(w, w_{Q_t}, D_i)$ denotes the distance between word w and word w_{Q_t} in document D_i. In our implementation the distance between two words, whithin a single document, is given by the minimum number of words that separate them. After a preliminary testing we decided to limit to 150 words the maximum word-word distance. A default distance of 300 is assigned to those words that exceed this limit. It is legitimate to believe that outside a certain window the semantic link between two words is unpredictable.

This distance metric could be furtherly improved (i.e. taking into account sentences, paragraphs, titles, punctuations, etc.) but it already provides a very informative tool.

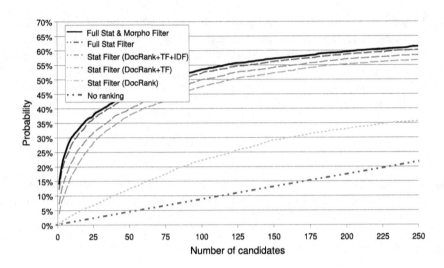

Fig. 8. Filtering perfomance. Probability of finding the correct answer as a function of the number of candidates that are taken into account. Looking further down in the candidate list, the probability of retrieving the target answer increases.

Other improvements could be obtained using a crossword-focused *idf* function (the *idf* values used here were obtained through a non-focused crawling session) or making use of the context in which each candidate appears. Figure 8 shows the contribution of all the elements used within the statistical filter. In a non ranked list the probability of finding the correct answer increases linearly with the number of candidates taken into consideration. If we rank the candidates for their TF value the probability increases for those words that are better placed in the list. It is easy to observe in figure 8 how the performances increase shifting from a basic filter to the full one which includes both the stastistical and morphological information.

3.4   The Morphological Filtering

The aim of this filter is to rank the candidates according to the morphological class they belong to. For this reason we made use of a Part-of-Speech (PoS) tagger, which associates a morphological class to each word of a sentence. Figure 9 shows the information flow of the morphological filter. The PoS tagger is used to tagged both the clue and each document related to it. Afterwards, the clue is processed by a multiclass classifier, which returns a weighted vector of the possible morphological classes the solution can belong to. Finally, for each word of the candidate list its morphological score is calculated by:

$$mf\text{-}score(w, C) = \sum_{i=0}^{\#\text{tags}} p(tag_i|w)score(tag_i, C) \tag{5}$$

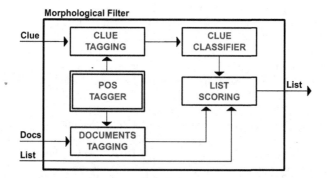

Fig. 9. Morphological Filter. A sketch of the internal architecture of the Morphological Filter.

$p(tag_i|w)$ is the information provided by the PoS-Tagger, $score(tag_i, C)$ is computed using the output of the classifier with the addition of a spread factor in order to enhance the impact of the classification.

With the attempt to maintain a strong language-independency we choosed an automatic trainable PoS tagger, called TreeTagger [12] [13], which is an extension of a basic Markov Model tagger. The TreeTager is based on two parts: a Lexicon and a Decision tree. Each word is first tagged using the Lexicon, which makes use also of a Prefix tree and a Suffix tree. This two trees are binary decision trees, generated by the training examples, which infer the possible tag of a word by examining, respectevly, its beginning or ending. Finally, a binary decision tree is used. This takes into account the tags of the k preceding words and returns a vector of the probable tags, based on the examples seen in the training corpus. We used 23 different classes to distinguish: articles, nouns and adjectives, verbs, adverbs, particles, interlocutory words, numbers, punctuation marks, abbreviations and others. A detailed list is given in table 4. At first, the TreeTagger was trained using an automatically extracted corpus form TUT [1]. The tagger was then used to tag a new corpus based on some CWDB's clues and documents from the web. This new corpus was corrected and added to the first one. Finally, the TreeTagger was retrained, obtaining an accurancy of about 93% on a cross validation test set.

The clue classifier was built using multiclass Kernel-based Vector Machine [15] [2]. First, a training set was created by extracting about 7000 clue-target pairs from the CWDB. Each clue was tagged by the TreeTagger and a feature vector $\bar{x} \in \mathbb{R}^n$ was then automatically generated for each example. The features extracted from each clue-answer pair were: the length of the target, the number of words in the clue, the number of capital letters in the clue, a set of the 250 most frequent clue-words and the probability tag vector accociated to each word of the clue. Finally, a target class $i \in \{1, \ldots, k\}$ was accociated to each example. We made use of 21 different target classes: almost all the morphological ones with the addition of name intials (IP) and non-semantic words (NS). A detailed list is shown in table 6.

Table 4. Morphological classes. This is the full list of the morphological classes used in our PoS Tagger. It differes from usual PoS tagging lists as the choise was to stress information relevant for finding the solution of a clue.

class	description	class	description
MS	Noun or Adj. or Pron., masc. sing.	AFP	Article, feminine plural
FS	Noun or Adj. or Pron., fem. sing.	AV	Adverb
MP	Noun or Adj. or Pron., masc. pl.	PART	Particle
FP	Noun or Adj. or Pron., fem. pl.	NUM	Number
NP	Proper Noun	EP	Interlocutory words
VS	Verb, cong. singular	ABBR	Abbreviation
VP	Verb, cong. plural	PC	Compound Words
VI	Verb, base form	SCRIPT	Script words in html doc.
VOTHER	Verb, other	SENT2	Punctuation, all the others
AMS	Article, masculine singular	SENT	Punctuation a the end
AFS	Article, feminine singular		of a sentence
AMP	Article, masculine plural	OTHER	all the rest

Our classifier is based on a multiclass Kernel-based Vector Machine, whose aim is to learn a linear function $H : X \rightarrow Y$ of the type $H(\bar{x}, M) = \langle M, \Phi(\bar{x}) \rangle$, where the predicted class is given by the function

$$f(\bar{x}) = \operatorname*{argmax}_{i \in \{1,\dots,k\}} H_i(\bar{x}, M) \tag{6}$$

$H_i(\bar{x}, M) = y_i$ is the i-th entry of the vector $\bar{y} = H(\bar{x}, M)$, corresponding to the score given to the class i. The goal is to minimize the empirical risk over all the training examples

$$\mathcal{R}(f) = \sum_t \Delta(y_t, f(\bar{x}_t)) \tag{7}$$

where $\Delta(y_t, \hat{y}_t)$ is the loss associated to the predicted class $\hat{y}_t = f(\bar{x}_t)$. $\Delta(y_t, \hat{y}_t) = 0$ if $y_t = \hat{y}_t$. Instead, $\Delta(y_t, \hat{y}_t) = pos\_loss + c \sum_{j:(y_j - y_t) > 0} (y_j - y_t)$ if $y_t \neq \hat{y}_t$, where pos_loss is the distance in positions of y_t from the first value $\hat{y}_t$ and c is a normalization parameter.

Using a cross validation test over the training set described above, we obtain with a linear kernel an accuracy of 54,30% on the predicted class. The accurancy is not very high as there are many clues where it is hard, also for humans, to determine the exact class of the solution. This ambiguity occurs mainly between the classes of these two subset: {MS,FS,NP} and {MP,FP} [11]. For the latter reason and taking into account that no candidate is pruned but just re-weighted, we considered as a more significant value the covarage of the classifier on the first

---

[11] For example, in some clues is not possible to determine the gender of the solution, such as ≺Ricopre i vialetti: **ghiaia, FS**≻ (≺It can cover a drive: **gravel**≻) or ≺Si cambiano ad ogni portata: **piatti, MP**≻ (≺You use different ones at each course: **plates**≻).

Table 5. Coverage. Here is reported the probability of finding the correct answer in the first k positions.

Table 6. Class accuracy. For each class it is given the percentage of examples inside the training set and the accuracy of the classifier.

Position	Coverage
1st pos	54.30%
2nd pos	73.01%
3rd pos	82.67%
4th pos	87.77%
5th pos	91.38%
6th pos	93.60%
7th pos	95.01%
8th pos	96.16%
9th pos	96.91%
10th pos	97.20%

class	P of ex.	acc.	class	P of ex.	acc.
MS	24.82%	50.23%	AV	1.22%	16.28%
NP	18.68%	68.17%	EP	1.01%	5.00%
FS	13.68%	32.36%	OTHER	0.89%	12.50%
MP	11.17%	65.12%	NUM	0.81%	38.71%
NS	9.04%	84.64%	AMS	0.36%	21.43%
FP	5.18%	18.99%	VS	0.16%	0.00%
ABBR	3.67%	67.18%	AMP	0.10%	20.00%
IP	2.86%	92.16%	AFP	0.06%	33.33%
VI	2.64%	67.39%	AFS	0.03%	33.33%
PC	2.33%	34.62%	VP	0.01%	0.00%
PART	1.28%	25.45%			

n predicted classes. As shown in table 5, the coverage increases very rapidly and it is equivalent to 91,38% if we look over the first 5 predicted classes. Thus, as the number of different target classes is large, this can be considered a very good result. In fact, the use of the output of the clue classifier causes an increment in the WSM performance.

Table 6 shows the occurence of each class in the data set, which should be similar to the one in the whole CWDB. No re-balacing has been made, as the learning algorithm, during each loop, process the "most violated" constraint using a cutting plane method. It can be seen also that there are several classes whose accurancy is high, such as IP, NS, VI, NP and MP.

Moreover, the two non-morphological classes (IP and NS) were introduced in order to better exploit the morphological classifier. A submodule (NI) was implemented which generates name initials [12] when two subsequent proper nouns are found in a sentence. The NS class, instead, is associated to all those clues where the solution does not generally belong to the dictionary, but it can be inferred from the clue itself [13]. At this moment, this type of clues is mainly covered by the rule-based module. In future, a specific module will be implemented which will generate appropriate solutions in a more machine learning fashion. This means by infering likely solutions from previously seen examples.

3.5   Estimating a Confidence on the Lists

After generating a candidate, each module has to estimate the probability that this list contains the correct answer. This information is then processed by the merger, in order to correctly join the lists produced by the modules.

---

[12] E.g., ≺*Iniziali di Celentano:* **ac, IP**≻ (≺*Name initials of Celentano:* **ac**≻). The celebrity we are talking about is Adriano Celentano, with name initials A.C.

[13] E.g., ≺*Trasformano la forza in norma:* **nm, NS**≻ (≺*they change the force in norm:* **nm**≻, it should be read ≺*they change the word 'forza' in 'norma':* **nm**≻).

The confidence estimator of the Web Search Module has been implemented using a standard MLP neural network. This was trained on a set of 2000 candidate lists, using a cross validation set of 500 examples. The main features used for the description of a candidate list example include: the length of the query, the idf values of its words, the length of the list and the scores of the candidates. The output target was set to 1 when the list contained the correct answer, 0 when this was absent. At the end of the training the estimator produced on the validation set an average square error of 0,08.

4   The Other Modules

The experience gained with the Proverb project led to conclude that there are a number of important modules which provide an important contribution to the overall performance of crosswords solving systems.

4.1   The Data-Base Modules

Three different DB-based modules have been implemented in order to exploit the 42973 clue-answer pairs provided by our crossword database. As a useful comparison, the CWDB used by Proverb contained around 3.5×10^5 clue-answer pairs.

CWDB-EXACT simply checks for an exact clue correspondence in the clue-entries. For each answer to a clue C the score-probability is computed using the number of occurrences in the record C. CWDB-PARTIAL employs MySQL's partial-match functions, query expansion and positional term distances to compute clue-similarity scores. The number of answer occurrences and the clue-similarity score are used to calculate the candidates probabilities. CWDB-DICTIO simply returns the full list of words with the correct length, using the number of total occurrences to rank the candidates. Finally, the confidence estimation of the CWDB lists is an entropy function based on the probabilities and occurrences of the candidates.

4.2   The Rule-Based Module

Italian crosswords often contain a limited set of answers that have no semantic relation with their clues, but that are cryptically hidden inside the clue itself. This especially occurs in two-letter and three-letter-answers. Some of the cryptic jokes that crossword authors apply are more or less standard. The rule-based module (RBM) has been especially designed to handle these cases. We have defined eighteen rules for two-letter words and five rules for the three-letter case.

For example, with a clue like ≺*Ai confini del mondo:* **mo**≻ [14] the RBM works as follows: pattern → *ai confini*; object → *mondo*; rule → extract first and last letter from the object. Hence, answer → **mo**.

---

[14] ≺*At the edge of the world:* **wd**≻, **wd** is the fusion of the first and the last letter of object *world*.

4.3 The Implicit Module

The goal of the implicit module is to give a score to sequences of letters. The implicit module is used in two ways: first, within the grid-filling algorithm, to guarantee that the slots that have no candidate words left during the solving process are filled with the most probable sequence of characters; second, as a list filter to rank the terms present in the dictionaries. To do so we used tetra-grams. The global score of a letter sequence results by the productory of all the inner tetragram probabilities. Following a data-driven approach the tetragram probabilities were computed from the CWDB answers.

4.4 The Dictionary Module

Dictionaries will never contain all the possible answers, being crosswords open to neologisms, acronyms, proper names and colloquial expressions. Nevertheless these sources can help to increment the global coverage of the clue-answering.

Two Italian dictionaries were used. The first one containing 127738 word lemmas, and the second one containing 296971 word forms. The output of this module is given by the list of terms with the correct length, ranked by the implicit module.

5 Merging the Candidate Lists

The merger module has been implemented in a very straightforward way (a more sophisticated version will be required in the future). The final score of each term w is computed as: $p(w) = c\sum_{i=0}^{m}(p_i(w) \times \text{conf}_i)$ where m is the number of modules used, conf_i is the confidence evaluation of module i, $p_i(w)$ is the probability score given by module i and c is a normalizing factor.

6 Filling the Crossword Puzzle

As demonstrated by [14] crossword solving can be successfully formalized as a Probabilistic-CSP problem. In this framework the slots of the puzzle represent the set of variables, the lists of candidates provide the domain of legal values for the variables. The goal is to assign a word to each slot in order to maximize the similarity between the final configuration and the target (defined by the cross-word designer). This similarity can be computed in various ways. We adopted the *maximum probability function*, described by the following equation.

$$\underset{\forall \text{sol}:v_1,\dots v_n}{\text{argmax}} \prod_{i=1}^{n} p_{x_i}(v_i) \tag{8}$$

where $p_{x_i}(v_i)$ is the probability that the value v_i is assigned to the variable x_i in the target configuration.

This means that given all the possible legal solutions we search for the one that maximizes the probability of the entire configuration.

A more efficient metric has been proposed in [14], the *maximum expected overlap* function[15]. We will include this feature in our further work.

Finding the maximum probability solution is an NP-complete problem that can be faced using heuristic search techniques as A*. In our implementation the path cost function is the product of the probabilities of the already assigned variables and the heuristic function is the product of the best remaining values of the unassigned variable. Taking the negative log probability, as in eq. 9 and 10, we transform the grid filling into a minimization problem that can be attacked using the classic A* cost function $f(X) = g(X) + h(X)$. Given d the number of already assigned variables in X, q the number of unassigned variables, $\#D_j$ the number of legal values for each unassigned variable x_j and v_j^k) the k-th legal value for x_j, we have the following:

$$g(X) = \sum_{i=1}^{d} - \log(p_{x_i}(v_i)) \tag{9}$$

$$h(X) = \sum_{j=1}^{q} - \log(\underset{k=1}{\overset{\#D_j}{\mathrm{argmax}}}(p_{x_j}(v_j^k)) \tag{10}$$

Due to the competition time restrictions and to the complexity of the problem the use of standard A* was discarded. For this reason we adopted as a solving algorithm a CSP version of WA* [11]. Our new cost function is given by:

$$f(X) = \gamma(d)(g(X) + wh(X)) \tag{11}$$

w is the weighting constant that makes A* more greedy, as in the classic definition of WA*, and $\gamma(d)$ represents an additional score, based on the number of assigned values d (the depth of the current node), that makes the algorithm more depth-first, which is preferable in a CSP framework. This depth score increases the speed of the grid-filling, but it also causes $f(X)$ to be non-admissible.

The grid-filling module works together with the implicit module in order to overcome the missing of a word within the candidates list. Whenever a variable x_i remains with no available values then a heuristic score is computed by taking the tetragram probability of the pattern present in x_i. The same technique is used when a slot is indirectly filled (by the insertion of a crossing word) with a term that is not present within the initial candidates list.

To produce a fast node consistency computation, whenever a variable is selected for expansion, we calculate the remaining legal words using the pointer technique proposed in [3].

7   Experimental Results

The whole crossword collection has been partitioned in five subsets. The first two belong to *La Settimana Enigmistica*, S_{ord}^1 containing examples of ordinary difficulty (mainly taken from the cover pages of the magazine) and S_{dif}^1 composed

---

[15] The aim here is to maximize the number of words that coincide with the target, and not the overall probability.

Table 7. Statistics of the test subsets. T^1_{ord} is the subset of "easy" crosswords with short answers, high number of blanks, limited number of clues. T^1_{dif} provides a tough challenge, having a high number of clues and long answers. T^2_{new} and T^2_{old} are extremely difficult because they contain a great quantity of socio-political references. T^3 is a miscellaneous of average difficulty.

	T^1_{ord}	T^1_{dif}	T^2_{new}	T^2_{old}	T^3
# Letters	160.7	229.5	156.6	141.1	168.5
# Blanks	69.4	37.5	29.8	31.3	37.8
Clues	59.7	79.5	59.5	61.4	50.5
Avg. Length	4.99	5.53	5.04	4.96	4.61
Target in 1-pos	40.3%	37.3%	37.3%	33.3%	31.2%

by crosswords especially designed for skilled cruciverbalists. An other couple belong to *La Repubblica*, S^2_{new} and S^2_{old} respectively containing crosswords that were published in 2004 and in 2001-2003. Finally, S^3 is a miscellaneous of examples from crossword-specialized web sites.

Sixty crosswords of the test set (3685 clues, avg. 61.4 each) were randomly extracted from these subsets in order to form the experimental test suite: T^1_{ord} (15 examples), T^1_{dif} (10 exs.), T^2_{new} (15 exs.), T^2_{old} (10 exs.) and T^3 (10 exs.). Some statistics about the test set are shown in table 7.

To securely maintain WebCrow within the 15 minutes time limit we decided to gather a maximum of 30 documents per clue. To download the documents, parse them and compute the statistical filtering an average of 8 minutes are required. An additional 35 secs are needed by the morphological filter. Thus, in less than 9 minutes WSM's work is completed. The other modules are much faster and the global list generation phase can be terminated in less than 10 minutes. To fulfill the competition requirements we limited the grid-filling execution time to 5 minutes. If a complete solution is not found within this time limit the best partial assignment is returned.

The results[16] that we obtained manifested the different difficulty inherent in the five subsets. Figure 10 reports WebCrow's performance on each example. On T^1_{ord} the results were quite impressive: the average number of targets in first position was just above 40% and the CSP module raised this to 80.0%, with 90.1% correct letters. In one occasion WebCrow perfectly completed the grid. With T^1_{dif} WebCrow was able to fill correctly 67.6% of the slots and 81.2% of the letters (98.6% in one case) which is more or less the result of a beginner human player. On T^2_{new} WebCrow performs with less accuracy averaging 62.9% (72% letters). On T^2_{old} (old crosswords), due to the constant refreshing of Web's information, the average number of correct words goes down to 61.3% (72.9% letters). The last subset, T^3, containes crosswords that belong to completely different sources, for this reason the contribution of the CWDB is minimal (the

---

[16] WebCrow has been implemented mainly in Java with some parts in C++ and Perl. The system has been compiled and tested using Linux on a Pentium IV 2GHz with 2GB ram.

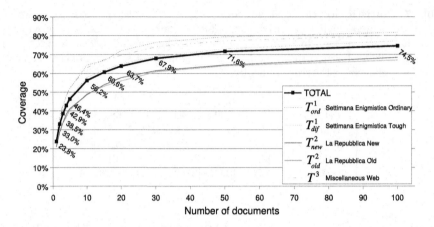

Fig. 10. The coverage of the WSM in relation to the number of documents used. The WSM can increase its coverage by using more documents for each clue. This sensibly slows the answering process.

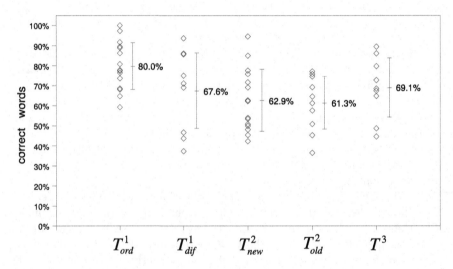

Fig. 11. WebCrow's performance on the five subsets. The average and the variance of the correct words are also reported.

coverage and the precision of CWDB-EXACT are more than halved). Nevertheless, the WSM still assures a good clue-answering and the solving module is able to reach 69.1% words correct and 82.1% letters correct.

Altogether, WebCrow's performance over the test set is of 68.8% (ranging from 36.5% to 100%) correct words and 79.9% (ranging from 48.7% to 100%)correct letters.

From preliminary tests we observed that allowing an extended time limit of 45 minutes and using more documents from the Web (i.e. 50 per clue) the system's performances increase by a 7% in average.

8 Conclusions

The discussion on crosswords solving and, particularly, the WebCrow project that is mainly described in this paper indicates very promising results. We-bCrow's overall architecture allows us to plug in several expert modules in order to increase the system's performances. The web-search approach has prooved to be consistent and we are confident that it could suitable to face a number of related real-world problems. Future research is mainly expected to b on multi-lingual crosswords so as to fully exploit the clue answering from the Web, which does not make linguistic assumptions. On the other hand, robust NLP system are likely to have a great impact on the answering of the clues. This can be done by adding several other list filters: stylistic, morpho-syntactical, lexical and logical. On the long run, WebCrow is expected to reach human performance at expert level and overcome humans in crosswords with multilingual clues.

Acknowledgments. We would like to thank Michael Littman for fruitful discussions and Shay Bushinsky for having inspired the human competition idea. We are also very grateful to Google Inc the financial support of the research under a Research Program Awards.

References

1. C. Bosco, V. Lombardo and D. Vassallo and L. Lesmo: Building a Treebank for Italian: a Data–driven Annotation Schema. Proceedings of LREC. (2002)
2. K. Crammer and Y. Singer: On the algorithmic implementation of multiclass kernel–based vector machines. Journal of Machine Learning Res. 2 (2002) 265–-292
3. M. L. Ginsberg, M. Frank, M. P. Halping and M.C. Torrance: Search lessons learned from crossword puzzles. Proceedings of the Eighth National Conference on Artificial Intelligence (AAAI–90). (1990) 210–215
4. M. L. Ginsberg: Dynamic Backtracking. Journal of Artificial Intelligence Research. 1 (1993) 25–46
5. G. A. Keim, Noam M. Shazeer and Michael L. Littman: PROVERB: the probabilistic cruciverbalist. AAAI '99: Proceedings of the sixteenth national conference on Artificial Intelligence. (1999) 710–717
6. C. Kwok, O. Etzioni and D. S. Weld: Scaling question answering to the web. ACM Trans. Inf. Syst. 19,3 (2001) 242–262
7. M. L. Littman: Review: computer language games. Journal of Computer and Games. 134 (2000) 396–404
8. M. L. Littman, Greg A. Keim and Noam M. Shazeer: A probabilistic approach to solving crossword puzzles. Journal of Artificial Intelligence. 134 (2002) 23–55
9. L. J. Mazlack: Computer construction of crossword puzzles using precedence relationships. Journal of Artificial Intelligence 7 (1976) 1–19

10. News–Nature: Program crosses web to fill in puzzling words. Nature 431 (2004) 620
11. I. Pohl: Heuristic search viewed as path finding in a graph. Journal of Artificial Intelligence. 1 (1970) 193–204
12. H. Schmid: Probabilistic Part–of–Speech Tagging Using Decision Trees. International Conference on New Methods in Language Processing. (1994)
13. H. Schmid: Improvements in Part-of-speech Tagging with an Application to German. Proceedings of the EACL SIGDAT Workshop. (1995)
14. N. M. Shazeer, Greg A. Keim and Michael L. Littman: Solving crosswords as probabilistic contraint satisfaction. AAAI '99: Proceedings of the sixteenth national conference on Artificial Intelligence. (1999) 156–152
15. I. Tsochantaridis, T. Hofmann, T. Joachims and Y. Altun: Support vector machine learning for interdependent and structured output spaces. ICML '04: Twenty–first international conference on Machine learning (2004)
16. E. M. Voorhees and D. M. Tice: Overview of the TREC–9 Question Answering Track. Proceedings of the Ninth Text REtrieval Conference (TREC–9). (2000)

Model-Based Diagnosis Through OBDD Compilation: A Complexity Analysis

Pietro Torasso and Gianluca Torta

Dipartimento di Informatica, Università di Torino, Italy
{torasso, torta}@di.unito.it

Abstract. Since it is known that Model-Based Diagnosis may suffer from a potentially exponential size of the search space, a number of techniques have been proposed for alleviating the problem. Among them, some forms of compilation of the domain model have been investigated. In the present paper we address the problem of evaluating the complexity of diagnostic problem solving when Ordered Binary Decision Diagrams are adopted for representing the normal and faulty behavior of the system to be diagnosed and the solution space. In particular we analyze the case of the diagnosis of static models that exhibit a directionality from inputs to outputs (an important example of this type of models is the class of combinatorial digital circuits). We show that the problem of determining the set of all diagnoses and of determining the minimum cardinality diagnoses can be solved in time and space polynomial with respect to the size of the OBDD encoding the domain model. These results hold regardless of the degree of system observability including whether observations are precise or uncertain. We then analyze the complexity of refining the set of diagnoses by making additional observations and by using a test vector for troubleshooting the system. In particular we show that in the latter case we lose the formal guarantee that the diagnosis can be performed in polynomial time with respect to the size of the compiled domain model.

1 Introduction

A problem recognized very early in the research on Model-Based Diagnosis (MBD) concerns the potentially exponential size of the search space and of the set of alternative diagnoses and consequently, the need for formalisms for a compact representation of both. Many alternative representation formalisms have been proposed (e.g. *kernel diagnoses* [13], *consequences* [11], *scenarios* [23]); most of them exploit the properties of the system structure or behavior in order to efficiently compute and encode diagnoses for restricted but practically relevant classes of systems.

Researchers in the MBD community have recently started to look at formalisms for the symbolic representation of the search and solution spaces in order to improve the efficiency of tasks such as assessment of system diagnosability and computation of diagnoses (e.g. [8], [20], [9], [10]).

O. Stock and M. Schaerf (Eds.): Aiello Festschrift, LNAI 4155, pp. 287–305, 2006.

Ordered Binary Decision Diagrams (OBDDs, see [6]) are a well-known mathematical tool used in several areas of computing (including AI, see e.g. [3], [15]) for efficiently representing and manipulating large state spaces. In [21] the relational model of a Discrete Event System is encoded as an OBDD $\mathcal{O}$; the authors show that the set of diagnoses can be characterized by a 1st order formula over the relations of the model and therefore, diagnoses can be obtained by manipulating $\mathcal{O}$ with standard OBDD operators.

However, the use of OBDDs is not a *per se* panacea for the MBD task. Firstly, the encoding of the system model (in particular the choice of the order of the system variables) must be done very carefully in order to avoid the explosion of the OBDD size. Secondly, once the system model has been encoded as an OBDD, the application of standard OBDD operators for the computation and extraction of diagnoses may lead to time and/or space intractability. Finally, some methods are needed to integrate the filtering and/or ordering of diagnoses based on preference criteria into the OBDD approach.

In the present paper, we address these issues in the context of the diagnosis of static models that exhibit a directionality from inputs to outputs. This class of system models is interesting from a computational point of view since it is possible to prove tractability results for diagnosis in terms of OBDD encoding, as we will discuss in the rest of the paper.

Moreover, static models with directionality have a practical relevance since they can capture the behavior of important classes of systems. First of all it is worth mentioning the class of combinatorial digital circuits (which has historically been one of the main test beds for MBD) whose behavior is defined in terms of flow information from inputs to outputs and does not involve the temporal dimension; other real-world systems can be translated into this formalism, in particular the ones that can be modeled via causal networks (e.g. a robotic arm [23] and an industrial plant [17]).

The class of static models with directionality obviously does not include dynamic systems (such as sequential digital circuits) as well as static systems whose model has to be expressed in terms of qualitative equations[1].

The paper is structured as follows. In section 2, we give formal definitions of the concepts of system model, diagnostic problem and diagnosis on which our work is based. In section 3, we first provide a short summary on OBDDs and we discuss how to encode a system description using OBDDs as well as heuristics for determining a suitable variable order. In section 4, we show that the problem of determining the set of all diagnoses and of determining the minimum cardinality diagnoses can be solved in a time and space polynomial with respect to the size of the OBDD encoding the domain model. Moreover, in sections 5 and 6, we exploit the flexibility of OBDDs in order to apply our techniques to ambiguous observations and to Context-Varying Diagnosis, where the system is observed with different test-vectors. Finally in section 7, we summarize the contributions of this paper.

---

[1] Let us consider for example a pipe in an hydraulic system; if the ends of the pipe are denoted as A and B we need an equation *flow(A) = flow(B)* and it is not possible to specify which of the two determines the value of the other.

2   Basic Definitions of Model-Based Diagnosis

The following definitions formalize the concepts of system description, diagnostic problem, diagnosis and preferred diagnosis.

Definition 1. *A* System Description *(SD) is a pair* $(\mathcal{SV}, DT)$ *where:*

- $\mathcal{SV}$ *is a set of discrete system variables partitioned in a subset* $\mathcal{SV}_{exo}$ *of exogenous variables and a subset* $\mathcal{SV}_{end}$ *of endogenous variables. Set* $\mathcal{SV}_{exo}$ *is further partitioned in subsets* INPUTS *(system inputs and commands) and* COMPS *(components), while* $\mathcal{SV}_{end}$ *is further partitioned into* OBS *(observables) and* INTVARS *(non-observables);* DOM(V) *is the finite domain of variable* $V \in \mathcal{SV}$. *In particular, for each* $C \in COMPS$, DOM(C) *contains a set of behavioral modes, one corresponding to the nominal mode (OK) and the others to faulty behaviors*
- *DT (Domain Theory) is an acyclic set of Horn clauses defined over* $\mathcal{SV}$ *representing the behavior of the system (under normal and abnormal conditions); variables in* COMPS *and* INPUTS *never appear in the head of a clause*

It is worth noting that the definition of *SD* is focused on systems whose behavior can be expressed in terms of input/output relations. Also note that we assume that *DT* contains a model not only for the nominal behavior but also for abnormal behaviors; in particular, for each behavioral mode of a component, the model must determine its outputs given its inputs.

Example 1. The class of systems captured by this definition includes devices with a relevant role in real-world applications such as combinatorial digital circuits. As an example, in Figure 1 we report the schema of circuit *c17* from the *ISCAS85* benchmark (for more details see e.g. [14]). While this circuit is very simple, it is worth noting that (as it happens in most of the ISCAS85 circuits) the components are not just the logical gates but also the connections, so the faults can occur also in the connections. Moreover, apart from nominal behavior, ISCAS85 foresees two kinds of faults for the components, that is *Stuck at 0* and *Stuck at 1* when the output of a component is 0 (or 1) independently of the value of the inputs.

In particular, we report the nominal and faulty behavior of one of the NAND gates of circuit *c-17* in Figure 2.   □

We now introduce the formal definitions of *Diagnostic Problem* and *Diagnosis* commonly adopted for *consistency-based diagnosis*.

Definition 2. *A* Diagnostic Problem *is a 3-tuple DP = (SD,* **OBS**, **INPUTS***) where SD is the System Description,* **OBS** *is an instantiation of* OBS *variables and* **INPUTS** *is an instantiation of* INPUTS *variables.*

Definition 3. *Let DP = (SD,* **OBS**, **INPUTS***) be a diagnostic problem. We say that an instantiation* $\mathbf{D} = \{C_1(bm_1), \ldots, C_n(bm_n)\}$ *of COMPS is a consistency-based diagnosis for DP iff:*

$$DT \cup \mathbf{INPUTS} \cup \mathbf{OBS} \cup \mathbf{D} \not\vdash \bot$$

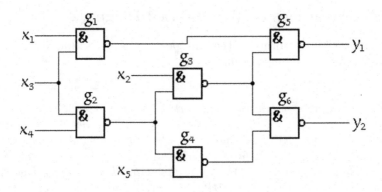

Fig. 1. Circuit c17 from the ISCAS85 benchmark

$$
\begin{array}{|l|}
\text{nandg(g1)} \wedge \text{ok(g1)} \wedge \text{X1(0)} \wedge \text{X3(0)} \Rightarrow \text{out(1)} \\
\text{nandg(g1)} \wedge \text{ok(g1)} \wedge \text{X1(0)} \wedge \text{X3(1)} \Rightarrow \text{out(1)} \\
\text{nandg(g1)} \wedge \text{ok(g1)} \wedge \text{X1(1)} \wedge \text{X3(0)} \Rightarrow \text{out(1)} \\
\text{nandg(g1)} \wedge \text{ok(g1)} \wedge \text{X1(1)} \wedge \text{X3(1)} \Rightarrow \text{out(0)} \\
\text{nandg(g1)} \wedge \text{sa0(g1)} \Rightarrow \text{out(0)} \\
\text{nandg(g1)} \wedge \text{sa1(g1)} \Rightarrow \text{out(1)}
\end{array}
$$

Fig. 2. Model of *g1* NAND gate

According to the above definition, a diagnosis assigns a behavioral mode to each component of the system to be diagnosed.

In most diagnostic systems, especially when the set of returned diagnoses can be very large, we are interested only in the preferred diagnoses, according to some particular preference criterion. In the following, we focus on a specific preference criterion involving the number of faults in the diagnoses

Definition 4. *Let* $\mathbf{D} = \{C_1(bm_1), \ldots, C_n(bm_n)\}$ *be a diagnosis for DP and* card($\mathbf{D}$) *be the number of assignments* $C_i(bm_i) \in \mathbf{D}$ *s.t.* $bm_i \neq OK$. *We say that* $\mathbf{D}$ *is a* minimum cardinality diagnosis *for DP iff* $\nexists\mathbf{D}$' *s.t.* $\mathbf{D}$' *is a diagnosis for DP and* card($\mathbf{D}$') < card($\mathbf{D}$).

3   Encoding the System Model into an OBDD

Before discussing the issues related to the encoding of the model of the system to be diagnosed, we report a short summary on the main characteristics of OBDDs that will be exploited in the remainder of the paper. For a much more detailed discussion on OBDDs, please refer to e.g. [6], [16].

3.1   Summary on OBDDs

An OBDD is a formalism for representing a Boolean function $\mathcal{F}(x_1, \ldots, x_n)$. Given an ordering of $x_1, \ldots, x_n$ an OBDD is a rooted DAG whose nodes include (at most) two terminal nodes labeled **0** and **1** and non-terminal nodes each labeled with one of the x_i variables.

Every internal node x_i has exactly two successors *low* and *high* (if x_j is successor of x_i in the DAG then x_i must precede x_j in the ordering).

Every path $\mathcal{P}$ from the root to node **1** can be viewed as an assignment to the variables involved in $\mathcal{P}$ ($x_i = 1$ if $high(x_i)$ is in $\mathcal{P}$ and $x_i = 0$ if $low(x_i)$ is in $\mathcal{P}$) which guarantees that the value of $\mathcal{F}$ is 1. The *size* $|\mathcal{O}|$ of an OBDD $\mathcal{O}$ is defined as the number of its nodes.

It is known that the OBDD of minimal size is unique for a given function $\mathcal{F}$ and a fixed variable order $\mathcal{VO}$. Variable order choice is a preminent issue when encoding a Boolean function as an OBDD: in [6], e.g., Bryant shows a theory whose OBDD encoding size varies from linear to exponential just because of different variable orders; moreover, improving a given variable order (w.r.t. the OBDD size) has been shown to be NP-complete [4].

Manipulations of Boolean functions can be mapped to operations on the OB-DDs which represent them. We denote with *build* the operator that, when applied to a Boolean function $\mathcal{F}$ and a variable order $\mathcal{VO}$, returns the OBDD representing $\mathcal{F}$ according to order $\mathcal{VO}$.

Binary logical operations can be performed on OBDDs $\mathcal{O}_1$ and $\mathcal{O}_2$ with the *apply* operator whose first argument is the binary operator *op* to be applied, while the *restrict* operator substitutes a constant to a variable in an OBDD.

The existential quantification of a variable B in an OBDD $\mathcal{O}$ consists in computing the restrictions $\mathcal{O}_1$ and $\mathcal{O}_0$ of $\mathcal{O}$ with $B = 1$ and $B = 0$ respectively and then applying the $\vee$ operator between $\mathcal{O}_1$ and $\mathcal{O}_0$.

Table 1 reports the time and space complexity of the main OBDD operators.

The following theorem due to Sieling and Wegener [19] relates the number of copies of each internal variable x_i of the OBDD with the encoded Boolean function f.

Theorem 1. (Sieling and Wegener) *Let $x_1, \ldots, x_n$ be the given variable order and let f be defined on $x_1, \ldots, x_n$. The reduced OBDD for f contains as many x_i-nodes (i.e., nodes labeled by variable x_i) as there are different subfunctions $f|_{x_1=a_1,\ldots,x_{i-1}=a_{i-1}}$, for $a_1, \ldots, a_{i-1} \in \{0, 1\}$, which depend essentially on x_i (function ϕ depends essentially on x_i if $\phi|_{x_i=0}$ is different from $\phi|_{x_i=1}$).*

According to the theorem, the number of copies of a variable x_i can be exponential in the number of variables that precede it in the variable order; for this reason the theorem provides an explanation of the exponential complexity of the *build* operator.

Table 1. OBDD operators and their complexity

op	time	output size
build($\mathcal{F}(x_1, \ldots, x_n)$, $\mathcal{VO}$)	$O(2^n)$	$\leq 2^n$
apply(op, $\mathcal{O}_1$, $\mathcal{O}_2$)	$O(|\mathcal{O}_1| \cdot |\mathcal{O}_2|)$	$\leq |\mathcal{O}_1| \cdot |\mathcal{O}_2|$
restrict($\mathcal{O}$, $x_i = b$)	$O(|\mathcal{O}|)$	$\leq |\mathcal{O}|$

However, the practical relevance of OBDDs stems from the fact that, in many cases, it is possible to find a variable order s.t. the size of the OBDD representing $\mathcal{F}(x_1, \ldots, x_n)$ is much smaller than 2^n; in the past few years a number of heuristic techniques for computing variable orders have been proposed (e.g. [1], [2]).

3.2   Encoding Functions and Sets over Multi-valued Variables

By definition an OBDD $\mathcal{O}$ represents a Boolean function $\mathcal{F}$ over Boolean variables $B_1, \ldots, B_m$. It is straightforward, however, to represent a Boolean function $\mathcal{M}$ over a set of multi-valued variables $\{V_1, \ldots, V_m\}$.

First, since OBDDs handle only Boolean variables (whose value is either 0 or 1), a generic multi-valued variable V with domain $DOM(V) = \{v_1, \ldots, v_k\}$ will be mapped to a set of Boolean variables $V_B = \{V_{v_1}, \ldots, V_{v_k}\}$ [2].

We need to explicitly enforce the fact that a multi-valued variable V assumes exactly one value; this can be expressed by a *completeness* formula (i.e. $V_{v_1} \vee \ldots \vee V_{v_k}$) and a set of *mutual-exclusion* formulas (i.e. $\sim(V_{v_i} \wedge V_{v_j})\ \forall i \neq j$).

Given a Boolean function $\mathcal{M}(V_1, \ldots, V_m)$ over multi-valued variables $\{V_1, \ldots, V_m\}$, let's consider function $\mathcal{M}_B(V_{1,B}, \ldots, V_{m,B})$ s.t.:

$$\mathcal{M}_B(V_1(v_1)_B, \ldots, V_m(v_m)_B) = 1 \Leftrightarrow \mathcal{M}(V_1(v_1), \ldots, V_m(v_m)) = 1$$

where $V_i(v_i)_B$ means that Boolean variable V_{i,v_i} is set to 1 and Boolean variables $V_{i,v_j}, v_j \in DOM(V_i), v_j \neq v_i$ are set to 0.

The OBDD $\mathcal{O}_{\mathcal{M}_B}$ which represents $\mathcal{M}_B$ also encodes the original function $\mathcal{M}$, since it is straightforward to map back and forth between instantiations of the set of multi-valued variables and instantiations of the associated Boolean variables.

From this result it directly follows that it is also possible to encode arbitrary *sets of instantiations*. Indeed, $\mathcal{M}$ implicitly defines a set $instset(\mathcal{M})$ as follows:

$$instset(\mathcal{M}) = \{\mathcal{I} = (V_1(v_1), \ldots, V_m(v_m)) | \mathcal{M}(\mathcal{I}) = 1\}$$

Given the OBDD $\mathcal{O}_{\mathcal{M}}$ which encodes $\mathcal{M}$, in order to compute $instset(\mathcal{M})$ it is sufficient to perform an exhaustive visit of all the paths in $\mathcal{O}_{\mathcal{M}}$ from the root to node $\mathbf{1}$. Note that due to completeness and mutual exclusion formulas we are guaranteed that each path contains exactly one assignment to each multi-valued variable.

Such a mechanism can be exploited for providing a complete enumeration of diagnoses once we have computed an OBDD which represents them. The enumeration has time complexity linear in the cardinality of the set of diagnoses.

---

[2] In general we'll denote with $\mathcal{S}_B$ the set of Boolean variables associated with the set $\mathcal{S}$ of multi-valued variables. The proposed encoding is clearly not the most efficient, especially for multi-valued variables with large ranges; all the discussions in this paper, however, apply to more efficient encodings as well.

3.3   Encoding the Domain Theory

Given a System Description according to Definition 1, the Domain Theory DT is a propositional logical theory expressed in terms of multi-valued variables $\{V_1, \ldots, V_m\} = \mathcal{SV}$. We can associate DT with a Boolean function $\mathcal{M}_{DT}$ over $\mathcal{SV}$ variables s.t.:

$$DT \cup \{V_1(v_1) \wedge \ldots \wedge V_m(v_m)\} \not\vdash \bot \Leftrightarrow \mathcal{M}_{DT}(V_1(v_1), \ldots, V_m(v_m)) = 1$$

i.e. an instantiation of $\mathcal{SV}$ is consistent with DT iff applying $\mathcal{M}_{DT}$ to it yields 1. As shown in the previous section, there exists an OBDD $\mathcal{O}_{DT}$ which encodes $\mathcal{M}_{DT}$; we will consider $\mathcal{O}_{DT}$ as the OBDD encoding of DT.

In order to build OBDD $\mathcal{O}_{DT}$ the formulas in DT first need to be rewritten by substituting the instances of multi-valued variables with the associated Boolean variables. Thus, a generic Horn clause in DT:

$$N_1(v_1) \wedge \ldots \wedge N_k(v_k) \Rightarrow M(u) \quad \text{becomes} \quad N_{1,v_1} \wedge \ldots \wedge N_{k,vk} \Rightarrow M_u$$

The OBDD $\mathcal{O}_{DT}$ is then built by putting in conjunction the mutual exclusion and completeness formulas for variables $V_1, \ldots, V_m$ with the transformed DT formulas by using the standard *apply* operator. It is worth noting that, due to the addition of the mutual exclusion and completeness formulas, $\mathcal{O}_{DT}$ encodes a propositional theory which is not Horn; in particular, the associated diagnostic problems fall in the class of *incompatibility abduction problems* that are known to be NP-hard [7].

3.4   Choosing an Order for the System Variables

As noted in section 3.1, one major concern in encoding any propositional theory as an OBDD regards the choice of variable order. An effective means to focus the search for a good order consists in taking into account the (implicit) structure of the Domain Theory (see e.g. [1], [2]).

In our approach, it is possible to make such a structure explicit by exploiting the directionality between inputs and outputs of the system. In particular we introduce the notion of dependency among system variables.

Definition 5. *Given $SD = (SV, DT)$ the associated* System Dependencies Graph $\mathcal{G}$ *is a DAG representing the causal structure of the system; the nodes of $\mathcal{G}$ are in one-to-one correspondence with the variables in $\mathcal{SV}$ and, whenever a formula $N_1(v_1) \wedge \ldots \wedge N_k(v_k) \Rightarrow M(u)$ appears in DT, nodes N_1 through N_k are parents of M in $\mathcal{G}$. We denote with $rev(\mathcal{G})$ the DAG obtained by inverting the direction of all edges in $\mathcal{G}$.*

We define a *family* to be a set of variables containing a variable V and its parents in $\mathcal{G}$.

The structure of $\mathcal{G}$ can be exploited in different ways to get a variable order. For our purposes, we have identified three criteria that relate the structure of $\mathcal{G}$, the order of multi-valued variables and the order of the corresponding Boolean variables.

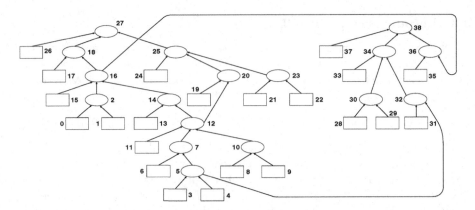

Fig. 3. System Dependencies Graph for circuit c17

- **directionality:** always index the parent variables in the families of $\mathcal{G}$ before the child variable
- **family vicinity:** index variables in the same family of $\mathcal{G}$ "as close as possible"
- **variable vicinity:** always index the Boolean variables representing the values in the domain of a multi-valued system variable in sequence (i.e. if $DOM(V) = \{v_1, \ldots, v_k\}$ then variables $V_{v_1}, \ldots, V_{v_k}$ have indexes i to $(i+k-1)$ in the order)

The first criterion plays an important role in the complexity results that will appear in the next section while the second corresponds to a guideline that is widely recognized as important for obtaining good variable orders ([1], [2]). Finally, the third criterion determines the order of the Boolean variables once the order of multi-valued variables has been fixed.

In [24] we have presented a simple strategy that satisfies the criteria introduced above. In particular, such a strategy orders the variables through a depth-first visit of $rev(\mathcal{G})$ taking into account that $rev(\mathcal{G})$ can be multiply-connected and multi-rooted.

Figure 3 reports the System Dependencies Graph $\mathcal{G}$ for the combinatorial digital circuit $c17$ from the ISCAS85 benchmark. Despite the fact that the circuit involves just 5 inputs, 6 logical NAND gates and 2 outputs, graph $\mathcal{G}$ is not so simple because also the connections between components are modeled and can fail.

Figure 3 reports also the variable order obtained with the strategy mentioned above; note in particular that the order satisfies the *directionality* condition. According to this variable order, the size of the OBDD encoding the model of $c17$ is just 721 nodes.

4   Computing Diagnoses

4.1   Computing the Complete Set of Diagnoses as an OBDD

Many approaches to MBD try to compute just preferred diagnoses in order to alleviate the computational cost of diagnosis (e.g. [12]) . On the contrary, in our

1 Function Diagnose($\mathcal{O}_{DT}$, **OBS**, **INPUTS**)
2    $\mathcal{O}_{DT,OBS}$:= DisregardVariables($\mathcal{O}_{DT}$, INTVARS)
3    $\mathcal{O}_{DIAG}$:= AssignVariables($\mathcal{O}_{DT,OBS}$, **OBS** $\cup$ **INPUTS**)
4   Return $\mathcal{O}_{DIAG}$
5 EndFunction

1 Function DisregardVariables($\mathcal{O}_{IN}$, ELIMVARS)
2    $\mathcal{O}_{TMP}$:= $\mathcal{O}_{IN}$
3   For Each $B \in$ ELIMVARS$_B$
4      $\mathcal{O}_{TMP}$:= apply($\vee$, restrict($\mathcal{O}_{TMP}$, B), restrict($\mathcal{O}_{TMP}$, $\sim B$))
5    $\mathcal{O}_{OUT}$:= $\mathcal{O}_{TMP}$
6   Return $\mathcal{O}_{OUT}$
7 EndFunction

1 Function AssignVariables($\mathcal{O}_{IN}$, KNOWNVARS)
2    $\mathcal{O}_{TMP}$:= $\mathcal{O}_{IN}$
3   ForEach $V(v) \in$ KNOWNVARS
4      $\mathcal{O}_{TMP}$:= restrict($\mathcal{O}_{TMP}$, V_v)
5    $\mathcal{O}_{OUT}$:= $\mathcal{O}_{TMP}$
6   Return $\mathcal{O}_{OUT}$
7 EndFunction

Fig. 4. Computation of Diagnoses

approach we show that it is computationally feasible (in most cases) to compute the set of diagnoses and then to extract the most preferred ones.

First of all, we show how the OBDD representing the set of all the diagnoses for a diagnostic problem can be computed by an appropriate sequence of logical operations involving the OBDD $\mathcal{O}_{DT}$ representing the Domain Theory. Figure 4 reports the diagnostic algorithm Diagnose().

Since we don't have any information about *INTVARS* variables, and they do not appear in the resulting diagnoses (which are defined just in terms of *COMPS*), the algorithm starts by filtering out the *INTVARS* variables from $\mathcal{O}_{DT}$ in order to obtain a new OBDD $\mathcal{O}_{DT,OBS}$.

This is accomplished by calling function DisregardVariables() which, given the set *ELIMVARS* of multi-valued variables to be disregarded, considers each Boolean variable $B \in$ ELIMVARS$_B$ and performs an existential quantification over it.

The subsequent call to AssignVariables() in Diagnose() has the effect of incrementally constraining $\mathcal{O}_{DT,OBS}$ with each piece of information provided by the inputs (i.e. **INPUTS**) and available observations (i.e. **OBS**), by using the standard OBDD operator *restrict*.

The resulting OBDD $\mathcal{O}_{DIAG}$ represents the set of instantiations of variables *COMPS* consistent with *DT*, **INPUTS** and **OBS**, i.e it represents the set of consistency-based diagnoses.

The following theorem states that the simple algorithm given above for computing diagnoses from $\mathcal{O}_{DT}$ is both correct and complete.

Theorem 2. *Let DP = (SD, **OBS**, **INPUTS**) be a diagnostic problem, and $\mathcal{O}_{DIAG}$ be the OBDD computed by algorithm* `Diagnose()`. *Then the set of instantiations of* COMPS *represented by* $\mathcal{O}_{DIAG}$ *contains all and only the consistency-based diagnoses for DP.*

Proof. From Section 3.3, we know that $\mathcal{O}_{DT}$ represents a Boolean function $\mathcal{M}_{DT}$ s.t.:

$$\mathcal{M}_{DT}(\mathcal{C},\mathcal{O},\mathcal{I},\mathcal{N}) = 1 \Leftrightarrow DT \cup \mathcal{C} \cup \mathcal{O} \cup \mathcal{I} \cup \mathcal{N} \not\vdash \bot$$

where $\mathcal{C}$ is an instantiation of *COMPS* variables, $\mathcal{O}$ is an instantiation of *OBS* variables, $\mathcal{I}$ is an instantiation of *INPUTS* variables and $\mathcal{N}$ is an instantiation of *INTVARS* variables.

Let $\mathcal{O}_{DT,OBS}$ be the OBDD obtained from $\mathcal{O}_{DT}$ through projection on *COMPS* $\cup$ *OBS* $\cup$ *INPUTS* variables, so that it represents the following Boolean function:

$$\mathcal{M}_{DT,OBS}(\mathcal{C},\mathcal{O},\mathcal{I}) \equiv (\exists \mathcal{N})(\mathcal{M}_{DT}(\mathcal{C},\mathcal{O},\mathcal{I},\mathcal{N}))$$

It follows that $\mathcal{M}_{DT,OBS}$ satisfies the equivalence:

$$\mathcal{M}_{DT,OBS}(\mathcal{C},\mathcal{O},\mathcal{I}) = 1 \Leftrightarrow \exists \mathcal{N}(DT \cup \mathcal{C} \cup \mathcal{O} \cup \mathcal{I} \cup \mathcal{N} \not\vdash \bot)$$

After we impose that *OBS* variables have value **OBS** and *INPUTS* variables have value **INPUTS** we obtain a new function $\mathcal{M}'_{DT,OBS}$ s.t.:

$$\mathcal{M}'_{DT,OBS}(\mathcal{C}) = 1 \Leftrightarrow \exists \mathcal{N}(DT \cup \mathcal{C} \cup \mathbf{OBS} \cup \mathbf{INPUTS} \cup \mathcal{N} \not\vdash \bot)$$

It is immediate to see that $\mathcal{M}'_{DT,OBS}(\mathcal{C})$ is the function encoded by OBDD $\mathcal{O}_{DIAG}$ returned by `Diagnose()`. In order to conclude that $\mathcal{M}'_{DT,OBS}$ is the characteristic function of the set of consistency-based diagnoses for *DP* there remains to show that:

$$\exists \mathcal{N}(DT \cup \mathcal{C} \cup \mathbf{OBS} \cup \mathbf{INPUTS} \cup \mathcal{N} \not\vdash \bot) \Leftrightarrow DT \cup \mathcal{C} \cup \mathbf{OBS} \cup \mathbf{INPUTS} \not\vdash \bot$$

(note that the right member of the equivalence is indeed exactly the definition of consistency-based diagnosis).

The implication from left to right is obvious, since the $\vdash$ relationship is monotonic. As for the implication from right to left, it is sufficient to note that, since the standard propositional derivation denoted with $\vdash$ is correct and complete, there must exist a consistent assignment to $\mathcal{SV}$ variables that assigns $\mathcal{C}$ to *COMPS*, **OBS** to *OBS*, **INPUTS** to *INPUTS* and some $\overline{\mathcal{N}}$ to *INTVARS*. Instantiation $\overline{\mathcal{N}}$ is then an example satisfying the existential quantification on the left side of the equivalence. □

As for the computational complexity of the diagnostic algorithm, we will report results concerning tractability of diagnosis. In particular, we first state two lemmas concerning the computational complexity of `DisregardVariables()` and `AssignVariables()` and then we give a result on the complexity of the `Diagnose()` function.

Lemma 1. *If the variable order satisfies directionality and variable vicinity, the time complexity of the call to* DisregardVariables() *is:*

$$O(|INTVARS| \cdot |\mathcal{O}_{DT}|^2)$$

Moreover, $|\mathcal{O}_{DT,OBS}| \leq |\mathcal{O}_{DT}|$.

Proof. We first show that, under the theorem's hypotheses, at each iteration of the ForEach loop in DisregardVariables() (Figure 4), the size of $\mathcal{O}_{TMP}$ does not increase.

Let $V \in INTVARS$, $DOM(V) = \{v_1, \ldots, v_k\}$ and assume that Boolean variables $V_B = \{V_{v_1}, \ldots, V_{v_k}\}$ have been assigned indexes $i, \ldots, i + k - 1$ (because of variable vicinity).

It is easy to see that, given a partial path $\mathcal{P}^{i-1}$ in OBDD $\mathcal{O}_{TMP}$ from the root up to the Boolean variable with index $(i - 1)$, all the extensions of $\mathcal{P}^{i-1}$ that lead to terminal node **1** must agree that some $V_{v_l} \in V_B$ has value **1** while each $V_{v_j} \in V_B$, $v_j \neq v_l$ has value **0** (in other words, all the extensions of $\mathcal{P}^{i-1}$ must agree that multi-valued variable V has value v_l).

This follows from the fact that directionality prescribes to order parent variables in the System Influence Graph $\mathcal{G}$ before their child variable and from our assumption that the model is deterministic.

From this fact, it is immediate to see that the existential quantification of $\{V_{v_1}, \ldots, V_{v_k}\}$ results just in the nodes labeled with $\{V_{v_1}, \ldots, V_{v_k}\}$ being removed from $\mathcal{O}_{TMP}$, without any further change, i.e. at each iteration the size of $\mathcal{O}_{TMP}$ does not increase.

From the above discussion it follows that each execution of the body of the ForEach loop in DisregardVariables() takes time $O(|\mathcal{O}_{DT}|^2)$, and then the execution of DisregardVariables() itself takes time $O(|INTVARS| \cdot |\mathcal{O}_{DT}|^2)$.

Moreover, since none of the operations performed by DisregardVariables() increases the size of $\mathcal{O}_{TMP}$, the size of OBDD $\mathcal{O}_{DT,OBS}$ must be smaller or equal to the size of $\mathcal{O}_{DT}$.   □

Lemma 2. *The time complexity of the call to* AssignVariables() *is:*

$$O(|INPUTS \cup OBS| \cdot |\mathcal{O}_{DT,OBS}|)$$

Moreover, $|\mathcal{O}_{DIAG}| \leq |\mathcal{O}_{DT,OBS}|$.

Theorem 3. *The time complexity of* Diagnose() *is:*

$$O(|INTVARS| \cdot |\mathcal{O}_{DT}|^2 + |INPUTS \cup OBS| \cdot |\mathcal{O}_{DT}|)$$

Moreover, $|\mathcal{O}_{DIAG}| \leq |\mathcal{O}_{DT}|$.

It is easy to see that Lemma 2 directly follows from the complexity of the basic OBDD operators (section 3.1). Theorem 3 just summarizes the results of the two lemmas.

This result provides us with upper bounds on time and space complexity of computation of the set of consistency-based diagnoses. This is a very relevant

```
1 Function MinCardDiagnose(𝒪_DT, OBS, INPUTS)
2    𝒪_{DT,OBS} := DisregardVariables(𝒪_DT, INTVARS)
3    𝒪_DIAG := AssignVariables(𝒪_{DT,OBS}, OBS ∪ INPUTS)
4    𝒪_PREF := apply(∧, 𝒪_DIAG, Filter[0]); k := 1
5    While (𝒪_PREF = 0)
6       𝒪_PREF := apply(∧, 𝒪_DIAG, Filter[k]); k := k+1
7    Return 𝒪_PREF
8 EndFunction
```

```
1 Function ComputeFaultCardinalityFilters(COMPS, 𝒱𝒪)
2    n := |COMPS|
3    Filter[0] := build(C_{1,OK} ∧ ... ∧ C_{n,OK}, 𝒱𝒪)
4    For k:=1 To n
5       Filter[k] := build(0)
6       For i:=1 To n
7          𝒪_i := restrict(FILTER[k-1], C_{i,OK})
8          𝒪_i := apply(∧, 𝒪_i, build(∼C_{i,OK}))
9          Filter[k] := apply(∨, Filter[k], 𝒪_i)
10   Return Filter[]
11 EndFunction
```

Fig. 5. Computation of Fault Cardinality Filters and Minimum Cardinality Diagnoses

result showing that *each* diagnostic problem can be solved in polynomial time provided that a compact encoding for the System Description can be found. Obviously this compact encoding is not always possibile (otherwise we would have found a polynomial algorithm for the NP-hard problem of MBD), however a good variable order in many cases produces a compact encoding.

The above results also say something relevant about the encoding of all possibile diagnoses (we know that in principle they may be exponential in the number of components): the size of the OBDD $\mathcal{O}_{DIAG}$ encoding the diagnoses is no larger than the size of the OBDD $\mathcal{O}_{DT}$ encoding the Domain Theory.

It is worth noting that for many devices the diagnostic problems always involve the same set of observable parameters *OBS*.

In this case we can exploit this regularity by executing `DisregardVariables()` only once off-line; as an important benefit, the on-line computational complexity becomes linear in $|\mathcal{O}_{DT}|$ (i.e. $O(|INPUTS \cup OBS| \cdot |\mathcal{O}_{DT,OBS}|)$).

4.2   Computing Preferred Diagnoses

As stated above, the client of a diagnostic system may be interested just in the preferred diagnoses, especially when the set of returned diagnoses can be very large. Preferred diagnoses can be efficiently computed from $\mathcal{O}_{DIAG}$ when the selected preference criterion is to minimize the number of faults (see Definition 4).

The basic idea consists in pre-compiling an OBDD *Filter[k]* representing the set of all assignments to *COMPS* involving k faults, for each $k = 0, \ldots, n$; by filtering the set of all the diagnoses for a specific diagnostic problem (i.e. $\mathcal{O}_{DIAG}$) with such OBDDs we can determine the set of diagnoses with k faults.

Figure 5 reports the diagnostic algorithm computing the minimum cardinality diagnoses. The algorithm is essentially the same as the one reported in Figure 4 but the computation of preferred diagnoses is added. In particular, the algorithm intersects $\mathcal{O}_{DIAG}$ with *Filter[k]* starting with $k = 0$ and stopping as soon as the result $\mathcal{O}_{PREF}$ is not empty.

The algorithm to be run offline for computing the complete set of *fault cardinality filters* is shown in Figure 5. OBDD *Filter[k]* represents all and only the instantiations of *COMPS* variables containing exactly k faulty components. OBDD *Filter[0]* represents the situation with no fault, i.e. all the components are in the *OK* mode. Intuitively, for each instantiation of *COMPS* represented in *Filter[k-1]*, *Filter[k]* substitutes the assignment of the *OK* mode to a component C_i with all the possible faulty behavioral modes of C_i.

For computational complexity we show results concerning the off-line computation of fault cardinality filters as well as results concerning the on-line computation of minimum-cardinality diagnoses based on the filters.

Theorem 4. *The time complexity for computing fault cardinality filters is polynomial in* $|COMPS|$. *The size of each filter* $Filter[k], k = 0, \ldots, |COMPS|$ *is* $O(|COMPS|^2)$.

Proof. Let's assume that the variables in *COMPS* are ordered as $(C_1, \ldots, C_n)$ according to $\mathcal{VO}$.

In order to prove that the size of *Filter[k]* is $O(|COMPS|^2)$, we first prove that, for each $C_i \in COMPS$, assuming that the first value of $DOM(C_i)$ in $\mathcal{VO}$ is fbm_i, *Filter[k]* contains at most $(k + 1)$ copies of the Boolean variable C_{i,fbm_i} encoding assignment $C_i(fbm_i)$.

Let $\mathcal{F}_k$ be the Boolean function represented by OBDD *Filter[k]*; moreover, let function $\mathcal{F}_k|_{C_1(bm_1)_B, \ldots, C_{i-1}(bm_{i-1})_B}$ be the restriction of $\mathcal{F}_k$ that assigns bm_j to $C_j, j = 1, \ldots, i - 1$.

The only contribution of $\mathcal{F}_k|_{C_1(bm_1)_B, \ldots, C_{i-1}(bm_{i-1})_B}$ to the value of $\mathcal{F}_k$ consists in the number of faults that are present in the following assignment: $C_1(bm_1), \ldots, C_{i-1}(bm_{i-1})$.

Such a number must clearly be between 0 and k and thus the number of copies of $C_{i,fbm}$ must be at most $k + 1$.

As for a variable C_{i,bm_i}, $bm_i \neq fbm_i$, it is immediate to see that the number of restrictions of $\mathcal{F}_k$ with constant assignments to all the Boolean variables preceding C_{i,bm_i}, $bm_i \neq fbm_i$ in $\mathcal{VO}$ now depends on two possibly independent factors:

- the number of faults that are present in the restriction (which must be in the range $0, \ldots, k$)
- whether multi-valued variable C_i has already been assigned a value in the restriction or not (either true or false)

It follows that the number of restrictions, and consequently of copies of node C_{i,bm_i}, is at most $2 \cdot (k + 1)$.

Since $k \leq |COMPS|$, the number of copies of *any* Boolean variable in *Filter[k]* is $O(|COMPS|)$ and the size of *Filter[k]* itself is $O(|COMPS|^2)$.

As for the time complexity of function `ComputeCardinalityFilters()`, we note that the body of the inner loop is executed $|COMPS|^2$ times.

Since the size of each $Filter[k]$ is $O(|COMPS|^2)$, it is easy to see that each execution of the inner body takes time $O(|COMPS|^4)$ (determined by the second *apply* operation). It follows that the execution of the function takes time polynomial in $|COMPS|$.   □

Corollary 1. *The time complexity of computing $\mathcal{O}_{PREF}$ starting from $\mathcal{O}_{DIAG}$ is $O(|\mathcal{O}_{DIAG}| \cdot |COMPS|^3)$.*

The theorem and its corollary ensure that both the off-line computation of fault cardinality filters is tractable and that the size of any computed filter is small despite the fact that some of the filters represent an exponential number of assignments to $COMPS$.

By combining the results of Theorem 3 and Corollary 1 we have the guarantee that the computation of preferred diagnoses is not only tractable but also efficient if the Domain Theory can be compactly encoded into an OBDD.

5   Dealing with Ambiguous Observations

In many domains of practical interest some parameters characterizing the system are continuous variables that are discretized into a set of qualitative values in order to describe the system behavior in a qualitative way. As an example, let us consider an electric circuit where the behavior is described in terms of qualitative deviations and (deviation of) the intensity I_x at a specific probe point x can take values into $\{0, -, --, +, ++\}$. In some cases these qualitative distinctions may not be directly captured by a sensor and therefore some form of ambiguity affects the observation; for example we may have an observation $(I_x, \{+, ++\})$ meaning that $I_x = + \vee I_x = ++$.

More generally, we assume that **OBS** consists of a list of pairs $(O, \{v_1, \ldots, v_k\})$ where O is a variable whose observed value is $v_1 \vee \ldots \vee v_k$.

In this situation each observation is just a weak constraint on the value of a variable O and such a constraint cannot be enforced via the *restrict* operator.

```
1 Function DiagnoseAmbiguousObs(𝒪_DT, OBS, INPUTS)
2    𝒪_DT,OBS := DisregardVariables(𝒪_DT, INTVARS)
3    𝒪_TMP := AssignVariables(𝒪_DT,OBS, INPUTS)
5    ForEach (O, {v₁, ..., v_k}) ∈ OBS
6       𝒪_O := build(0)
7       ForEach v_i ∈ {v₁, ..., v_k}
8          𝒪_O := apply(∨, 𝒪_O, build(O_{v_i}))
9       𝒪_TMP := apply(∧, 𝒪_TMP, 𝒪_O)
10   𝒪_DIAG := DisregardVariables(𝒪_TMP, OBS)
11   Return OBDD_DIAG
12 EndFunction
```

Fig. 6. Diagnosis with Ambiguous Observations

Function `Diagnose()` has to be revised and the new version `DiagnoseAmbigu-ousObs()` is reported in Figure 6. The main difference concerns the enforcement of the constraints represented by observations: in the new version we have to build an OBDD $\mathcal{O}_O$ for capturing the ambiguous reading of each observation O.

Observable variables are still present in $\mathcal{O}_{TMP}$ after intersecting it with all the $\mathcal{O}_O$s and therefore function `DisregardVariables()` has to be invoked.

Polynomial complexity is guaranteed also for diagnosis with ambiguous observations, through an extension of Corollary 3.

6   Dealing with Context-Varying Diagnostic Problems

An important way for discriminating among alternative diagnoses consists in observing the behavior of the system to be diagnosed under different contextual conditions (e.g. for isolating faults in digital circuits as well as for debugging software). With reference to the framework of [5] the change of inputs over time is the simplest class of temporal phenomena i.e. *context-varying*. In fact, discrimination among diagnostic hypotheses is possibile by assuming that in the time window $[1, \ldots, w]$ the behavioral modes of the components of the system do not change over time. Let us suppose that we have at disposal the pairs **INPUTS**$_i$, **OBS**$_i$ for any time instant i in the time window $[1, \ldots, w]$.

We can simply compute the diagnoses over the test vector according to the following expression:

$$\mathcal{O}_{DIAG} = apply(\wedge, \mathcal{O}_{DIAG,w}, apply(\ldots, apply(\wedge, \mathcal{O}_{DIAG,1}, \mathcal{O}_{DIAG,2})\ldots))$$

where $\mathcal{O}_{DIAG,i}$ is an OBDD representing instantaneous diagnoses at time i computed according to algorithm `Diagnose()` (Figure 4).

The extension to context-varying diagnosis seems quite strightforward: we have just to perform instantaneous diagnosis at w different time points and then intersect the results. Unfortunately, although we know that the computation of instantaneous diagnoses at each time point is polynomial in the size of $\mathcal{O}_{DT}$, the complexity of context-varying diagnosis is not guaranteed to be polynomial as stated in the following theorem.

Theorem 5. *Let DP = (SD, $\overline{OBS}$, $\overline{INPUTS}$) be a Context-Varying Diagnostic Problem, with $\overline{OBS}$ = (OBS$_1$, ..., OBS$_w$) and $\overline{INPUTS}$ = (INPUTS$_1$, ..., INPUTS$_w$).*

Moreover, let V_v be the Boolean variable representing value v of variable $V \in SV$ and let $WD_i(V_v)$ (resp. $WD(V_v)$) be the number of copies of V_v in OBDD $\mathcal{O}_{DIAG,i}$ (resp. $\mathcal{O}_{DIAG}$).

Then, if we define $WD_{max}(V_v) = max_{i=1,\ldots,w} WD_i(V_v)$, the following holds:

$$WD(V_v) = O(min((WD_{max}(V_v))^k, (D_{max})^{|COMPS|}))$$

where $k \leq w$ is the number of different contexts in $\overline{INPUTS}$ and D_{max} is the size of the largest domain of a component, i.e. $D_{max} = max_{i=1,\ldots,n}(|DOM(C_i)|)$.

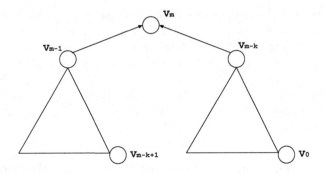

Fig. 7. Schema of $\mathcal{G}$ for the proof of theorem 5

Proof. Let $\mathbf{INPUTS}^1$, ..., $\mathbf{INPUTS}^k$ be the different contexts in $\overline{\mathbf{INPUTS}}$. By definition of Context-Varying Diagnosis we have that:

$$\mathcal{O}_{DIAG} = apply(\wedge, \mathcal{O}_{DIAG,w}, apply(\ldots, apply(\wedge, \mathcal{O}_{DIAG,1}, \mathcal{O}_{DIAG,2})\ldots))$$

Since $\wedge$ is commutative and associative, and since necessarily $\mathcal{O}_{DIAG,i} = \mathcal{O}_{DIAG,j}$ when $\mathbf{INPUTS}_i = \mathbf{INPUTS}_j$ (because of determinism), we can rewrite the expression above as:

$$\mathcal{O}_{DIAG} = apply(\wedge, \mathcal{O}_{DIAG}^k, apply(\ldots, apply(\wedge, \mathcal{O}_{DIAG}^1, \mathcal{O}_{DIAG}^2)\ldots))$$

where $\mathcal{O}_{DIAG}^i$ encodes the instantaneous diagnoses when $INPUTS$ has value $\mathbf{INPUTS}^i$.

Recalling that the maximum number of copies of Boolean variable V_v in each OBDD $\mathcal{O}_{DIAG,i}$ is bounded by $WD_{max}(V_v)$ and that all the OBDDs $\mathcal{O}_{DIAG}^i$ are defined exactly on the same variables, it is easy to see (by taking into account the way *apply* works [6]) that the maximum number of copies of V_v in $\mathcal{O}_{DIAG}$ can't be larger than $(WD_{max}(V_v))^k$.

At the same time, since $\mathcal{O}_{DIAG}$ encodes assignments to $COMPS$ variables, the maximum number of copies of V_v can't exceed the maximum number of such assignments (i.e. $(D_{max})^{|COMPS|}$) because of Theorem 1. Therefore, the claim of the theorem is demonstrated.

In order to show how this upper bound can actually be reached, let's consider a simple case where $rev(\mathcal{G})$ is a directed tree. Figure 7 shows a schematic vision of the System Influence Graph $\mathcal{G}$ where some of the nodes in $\mathcal{G}$ that are relevant for our discussion are showed as circles and labeled with the associated variable names.

We assume that the System Variables have been ordered as $(V_1, \ldots, V_m)$ respecting the directionality condition stated in section 3.4 (note e.g. in figure 7 that the root node of $rev(\mathcal{G})$ is V_m while the rightmost leaf is V_0). Moreover, for each variable $V \in SV$ s.t. $DOM(V) = \{v_1, \ldots, v_l\}$ we assume that the corresponding Boolean variables $V_B = \{V_{v_1}, \ldots, V_{v_l}\}$ have been ordered as $(V_{v_1}, \ldots, V_{v_l})$ (respecting the variable vicinity condition).

We further assume that variable V_{m-k+1} in Figure 7 belongs to $COMPS$ and focus on the number of copies of V_{m-k+1,v_1} (i.e. the Boolean variable that is *true* when value v_1 is assigned to variable V_{m-k+1})[3]. By applying arguments similar to the ones used in the proof of Theorem 4, it is not difficult to see that the number of copies of V_{m-k+1,v_1} in $\mathcal{O}_{DT}$ can be up to $|DOM(V_{m-k})|$. Intuitively, the assignments to variables V_0 to V_{m-k} are *summarized* by the value assigned to V_{m-k}, that is the only one to be "remembered" in order to determine, together with the value of variable V_{m-1}, the value of the root, i.e. variable V_m.

Thanks to Theorem 1, we know that the number of copies of V_{m-k+1,v_1} in $\mathcal{O}_{DT,OBS}$ is the same as in $\mathcal{O}_{DT}$; therefore, also the number of copies of V_{m-k+1,v_1} in $\mathcal{O}_{DIAG,i}$ is up to $|DOM(V_{m-k})|$.

Let's consider the OBDD $\mathcal{O}_{TMP}$ obtained by intersecting $\mathcal{O}^1_{DIAG}$ and $\mathcal{O}^2_{DIAG}$. The set of assignments to variables V_0 to V_{m-k-1} that cause V_{m-k} to take some value v may be different in OBDDs $\mathcal{O}^1_{DIAG}$ and $\mathcal{O}^2_{DIAG}$. This means that, in OBDD $\mathcal{O}_{TMP}$, for each assignment $\mathcal{I}$ to variables V_0 to V_{m-k} we may need to "remember":

- the value assigned by $\mathcal{I}$ to V_{m-k} when $INPUTS$ has value **INPUTS**^{1}
- the value assigned by $\mathcal{I}$ to V_{m-k} when $INPUTS$ has value **INPUTS**^{2}

This leads to a number of copies of V_{m-k+1,v_1} that is up to $|DOM(V_{m-k})|^2$. It is easy to generalize this result to the intersection of $\mathcal{O}^1_{DIAG}, \ldots, \mathcal{O}^k_{DIAG}$ where the number of copies of V_{m-k+1,v_1} can grow up to $|DOM(V_{m-k})|^k$ (provided it is smaller than $(D_{max})^{|COMPS|}$, as explained above). □

The theorem tells us that, even if the size of $\mathcal{O}_{DT}$ is not exponential in $|COMPS|$, the size of $\mathcal{O}_{DIAG}$ may be exponential in $|COMPS|$ in case enough different inputs are provided in the time window $[1, \ldots, w]$ associated with the diagnostic problem.

7 Discussion and Conclusions

In order to handle the computational complexity of the MDB task (and the potentially exponential number of solutions to diagnostic problems), in the MBD community there is an increasing interest regarding the adoption of symbolic methods for encoding both the domain knowledge as well as the solution space of diagnostic problems. So far, most of the attention has been drawn by OBDDs (see e.g. [21], [18]), because of the maturity of the theoretical analysis on OBDDs and the availability of efficient tools implementing the standard operators.

In the present paper, starting from the preliminary results reported in [22] and [24], we have proposed possible solutions to the issues that arise when using OBDDs for the diagnosis of static systems, from the system encoding up to the presentation of preferred diagnoses.

The main contribution of the paper concerns the analysis of both space and time complexity of the diagnostic algorithm. These theoretical results provide

---

[3] If $V_{m-k+1} \notin COMPS$ the proof would just be slightly more complicated.

upper bounds on the time and space needed for solving diagnostic cases; such bounds are polynomial in the size of the OBDD encoding the system model and hold for any diagnostic case independently of the number of faults and the available observations. We have also shown that similar results hold when the observations are ambiguous.

The second major contribution of the paper concerns the result on computational complexity of context-varying diagnosis. In particular, we have shown that there is no guarantee that the space and time upper bounds are polynomial in the size of O_{DT}. This negative result is quite important because it shows that the tractability properties that hold for the atemporal diagnoses cannot be extended to any system where some of the parameters change over time. In fact, the case of Context-Varying Systems is just the simplest class in the temporal diagnosis ontology proposed in [5].

All the positive complexity results stated in the paper depend on the size of the OBDD encoding the system model. While it is often possible, by carefully selecting the variable order, to get compact encodings even for large systems, there are cases where the size of the encoding is too large to be practically manageable.

However, this does not necessarily mean that the methods described in this paper are not applicable. In a recent paper [25] we have shown how a complex system can be partitioned into a set of subsystems that can be encoded separately and how global diagnoses can be computed from diagnoses local to each subsystem. In this way we have been able to diagnose very complex digital circuits involving hundreds of components.

References

1. F. Aloul, I. Markov, K. Sakallah: Faster SAT and Smaller BDDs via Common Function Structure. Proc. Int. Conf. on Computer Aided Design (2001)
2. F. Aloul, I. Markov, K. Sakallah: FORCE: a Fast and Easy-to-Implement Variable-Ordering Heuristic. Proc. Great Lakes Symposium on VLSI (2003) 116-119
3. Bertoli, P., Cimatti, A., Roveri, M., Traverso, P.: Planning in Nondeterministic Domains under Partial Observability via Symbolic Model Checking. Proc. IJCAI01 (2001) 473-478
4. Bollig, B., Wegener, I.: Improving the Variable Ordering of OBDDs is NP-complete. IEEE Transactions on Computers **45**(9) (1994) 932–1002
5. Brusoni, V., Console, L., Terenziani, P., Theseider Dupré, D.: A Spectrum of Definitions for Temporal Model-Based Diagnosis. Artificial Intelligence **102** (1998) 39–79
6. Bryant, R.: Symbolic boolean manipulation with Ordered Decision Diagrams. ACM Computing Surveys **24** (1992) 293-318
7. Bylander, T., Allemang, D., Tanner, M., Josephson, J.: The Computational Complexity of Abduction. Artificial Intelligence **49**(1-3) (1991) 25-60
8. Cimatti, A., Pecheur, C., Cavada, R.: Formal Verification of Diagnosability via Symbolic Model Checking. Proc. IJCAI (2003) 363–369
9. Console, L., Picardi, C., Ribaudo, M.: Process algebras for system diagnosis. Artificial Intelligence **142**(1) (2002) 19-51

10. Cordier, M.-O., Largouet, C.: Using model-checking techniques for diagnosing discrete-event systems. Proc. DX01 (2001) 39-46
11. Darwiche, A.: Model-based diagnosis using structured system descriptions. Journal of Artificial Intelligence Research **8** (1998) 165-222
12. de Kleer, J.: Using Crude Probability Estimates to Guide Diagnosis. Artificial Intelligence **45**(3) (1990) 381-391
13. de Kleer, J., Mackworth, A., Reiter, R.: Characterizing Diagnoses and Systems. Artificial Intelligence **56**(2-3) (1992) 197-222
14. ISCAS 85 Benchmark `http://www.visc.vt.edu/~mhsiao/iscas85.html`
15. Jensen, R. M.,Veloso, M. M.: Obdd-based universal planning: Specifying and solving planning problems for synchronized agents in nondeterministic domains. Lecture Notes in Computer Science **1600** (1999) 213-248
16. Meinel, C., Theobald, T.: Algorithms and Data Structures in VLSI Design: OBDD - Foundations and Applications. Springer-Verlag, Berlin, Heidelberg, New York (1998)
17. Pogliano, P., Riccardi, L.: Modeling Process Diagnostic Knowledge through Causal Networks. Lecture Notes in Computer Science **992** (1995) 323-334
18. Schumann, A., Pencolé, Y., Thiébaux, S.: Diagnosis of Discrete-Event Systems using Binary Decision Diagrams. Proc. 15th Int. Work. on Principles of Diagnosis (DX04) (2004) 197-202
19. Sieling, D., Wegener, I.: NC-algorithms for Operations on Binary Decision Diagrams. Parallel Processing Letters **3**(1) (1993) 3-12
20. Struss, P., Rehfus, B., Brignolo, R., Cascio, F., Console, L., Dague, P., Dubois, P., Dressler, P., Millet, D.: Model-based Tools for the Integration of Design and Diagnosis into a Common Process - A Project Report. Proc. DX02 (2002)
21. Sztipanovits, J., Misra, A.: Diagnosis of discrete event systems using Ordered Binary Decision Diagrams. Proc. DX96 (1996)
22. Torasso, P., Torta, G.: Computing Minimum-Cardinality Diagnoses Using OBDDs. Lecture Notes in Artificial Intelligence **2821** (2003) 224-238
23. Torasso, P., Torta, G.: Compact Diagnoses Representation in Diagnostic Problem Solving. Computational Intelligence **21**(1) (2005) 27–68
24. Torta, G., Torasso, P.: The Role of OBDDs in Controlling the Complexity of Model Based Diagnosis. Proc. Int. Work. on Principles of Diagnosis (2004) 9-14
25. Torta, G., Torasso, P.: On the use of OBDDs in Model-Based Diagnosis: an Approach Based on the Partition of the Model. Knowledge-Based Systems **19** (2006)

Examples of Integration of Induction and Deduction in Knowledge Discovery

Franco Turini, Miriam Baglioni, Barbara Furletti, and Salvatore Rinzivillo

Dipartimento di Informatica,
University of Pisa, Italy
{turini, baglioni, furletti, rinziv}@di.unipi.it

Abstract. The use of classification trees in two quite different application areas –business documents on one side and geographic information
systems on the other– is presented. What is in common between such
so different applications of the classification techniques based on trees is
the need of complementing the straightforward use of induction with the
exploitation of some form of deductive, or better to say expert, knowledge. When working on business documents, the expert knowledge, in
the form of rules elicited from human experts, is used to improve the
construction of the classification tree by complementing the inductive
knowledge coming from the examples in the choice of the next node
to add to the tree. When working on geographic information systems,
the expert knowledge, in the form of specifying which are the spatial
relationships among the geographic objects, is used to extract the information from the GIS in a form that can be then processed in an inductive
style.

1   Introduction

Classification is one of the most useful techniques in knowledge discovery. It
allows one to construct a classifier via the analysis of already classified examples,
and then to use it to assign a new observation to a class. Several models for
classification have been proposed, and classification trees are among the most
successful. In the next subsection a description of the construction process for a
classification tree is given.

In our research experience, especially when looking at applications of knowledge discovery techniques to real problems, we have found the model of classification trees extremely useful. At the same time, as soon as we have approached
applications, we have realized that plain induction techniques, like the ones embedded in the basic algorithms for the construction of classification trees, are
not sufficient. In fact, such applications do not offer data simply organized as a
table in which the columns are the attributes common to each observation and
the rows are the single observations.

The solutions we propose follow a common strategic approach, that is the
integration of the basic induction techniques with expert knowledge. In a way,
we propose a reconciliation between the basic inductive approach emerging from

O. Stock and M. Schaerf (Eds.): Aiello Festschrift, LNAI 4155, pp. 307–326, 2006.

the machine learning field and the knowledge based deduction approach emerging from the expert systems field.

The need for approaching the construction of classification trees in such a more complex way came out naturally when dealing with complex data and knowledge.

The first opportunity is given by the need of extending the classification approach to knowledge stored in a geographical information system. Geographic Information Systems, GIS from now on, contains geo-referenced information, i.e., in database terms, the values of the attributes are bound to some geographic entity. The point is that the geography may be different for each of the attributes and the relationships among the geographies of the different attributes may be interpreted according to different perspectives. Here it comes the need of using some deep knowledge of the application area in order to extract the data for the induction step.

The second opportunity is given by an application in the business area. The idea is to classify company plans for innovating their products and processes in order to foresee whether they may be successful or not. The construction of the classifier is based, as usual, on existing examples for which the success/failure is known. However, it came out quite clearly that the information embedded in the plans, although quite complex, was not sufficient for constructing a well working classifier. The information in the plans is somewhat *contextual*, and in order to exploit it in an optimal way it is necessary to complement it with a general understanding of the rules - the general rules - that underlie the innovation process. Such rules have to be elicited from experts of the field. We have actually interviewed colleagues in the Business School of our University to that purpose. A critical point has been how to represent the extracted knowledge and how to use it. Our solution has been to use Bayesian Clausal Maps for representing the knowledge and the dynamic extraction of rules for affecting the construction of the tree during the construction of the tree itself.

The rest of the paper is organized as follows. Section 2 contains some background material on the classification process and on classification trees. Section 3 deal with spatial classification, with special attention to the phase of preparing the data for the induction step. Section 4 deals with the second approach, in which the expert knowledge is used to drive a better construction of classification trees in the area of assessment of business plans. The conclusions outline some of the work going on that builds on the approach described here.

2   A Brief Overview on Classification Process

The aim of a classifier is to create a model capable of assigning a class to transactions according to the values of (some of) their attributes. Usually, the input is a table of tuples, where one of the columns is chosen as the *class attribute*. The task is to build a model for predicting the value of the *class attribute*, knowing only the values of the others.

Many methods have been proposed in the literature for classification, such as Bayesian classifiers [5,18,12], decision tree[15,14,2,11,10], and neural networks [17,8]. We take into consideration *decision tree* models for their understandability, and their robustness to noisy data, since this property is crucial in our application contexts.

A decision tree is a tree data structure consisting of *decision nodes* and *leaves*, where each decision node denotes a test over one of the attributes and the leaf nodes represents one of the possible classification values. Unknown samples are classified by testing their attributes against the decision nodes, and they are assigned to the class corresponding to the reached leaf node.

The construction of a classification requires two separate phases.

- *Learning phase.* The tree is built using a training set, starting from a root node and recursively splitting the data according to a statistical measure to grow the subtrees of the root node. The decision tree classifiers proposed in the literature can be distinguished according to the statistical criterion used for dividing the data. For example CART [2] uses the *Gini index*, whereas *ID3* [14] and *C4.5* [15] use the *entropy* to measure the (im)purity of the samples.
- *Test phase.* The built tree is used to classify a set of known samples (i.e. the *test set*) and, possibly, it is re-structured to improve prediction accuracy and speed. The revision is based on statistical measures to remove the less reliable branches. Such pruning task is performed using two main techniques:
    - *stopping methods*: a tree is pruned by stopping its construction. For example, a node can be prevented to be split if the value of the attribute is below a threshold.
    - *post-pruning methods*: the tree is pruned after it has been fully constructed. Some branches are removed by replacing a split node by a leaf. This usually happens, for example, when the prediction error of the pruned branch is not worse than the error of the unpruned tree.

Example 1 (ID3 Example). Consider the well known example presented in [14]. The input consists of a table containing a set of tuples that describe the environmental conditions suitable for playing (or not playing) tennis. In Table 1 the tuples have been ordered according to the attribute *Outlook*. Using the ID3 learning algorithm we may obtain the decision tree in Figure 1.

When a sample exhibits the set of attribute/value pairs in a path from the root to a leaf, it is classified according to the class associated with the leaf. For example the sample { *Outlook=Rainy, Wind=Strong*} is classified as *No*: the first attribute selects the rightmost branch of the tree (*Rainy*); the second attribute selects the left branch of the *Wind* decision node.

The decision tree for the example above is obtained by recursively splitting the training set until all the samples (or the majority of the samples) in each partition belong to the same class.

Table 1. Training set of the *Play Tennis* example

Outlook	Temperature	Humidity	Wind	Class
Overcast	Hot	High	Weak	Yes
Overcast	Cool	Low	Strong	Yes
Overcast	Mild	High	Strong	Yes
Overcast	Hot	Low	Weak	Yes
Rainy	Cool	Low	Strong	No
Rainy	Mild	High	Strong	No
Rainy	Mild	High	Weak	Yes
Rainy	Cool	Low	Weak	Yes
Rainy	Mild	Low	Weak	Yes
Sunny	Hot	High	Weak	No
Sunny	Hot	High	Strong	No
Sunny	Mild	High	Weak	No
Sunny	Cool	Low	Weak	Yes
Sunny	Mild	Low	Strong	Yes

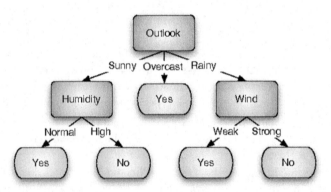

Fig. 1. A decision tree obtained from the training set in Table 1

2.1   Selecting Best Split

The choice of the splitting attribute for each decision node is crucial for the quality of the final decision tree. In fact, we aim at building compact decision trees, by choosing at each decision node the attribute that reduces the "impurity" of the samples the most. Moreover, the ID3 algorithm (like most decision tree learning methods) adopts a greedy strategy: an attribute is chosen on the basis of local measures and the alternative cases are not explored.

The criterion for the choice of the splitting attribute is based on the *entropy* measure of the distribution of samples. At each decision node, the algorithm chooses a possible attribute test with, say, n outcomes. The set S of the training samples can be partitioned into subsets $S_1, \ldots, S_n$. The only information available to evaluate the test attribute is the distribution of the classes in each of the subsets. The ID3 algorithm uses a statistical criterion, the *information gain*, to quantify how well a test attribute splits the samples. From the information theory, the entropy is used to estimate the information brought by each subset, by considering the probabilities that each class occurs in each partition.

If S is a set of samples and C_i is one of the possible classification values, we denote with $freq(C_i, S)$ the number of samples S that belong to class C_i. The probability that a randomly picked sample in S will belong to a class C_j is given by:

$$\frac{freq(C_i, S)}{|S|}.$$

Then, the information brought by this event is given by :

$$-\log_2\left(\frac{freq(C_i, S)}{|S|}\right).$$

To determine the expected information conveyed by a set of samples S, the information bound to the classes is summed up in proportion to their probabilities:

$$info(S) = -\sum_{i=1}^{n} \frac{freq(C_i, S)}{|S|} \times \log_2\left(\frac{freq(C_i, S)}{|S|}\right)$$

where n is the number of the possible classifications. Intuitively, $info(S)$ gives the average information needed to identify the class of a sample in S. For example, if S contains only samples belonging to a single class C_k then $info(S)$ will be zero, thus expressing the fact that each sample is univocally labeled by C_k.

When the samples in S are splitted according to a test attribute T with l outcomes, the expected information of the splitting is measured as the weighted sum of the information in each subset:

$$info_T(S) = \sum_{j=1}^{l} \frac{|S_i|}{|S|} \times info(S_i)$$

where S_i is the ith subset of the partition of S.

The information gained by partitioning S by means of the test attribute T is given by

$$gain(T) = info(S) - info_T(S).$$

The *gain criterion* is based on this measure and it selects the attribute that maximizes this quantity. The strategy of choosing an attribute that maximizes the information gain (and then minimizes the entropy) is justified by the observation that the splitting divides the data into smaller subsets, where the uniformity of samples may increase.

The C4.5 algorithm works essentially as the ID3 does, but the splitting criterion is based on the *information gain ratio*. The information gain ratio is the gain normalized by the information due to the split of S on the basis of the testing attribute:

$$gainRatio(T) = \frac{gain(T)}{-\sum_{j=1}^{l} \frac{|S_i|}{|S|} \times log_2 \frac{|S_i|}{|S|}}$$

Example 2 (Information Gain). In the training set in Table 1 there are 14 cases, nine of which are positive (*Yes*) and five are negative (*No*). The initial information of the samples is:

$$info(S) = -\frac{9}{14} \times log_2 \frac{9}{14} - \frac{5}{14} \times log_2 \frac{5}{14} = 0,9403$$

Once the samples have been divided according to the *Outlook* attribute we obtain three subsets (one for each value of the attribute). The entropy of the samples in each partition is:

$$info(S|Outlook=Overcast) = 0,0000$$
$$info(S|Outlook=Rainy) = 0,9709$$
$$info(S|Outlook=Sunny) = 0,9709$$

Then the expected entropy of the split is:

$$info_{Outlook}(S) = \frac{4}{14} \times 0,0000 + \frac{5}{14} \times 0,9709 + \frac{5}{14} \times 0,9709$$
$$= 0,6935$$

The information gain given by the attribute *Outlook* is:

$$gain(Outlook) = 0,9403 - 0,6935 = 0,2468$$

The attributes *Humidity*, *Wind*, and *Temperature* have a higher entropy (and hence a lower gain) than the attribute *Outlook*. Thus *Outlook* is selected as the best attribute for splitting and the root node is created accordingly. The growth of the tree will continue in each of the three partitions to create the subtrees. Notice that the partition relative to the value *Overcast* contains only *Yes* samples. In this case a leaf is created to represent the class of the samples (see Figure 1).

3   Spatial Classification

In a GIS application, a spatial dataset consists of a set of layers, where each layer brings the information on a particular aspect of the real world. What characterizes a geographic region is the union of all the pieces of information in all the layers. This way of organizing spatial data raises a new challenge in defining a spatial transaction. In fact, a transaction is a tuple of attributes brought together by all the layers and associated with a representative geometry (i.e. the geometry where the tuple holds). In general, one of the available layers is chosen as the *reference layer*, and each feature in this layer is used to select the features in the other layers.

3.1   Spatial Transactions

Formally, let $\mathcal{L} = \{L_1, L_2, \ldots, L_n\}$ be a set of layers, L_r be a reference layer, and S_R be a set of spatial relations. Each layer has a set of non-spatial attributes that describe the state of each object in the layer. For the clarity of presentation, we assume that each layer L_i has only one categorical attribute $attr_{L_i}$. For each object $o \in L_i$ the value of the attribute $attr_{L_i}$ is given by the term $o.attr_{L_i}$.

Dimensionally Extended 9-intersections Model. In order to rigorously define the spatial relation between two geometries, we adopt the *Dimensionally Extended 9-intersections Model* [4,?]. Given two objects embedded in a topological space, three sets of points are determined for each object: the interior (denoted by °), the boundary (denoted by δ), and the exterior, or complement (denoted by ⁻).

It is possible to determine the spatial relation between two geometries A and B by considering all the possible intersections, actually 9 intersections, among the three sets of points associated to both A and B. For each intersection we consider the dimension of the intersection itself. In particular, given a set S, the dimension of S is given by the function $dim(S)$:

$$dim(S) = \begin{cases} - & \text{if } S = \varnothing \\ 0 & \text{if } S \text{ contains at least a point and no lines or areas} \\ 1 & \text{if } S \text{ contains at least a line and no areas} \\ 2 & \text{if } S \text{ contains at least an area} \end{cases}$$

The spatial relation $R(A, B)$ is given by the following *9-intersection matrix*:

$$R(A, B) = \begin{pmatrix} dim(\delta A \cap \delta B) & dim(\delta A \cap B°) & dim(\delta A \cap B^-) \\ dim(A° \cap \delta B) & dim(A° \cap B°) & dim(A° \cap B^-) \\ dim(A^- \cap \delta B) & dim(A^- \cap B°) & dim(A^- \cap B^-) \end{pmatrix}$$

Materializing Spatial Transactions. The selection of the transactions within a GIS dataset can be performed by the application of some basic GIS operations. In this section we present the basic operations involved for materializing a set of spatial transactions.

Select by value. Given a layer $L = \{f_1, f_2, \ldots, f_n\}$ and a value v of the categorical attribute of L, the operation selects all the features in L whose value is v.

$$select(L, v) = \{f_i \in L | f_i.attr_L = v\}$$

Group by value. The natural extension of the *select-by-value* operation is the clustering of features in the layer according to all the possible values of the categorical attribute.

$$group\text{-}by\text{-}value(L) = \{select(L, v_1), \ldots, select(L, v_l)\},$$

where $v_1, \ldots, v_l$ are all the possible values of the categorical attribute. A typical example of this operation is presented by many GIS viewer applications: each feature in the layer is colored according to the value of one of its attributes.

Select by relation. The operations presented so far consider only the non-spatial attributes of the layer. We consider now the spatial extension of the features by exploiting it for selecting a set of objects in the neighborhood of a given feature:

$$select\text{-}by\text{-}relation(f, L, R) = \{f_i \in L | R(f, f_i) holds\}$$

Although it is not necessary, we assume that the reference feature f does not belong to the layer L, by considering layers that do not contain self-overlapping features.

Layer join. The *select-by-relation* operation can be extended to the case of two layers by iterating the selection process for each feature in the reference layer. In details:

$$layer\text{-}join(L_1, L_2, R) = \{\langle f_1, F_2\rangle | F_2 \neq \varnothing,$$
$$F_2 = select\text{-}by\text{-}relation(f_1, L_2, R), \forall f_1 \in L_1\}$$

The result is a set of pairs, where each pair contains a reference feature of the first layer and the set of features selected in the second layer. When the selection-by-relation is empty, the pair is not included in the resulting set.

To generate the set of spatial transactions, the operation can be easily extended to the case of $n + 1$ layers, where the first layer is used as reference layer:

$$layer\text{-}join(L_r, L_1, \ldots, L_n, R) = \{\langle f_{ri}, F_{1i} \cup \cdots \cup F_{ni}\rangle | F_{ij} \neq \varnothing,$$
$$F_{ij} = select\text{-}by\text{-}relation(f_{ri}, L_{ji}, R), \forall f_{ri} \in L_r, j = 1, \ldots, n\}$$

So far we have considered a generic set of spatial relations to drive the extraction of the spatial transactions. However, the set of relations can be organized in a hierarchy. In this way, the spatial transactions can be generated at different levels of details. For example, when aggregating specific relations into a coarser relation, the select-by-relation operation may return a single selection rather than a multi selection (since many objects selected by similar relations will form a unique multi-object). The opposite process, i.e. the use of more specific relations, produces a large number of multi selections. The generation of spatial transactions can also be driven by the hierarchies of concepts. Like the spatial relations, the attribute values of each layer may be used to group similar objects together. For example, the hierarchy of a road layer may consider the object at different layer of details: the single road (e.g., A11, A4), the road type (e.g. highway, motorway), and so on.

The choice of the relation(s) in the above operations is a matter of expert choice. In the experiments performed so far, such choices have been performed directly by the data miner during the process of data preparation for the induction phase. One of the objective we are pursuing now is the derivation of such choices from background knowledge coded in a domain ontology.

3.2   Spatial Decision Trees

Our goal is to build a decision tree capable of assigning an area to a class, given the values of the other layers with respect to the area. Like in transaction classification, we follow two steps: first, we build a model from a set of samples, namely a *training set*; then, we use the model to classify new (unseen) areas. The training set is determined by the spatial transactions extracted from the dataset.

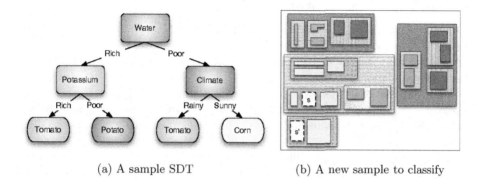

(a) A sample SDT   (b) A new sample to classify

Fig. 2. A possible spatial decision tree

Definition 1 (Spatial Decision Trees). *A* Spatial Decision Tree *(SDT) is a rooted tree where* (i) *each internal node is a decision node over a layer,*(ii) *each branch denotes an outcome of the test and* (iii) *each leaf represents one of the class values. A decision node n_i is associated with a layer L_i and with the attribute X_i of the layer. The outcoming edges are labeled with the possible values of X_i.*

3.3   SDT Classification

An area A is classified by starting at the root node, testing the layer associated with this node and following the branch corresponding to the test result. Let $x_1, x_2, \ldots, x_m$ be the labels of the m edges of the root node. If A is related to an object of type x_j in the layer associated with the root node, then the edge labeled with x_j is followed. This testing process is repeated recursively starting from the selected child node until a leaf node is reached. The area A is classified according to the value in the leaf. When the query region A relates to several areas with distinct values, then all the corresponding branches are followed. The area A is split according to the layer values and each portion is classified independently.

Example 3. In Figure 2(a) a spatial decision tree for the example in Figure 2(b) is presented. This decision tree classifies areas according to whether they are suitable for a type of crop rather than another. In particular, in this example we have three kind of crops: *Corn, Tomato* and *Potato*. Given a new instance **s** (marked with **s** in Figure 2(b)), we test **s** starting from the layer associated with the root node, i.e. the *Water* layer. Since **s** overlaps –the kind of relation chosen in this case– a water region whose value is *Poor*, the corresponding branch is followed and the node associated with the *Climate* layer is selected. Thus, **s** is tested against the features in the *Climate* layer: in this case it overlaps a *Sunny* region, so the class *Corn* is assigned to the instance **s**.

3.4   SDT Learning Algorithm

Spatial transactions are extracted from the set of layers by first selecting a reference layer, i.e. the layer with the attribute chosen as the classifier, then by relating the objects in the other layers via the spatial relation *layer-join* discussed in Section 3.1.

Following the basic decision tree learning algorithm [14], our method [16] employs a top-down strategy to build the model. Initially, a layer is selected and associated with the root node, using a statistical test to verify how well it classifies all samples. Once a layer has been selected, a node is created and a branch is added for each possible value of the attribute of the layer. Then, the samples are distributed among the descendant nodes and the process is repeated for each subtree.

The crucial point of the algorithm is the selection of the split layer for the current node. In Section 3.6 a strategy based on the notion of entropy is presented to quantify how well a layer separates samples. As we see later, the use of the spatial measure (i.e. the aggregate *area*) is crucial here. Once a layer is selected for a test node, the spatial transactions are partitioned according to the layer itself and the associated spatial relation. In Section 3.5 we show how to compute this partition.

3.5   Splitting Spatial Transactions

We aim at grouping spatial transactions according to the categorical attribute of the layers. We select a layer L_i and we split the transactions according to this layer. In general, if layer L_i has q possible values then it can split the samples in $q + 1$ subsets, i.e. a subset for each value $v_j, j = 1, 2, \ldots, q$, and a special subset corresponding to none of these values (termed $\neg L(C)$). The choice of the attribute to use for the split is crucial for the quality of the learned model. In the next section we discuss how to choose such an attribute among all the candidates. An example of splitting the transactions according to the *Water* layer and the *overlap* relation is presented in Figure 3.

3.6   Selecting the Best Split

In this section we introduce a statistical measure, the *spatial information gain*, to select a layer that classifies training samples better than the others. The information gain is based on the notion of *entropy*.

Spatial Information Gain. We present now the method to compute the entropy for a layer L, with respect to the class label layer S. First, we evaluate the entropy of the samples, i.e. the information needed to identify the class of a spatial transaction. While in tuple-transaction the frequency of a sample is expressed as a ratio of transaction occurrences, we use here the spatial measure of the samples. In order to maintain the presentation clear, we assume that all the samples are polygonal object and we use the aggregate *area* to measure their extents. The method is easily extensible to other spatial dimensions (i.e. lines, points).

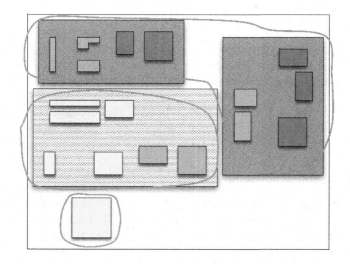

Fig. 3. Samples splitted according to the *Water* layer

Thus, given a set of spatial transactions L, we denote with $mes(L)$ the sum of the areas of all polygons associated with the transactions in L. If the set of samples S has l distinct classes (i.e. $c_1, c_2, \ldots, c_l$) then the entropy for S is:

$$H(S) = -\sum_{i=1}^{l} \frac{mes(S_{c_i})}{mes(S)} log_2 \frac{mes(S_{c_i})}{mes(S)} \qquad (1)$$

Given a layer L with values $v_1, v_2, \ldots, v_q$, we split the spatial transactions according to the values of this layer, as showed in Section 3.5. We can figure out the result of the splitting as a set of transaction $L(v_i, S)$ for each possible value v_i in L and, possibly, $\neg L(S)$. From equation (1) we can compute the entropy for samples in each set $L(v_i, S)$. The expected entropy value for splitting is given by:

$$H(S|L) = \frac{mes(\neg L(S))}{mes(S)} H(\neg L(S)) + \sum_{j=1}^{q} \frac{mes(L(v_j, S))}{mes(S)} H(L(v_j, S)) \qquad (2)$$

The set $\neg L(S)$ represents the transactions that can not be classified by the layer L (i.e. the samples not related with the layer L).
The *spatial information gain* for layer L is given by:

$$\text{Gain}(L) = H(S) - H(S|L) \qquad (3)$$

Clearly, the layer L that presents the highest gain is chosen as the *best split*: we create a node associated with L and an edge for each value of the layer. The samples are split among the edges according to each edge value. The selection process is repeated for each branch of the node by considering all the layers except L.

4   Classification of Business Plans

In real applications is not always enough to use the data coded into tables to extract a model that well describes the problem at hand. Quite often there is other knowledge around, that can usefully complement the one hidden inside the examples.

The problem we are addressing is how to build a classification tree taking into account a set of examples (table) and the knowledge proper of the domain of interest, the one owned by the experts in the field (background knowledge).

4.1   Background Knowledge

The background knowledge is represented by a network of dependency relationships via Bayesian Causal Maps (BCM). A BCM is a directed graph that connects concepts via a cause-effect relation. In this kind of graph, arcs connecting related concepts are associated to a probability measure of the strength with which these concepts are related. Bayesian Causal Maps (BCMs) are obtained by merging Causal Maps [6,9], which are used to represent the human way of thinking, and Bayesian Networks [12]. Causal maps include relationships expressed in the form of believes, values and perceptions held by individuals. Causal Maps are useful to represent a simplification of the reality, to highlight the most important elements to solve a specific problem, and to identify each possible alternative. These points are useful when we are interested in pointing out the cause-effect relationships among the variables belonging to the application domain. Causal Maps are also called Dependence Maps (D-MAP) because they guarantee that connected concepts are dependent. A Bayesian network model can be seen both as a qualitative and a quantitative model. At the qualitative level, the model is represented as a directed acyclic graph in which each node represents a variables and each arc represents a probabilistic dependence. The lacking of an arc from a node to another node means that the two nodes are conditionally independent. This is the reason why Bayesian Networks are also called Independence Maps (I-Map). At the quantitative level the influence relations among the variables are expressed as conditional probability distributions.

Our aim is now to use this knowledge to drive the construction of a classification tree. To this purpose, we extract from the map a set of rules, which we call *domain rules*, and we use those rules to change the probability measures used in the entropy computation.

4.2   Domain Rules

Starting from the representation of domain knowledge as a BCM, we are able to identify and extract rules such as: $L \xrightarrow{p_i} R$, which means that when the left hand side L holds, then the right hand side R holds with a probability p_i. The left hand side of a rule can be:

- an element `Attribute-relation-value`, e.g. $A = a$, which refers to a direct connection (Fig. 4(a)) (*simple rule*);

- more than one element `Attribute-relation-value`, each of which refers to a direct connection (Fig. 4(c)), e.g. $A = a \cup C = c$ (*complex rule*);
- an element `Attribute-relation-value` which refers to an indirect connection (Fig. 4(b)), e.g. $A = a\|_{C=c}$ (*indirect simple rule*);
- more than one element `Attribute-relation-value`, each of which refers to a combination of direct and indirect connections (Fig. 4(d)), e.g. $A = a\|_{D=d} \cup C = c$ (*indirect complex rule*).

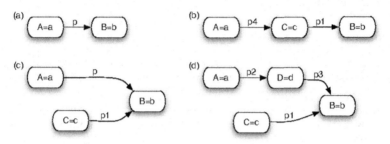

Fig. 4. *Fragments of BCMs*

The right hand side of the rule always contains a single element *Attribute − relation − value*. Figure 4 shows one example for each of the cases reported above. Figure 4(a) shows the fragment of a BCM from which it is possible to derive the simple implication $A = a \xrightarrow{p} B = b$ that states that when A is equal to a then B is equal to b with a likelihood of p. Figure 4(b) shows a fragment of a BCM from which it is possible to extract the extended simple implication $A = a\|_{C=c} \xrightarrow{p_x} B = b$. This rule states that it is possible from $A = a$ to say something about B being equal to b only passing through $C = c$. This is obviously possible only when we do not know anything about the value associated to C. Suppose to know that $C = c$ then from $A = a$ we cannot deduce anything about C because C is already known. The same thing would have happened if we had known that C was equal to c_1. Hence extended implications are a way of connecting attributes not directly connected. Then the likelihood associated to these rules will be dependent on the likelihood of both the connections and it will be smaller than that associated to each connection (e.g. p_4 and p_1 with respect to Figure 4(b)).

From the fragment of Figure 4(c) it is possible to extract the complex implication $A = a \cup C = c \xrightarrow{p_y} B = b$. In this case we have a right hand side ($B = b$) that directly depends on two other elements. These kinds of rules can be extracted only if the value of both the attributes in L (A and C) are known to be equal to that reported in the BCM for these attributes. If only one of the attributes verifies this requirement, a simple implication will be extracted. The likelihood associated to complex implications will be greater than that of each connection starting from a node in L.

Figure 4(d) shows a fragment of a BCM from which it is possible to extract the indirect complex implication $A = a\|_{D=d} \cup C = c \xrightarrow{p_z} B = b$. In this kind of rules the concepts expressed for extended and complex implications are merged. Roughly speaking they are complex implications on the right hand side of an extended rule.

From the things said so far, it follows that the map is a dynamic entity, and that the rules that can be extracted depend both on the attribute value associated to the current sub-tree root node and on the path followed from the root to reach the sub-tree root node. For any further detail in the extraction of domain rules, refer to [1].

4.3   Probability Computation

The likelihood associated to simple implications is, trivially, the value of the influence relationship between the nodes involved in the rule. According to Figure 4(a) the likelihood is p.

When dealing with extended implications we have to consider that this kind of rules can be extracted only if we know nothing about "the node in the middle" of the implication. Hence the likelihood associated to this kind of rules is the product of the influence relationships, which is $p_4 \times p_1$ according to Figure 4(b). More complex tasks are related to the computation of complex rules probabilities. Consider, for example, Figure 4(c). What we want to determine is the probability that $B = b$, knowing that $A = a$ and $C = c$. Hence we have to consider the independent events $E_1 = (B = b|A = a)$ and $E_2 = (B = b|C = c)$. Since both of the events have to be considered (the map tells us that the truth of $B = b$ depends on both of them) we have to compute the probability of the union of the events: $P(E_1 \cup E_2) = P(E_1) + P(E_2) - P(E_1 \cap E_2)$. Being E_1 and E_2 independent in the current instantiation of the map, we can compute the intersection of the events as the product of the probabilities associated to the events, obtaining $p + p_1 - p \times p_1$ as the likelihood associated to this complex rule. This way of reasoning can be extended to more than two complex implications.

For indirect complex implications both the computation of the indirect rule probability and the computation of complex rules probability are merged. According to Figure 4(d) the likelihood associated to the indirect complex rule is $(p_2 \times p_3) + p_1 - (p_2 \times p_3) \times p_1$.

The probability associated to these rules will be used to modify the entropy estimation, which drives the choice of the next node during the construction of the tree.

4.4   Attribute Selection and Domain Rules

The probability expressed by the rule will be used to replace the coefficient $\frac{|S_i|}{|S|}$ in the equation for the entropy computation. Consider the Table 2, related to the subset we are investigating to find the tree next best node, and the rule $R : A = a \xrightarrow{p} C = c_j$.

We replace the coefficient $\frac{|S_j|}{|S|}$ (where $1 \leq j \leq n$ and j denotes the attribute value specified in the rule) with the rule probability. This value, in fact, means

Table 2. A table fragment

Table 2. A table fragment

A	B	C	...
a	b	c	...
a	b	c	...
a	b	c	...
a	b	c	...
a	b1	c1	...
a	b1	c1	...
a	b1	c1	...
a	b2	c1	...
a	b3	c2	...
a	b	c2	...

that there is a different distribution for the attribute values c_j with respect to the one expressed in the data set. The remaining values ($\frac{|S_i|}{|S|}$ where $i \neq j$), that do not occur in the rule, are modified according to the data set values. By using the rule R, the formula to compute the expected information of the splitting of S with respect to a chosen attribute becomes:

$$Info_C(S) = p\Big(Info(C = c_j)\Big) + (1 - p) \sum_{i=1,i\neq j}^{n} \Big(\frac{|C = c_i|}{|S| - |C = c_j|}\Big) Info(C = c_i)$$

When computing the entropy, S identifies the subset of the data set in which we are considering the attribute C, while $C = c_j$ identifies the value of the attribute C for which the rule holds. Hence, we compute the entropy associated to the attribute C by modifying the coefficients associated to its values. The coefficient associated to the value c_j becomes p (the probability associated to the rule), while the ones related to the other values for C are computed according to the probability measure $(1 - p)$ proportionally distributed on the remaining instances.

Example 4. Consider the following instantiation of the rule R

$$R: A = a \xrightarrow{p=0.3} C = c$$

and the Table 2 as our running example. We compute $(1 - 0.3)\frac{4}{6}$ for the value $c1$[1], and $(1 - 0.3)\frac{2}{6}$ for the value $c2$ of variable C. Hence the formula becomes:

$$Info(C|A = a) = 0.3 Info(C = c) + 0.7\frac{4}{6} Info(C = c1) + 0.7\frac{2}{6} Info(C = c2)$$

4.5   A Concrete Example

In order to keep the example manageable, we refer to a concrete but very simple case. Consider the set of supermarket transactions taken from the file coop.arff, and the BCM in Figure 5:

---

[1] Value c1 occurs four times and there are six values for C not equal to c.

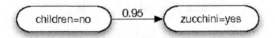

Fig. 5. Bayesian Causal Map

```
DrC45 unpruned tree
-------------------

yogurt = yes: yes (6.0)
yogurt = no
|   detergent = yes: no (3.0)
|   detergent = no
|   |   chicken = yes: no (7.0/1.0)
|   |   chicken = no
|   |   |   zucchini = yes: no (4.0/1.0)
|   |   |   zucchini = no
|   |   |   |   salad = yes
|   |   |   |   |   water = yes
|   |   |   |   |   |   milk = yes: no (2.0)
|   |   |   |   |   |   milk = no
|   |   |   |   |   |   |   corn = yes: yes (3.0/1.0)
|   |   |   |   |   |   |   corn = no
|   |   |   |   |   |   |   |   biscuits = yes: yes (3.0/1.0)
|   |   |   |   |   |   |   |   biscuits = no: no (4.0)
|   |   |   |   |   water = no: no (2.0)
|   |   |   |   salad = no: yes (2.0)
```

Fig. 6. Tree built using domain rules

```
J48 unpruned tree
-----------------

yogurt = yes: yes (6.0)
yogurt = no
|   detergent = yes: no (3.0)
|   detergent = no
|   |   chicken = yes: no (7.0/1.0)
|   |   chicken = no
|   |   |   beer = yes: no (2.0)
|   |   |   beer = no
|   |   |   |   salad = yes: yes (3.0/1.0)
|   |   |   |   salad = no
|   |   |   |   |   water = yes
|   |   |   |   |   |   corn = yes: yes (3.0/1.0)
|   |   |   |   |   |   corn = no: no (8.0/3.0)
|   |   |   |   |   water = no: no (4.0)
```

Fig. 7. Tree built in the usual way

From the map we extract the rule $R : chicken = no \xrightarrow{p=0.95} zucchini = yes$, if *chicken* is selected as root of current (sub)tree.

The tree obtained by the application of the domain rule is shown in figure 6 and as one can notice, differs from the one produced by applying the C4.5 algorithm[2], shown in picture 7, starting from the fourth node from the root.

Considering the execution of C4.5 algorithm, the attribute "beer" results to be the attribute with lower entropy than the others, and its entropy value is the following:

---

[2] Weka J48 implementation of the algorithm.

$$Info(beer|chicken = n) = 0 + \frac{18}{20}(-\frac{7}{18}log_2\frac{7}{18} - \frac{11}{18}log_2\frac{11}{18}) = 0.88.$$

By using the domain rule, the node below on the path where "chicken = n", becomes "zucchini" instead of "beer". In fact its entropy value changes and decreases from

$$Info(zucchini|chicken = n) = \frac{4}{20}(-\frac{1}{4}log_2\frac{1}{4} - \frac{3}{4}log_2\frac{3}{4})+$$
$$+\frac{16}{20}(-\frac{6}{16}log_2\frac{6}{16} - \frac{10}{16}log_2\frac{10}{16}) = 0.93$$

to

$$Info_R(zucchini|chicken = n) = 0.95(-\frac{1}{4}log_2\frac{1}{4} - \frac{3}{4}log_2\frac{3}{4})+$$
$$+0.05(-\frac{6}{16}log_2\frac{6}{16} - \frac{10}{16}log_2\frac{10}{16}) = 0.82,$$

The gain ratio with respect to the attribute *beer* is 0.13681, and the one with respect to *zucchini* considering the domain rule is 0,2864.

Consider now the subset of instances related to the subtree, the root of which is labeled by the attribute chosen by the application of the domain rule R. If there are no rules related to any attribute in that subset, the computation goes on as usual, by calculating the entropy without using external knowledge, and having the node chosen by the algorithm itself. If there are rules related to the attributes in that subset, the computation will take into account these rules. It is worth noting, that the same choice could be made by both the algorithms, also in the presence of domain rules.

As discussed in the introduction, the application that lead us to the modification of the algorithm for the construction of classification trees by the exploitation of expert knowledge was the need of classifying innovation plans. In the discussions with our colleagues of the economic school it quickly came out that the data of the innovation plans were really significant only when interpreted in the context of general knowledge about innovation in companies.

For business documents the experts supplied six macro categories.

1 Company managers and management systems: it contains information about the company managers (i.e. the age, the level of studies and so on) and about the company organizational systems (i.e. the existence of a budgeting process, the existence of a reporting systems and so on).

2 Commercial dimension: it contains information about the reference markets, the relationship between the company and the customers and the *know how* of the person in charge of sales.

3 Technical-productive dimension: it contains information on production agreements, patents, employer skills, skills of the person in charge of production and product design, and the degree of the equipment novelty.

4 Environment caring: it contains information about the geographical concentration and the infrastructures of the company.

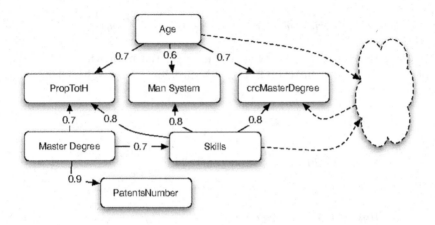

Fig. 8. Fragment of the BCM provided by the experts

5 Competitors analysis: it contains information about providers, customers,
  potential customers, and competitors.
6 Economical-financial analysis: it contains information about the financial
  and economic structure, and the income of the company.

The economic model that forms the domain knowledge (BCM), consists of a
set of weighted influence relations among items. Each relation regulates the
dependency between items, and the weight (probability measure) expresses the
strength of the relation.

A typical example of general knowledge elicited from the experts is represented
by the fragment of map in fig. 8, where

Age. The attribute **Age** contains information regarding the age of all the managers;
PropTotH. Contains the percentage of the working hours devoted to meetings,
  product quality checking, and so on;
ManSystem. The attribute **ManSystem** contains information regarding the
  company management system;
crcMasterDegree. The attribute **crcMasterDegree** indicates if the person
  in charge of commerce has got a master degree;
MasterDegree. The attribute **MasterDegree** refers to the percentage of the
  company manager with a master degree;
Skills. The attribute **Skills** contains information regarding the experience (both
  practical and theoretical) of the management;
PatentsNumber. The attribute **PatentsNumber** contains information regarding the number of patents owned by the company.

5   Conclusions

Classification is one of the most useful techniques for solving application problems, where a new situation has to be understood. The basic solution consists

in constructing a classifier by inducing a model (e.g. a classification tree) out of a number of already classified cases, and to use it for classifying new cases. The classification methods work well in well structured domains, where the cases can be neatly described via a number of well understood attributes. As soon as one attempts to apply classification methods to domains where the information is not that neatly organized, matters become more complex. This paper reports on two experiences, where such difficulties clearly came out and on how we addressed them. Apart from the specific solutions we found for addressing business documents on one side and geographical information systems on the other, we maintain that the general lesson we learned is that pure induction out of known examples is not enough in such complex applications. The way out we found is the exploitation of some form of expert knowledge, that, by enriching the information of the examples with background knowledge, allows us a better classification. We found that domain knowledge can be exploited in different phases of the construction of the classifier. When dealing with geographical information systems we found it useful to exploit it in the preparation of data, while in the context of business documents we found it useful to drive the construction of the tree. What is still unsatisfactory is the way expert knowledge is represented and our current goal is to find a methodology for addressing this aspect. The solution we are looking at, also in the context of three european projects we are working in (GeoPKDD [7], BRITE [3], MUSING [13]), is to derive the technical representation of knowledge from general ontologies, describing the application context. For example, the knowledge about the proper relations we need to use for extracting transactions from GIS can be driven by general descriptions of the geographic objects and their hierarchies. The knowledge behind either a balance sheet or a business plan can be coded, and it is being coded within the aforementioned projects, in proper ontologies. Bayesian causal networks can then be automatically extracted. In summary, the lesson we learned is that both the expert system approach, that underlies current ontological research, and the machine learning approach, that underlies current data mining approach, can contribute, when properly integrated, to the solution of complex application problems.

References

1. M. Baglioni, B. Furletti, and Turini F. Drc4.5: Improving c4.5 by means of prior knowledge. In *Proocedings of the 2005 ACM Symposium on applied computing*, pages 474–481, 2005.
2. L. Breiman, J. Friedman, R. Olshen, and C. Stone. *Classification and Regression Trees*. Wadsworth and Brooks, 1984.
3. BRITE, http://www.briteproject.net/.
4. Eliseo Clementini, Paolino Di Felice, and Oosterorn Oosterorn. *A small set of formal topological relationships suitable for end-user interaction*. SSD '93, Singapore, Springer Verlag LNCS 692, 1993.
5. R. O. Duda and P. E. Hart. *Pattern Classification and Scene Analysis*. Wiley, New York, 1972.

6. C. Eden, F. Ackermann, and S. Copper. The analysis of cause maps. *Journal of Management Studies*, 29(3):309/323, 1992.
7. GeoPKDD, http://geopkdd.isti.cnr.it/.
8. Jiawei Han and Micheline Kamber. *Data Mining: Concepts and Techniques*. Morgan Kaufmann Publishers, San Francisco, 2001.
9. B. Kemmerer, S. Mishra, and P. P. Shenoy. Bayesian causal maps as decision aids in venture capital decision making: Methods and applications. In *In proceedings of the Accademy of Management Conference*, 2002.
10. W.-Y. Loh and Y.-S. Shih. Split selection methods for classification trees. *Statistica Sinica*, 1997.
11. W. Y. Loh and N. Vanichsetakul. Tree-structured classification via generalized discriminant analysis. *Journal of the American Statistical Association*, 83:715–728, 1988.
12. Tom M. Mitchell. *Machine Learning*. McGraw-Hill, New York, 1997.
13. MUSING, http://musing.metaware.it/.
14. J R Quinlan. Induction of decision trees. *Machine Learning*, 1(1), 1986. QUINLAN86.
15. J. Ross Quinlan. *C4.5: Programs for Machine Learning*. Morgan Kaufmann, 1992.
16. Salvatore Rinzivillo and Franco Turini. Classification in geographical information system. In *8th European Conference on Principles and Practice of Knowledfe Discovery in Databases*, 2004.
17. D. E. Rumelhart, Geoffrey E. Hinton, and R. J. Williams. Learning internal representations by back-propagating errors. In Rumelhart D. E., editor, *Parallel Distributed Processing: Explorations in the Microstructure of Cognition*, Cambridge, MA, 1986. Bradford Books.
18. S. M. Weiss and C. A. Kulikowski. *Computer Systems That Learn, Classification and Prediction Methods from Statistics, Neural Networks, Machine Learning and Expert Systems*. Morgan Kaufmann, San Mateo, CA, 1991.

SharedLife: Towards Selective Sharing of Augmented Personal Memories

Wolfgang Wahlster, Alexander Kröner, and Dominik Heckmann

German Research Center for Artificial Intelligence (DFKI) GmbH
Stuhlsatzenhausweg 3, 66123 Saarbrücken, Germany
`firstname.lastname@dfki.de`
`http://www.dfki.de/specter/`

Abstract. The rapid deployment of low-cost ubiquitous sensing devices
– including RFID tags and readers, global positioning systems, wireless
audio, video, and bio sensors – makes it possible to create instrumented
environments and to capture the physical and communicative interaction
of an individual with these environments in a digital register. One of the
grand challenges of current AI research is to process this multimodal
and massive data stream, to recognize, classify, and represent its digital
content in a context-sensitive way, and finally to integrate behavior understanding with reasoning and learning about the individual's day by
day experiences. This augmented personal memory is always accessible
to its owner through an Internet-enabled smartphone using high-speed
wireless communication technologies. In this contribution, we discuss how
such an augmented personal memory can be built and applied for providing the user with context-related reminders and recommendations in
a shopping scenario. With the ultimate goal of supporting communication between individuals and learning from the experiences of others, we
apply this novel methods as the basis for a specific way of exploiting
memories – the sharing of augmented personal memories in a way that
doesn't conflict with privacy constraints.

1 Introduction

The rapid deployment of low-cost ubiquitous sensing devices – including RFID
tags and readers, global positioning systems, wireless audio, video, and bio sensors – makes it possible to create instrumented environments and to capture
the physical and communicative interaction of an individual with these environments in a digital register. One of the grand challenges of current AI research
is to process this multimodal and massive data stream, to recognize, classify,
and represent its digital content in a context-sensitive way, and finally to integrate behavior understanding with reasoning and learning about the individual's
day by day experiences. If we add the clickstream history, bookmarks, digital
photo archives, email folders, calendar, blog and wiki entries of an individual,
we can compile a comprehensive infrastructure that can serve as his augmented
memory. This personal memory is always accessible to its owner through an

O. Stock and M. Schaerf (Eds.): Aiello Festschrift, LNAI 4155, pp. 327–342, 2006.

Internet-enabled smartphone using high-speed wireless communication technologies. We have realized a broad range of augmented memory services in our system SPECTER (see, for instance, [15], [25], [21], [20], and [3]).

Ever since ancient times, storytelling has been a way of passing on personal experiences. The selective sharing of personal augmented memories is the modern counterpart of storytelling in the era of mobile and pervasive internet technology. In our SHAREDLIFE project, we are creating augmented episodic memories that are personal and sharable. The memory model does not aim at a simulation of human memory. Instead we are realizing an augmented memory in an unintrusive way, that may contain perceptions noticed by SPECTER but not by the user.

Although some researchers believe that it is feasible to store a whole human lifetime permanently, we are currently concentrating on a less ambitious task. We try to record and understand an individual's shopping behavior for a few days and share relevant experiences with others in a way that doesn't conflict with his privacy constraints. Dealing with shopping experiences is a limited, but meaningful task against which we can measure progress on our augmented memory research.

2   Related Work on Augmented Memories and Knowledge Sharing

The building of augmented personal memories in instrumented environments for the purpose of extending the user's perception and recall has been studied for more than 10 years (see, e.g., [17]; [9]). While this research has focused on user interface design for the retrieval of memories (among others, [8]; [1]), other research has looked into ways of processing the contents of such memories so as to increase their accessibility to their owner (see, e.g., [12]; [7]).

The exploitation of augmented memories has been researched for diverse scenarios. For instance, work conducted in the E-NIGHTINGALE project shows how automatically created nursing records may help to avoid medical accidents in hospitals (cf. [16]). How RFID technology and Web mining can be applied to support the user with everyday activities is discussed in [19]. In the project LIVING MEMORY [23], records of people's activities and access to community-related information are automatically processed in support of community-related behavior in relatively complex ways. An example of how memories can support social matching is offered by the system AGENTSALON (see, e.g., [24]). The system uses experience logs of participants in an academic conference in order to stimulate conversation via rather extraordinary means, involving animated characters.

3   Personalized Assistance in Mixed-Reality Shopping

Today, the retail industry introduces sensor networks based on RFID technology for advanced logistics, supply chain event management, digital product memories, innovative payment systems, and smart customer tracking, so that

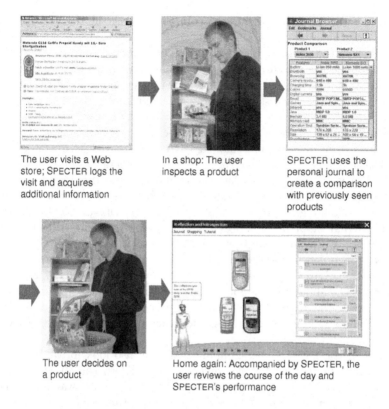

The user visits a Web store; SPECTER logs the visit and acquires additional information

In a shop: The user inspects a product

SPECTER uses the personal journal to create a comparison with previously seen products

The user decides on a product

Home again: Accompanied by SPECTER, the user reviews the course of the day and SPECTER's performance

Fig. 1. A typical action sequence from a shopping scenario explored in SPECTER

shops turn into instrumented environments providing ambient intelligence. Instrumented shopping environments like the METRO future store or the experimental DFKI Cybershopping mall support mixed-reality shopping, which augments the usual physical shopping experience with personalized virtual shopping assistance known from some online shops. Currently, our DFKI installation includes three small shops with instrumented shelves: a grocery store, a camera and phone shop, and a CD shop.

Up to now, such instrumented shopping environments provide more benefits to the supplier than to the customer. Our research is aimed at exploiting the networked infrastructure for more personalized shopping assistance like digital shopping list support, automatic comparison shopping, cross- and up-selling, proactive product information, and in-shop navigation. The combination of advanced plan-recognition techniques with augmented memory retrieval is a prerequisite for the generation of user-adaptive cross-selling and up-selling recommendations. For example, the system, recognizing that the customer is picking up the usual ingredients for lasagna, may recommend a discounted Italian red wine from Tuscany — a wine similar to one the customer enjoyed some time before but has since forgotten. Such personalized services make sure that shoppers get the best value. In this way, they are compensated for the risk of losing some privacy in instrumented shops.

Fig. 1 illustrates the use of augmented memory functions for automatic comparison shopping in a mixed-reality environment. The user wants to buy a new cell phone. At home, he searches the internet for new models. Our augmented memory service tracks the user's browsing behavior and stores the result in his personal journal. When the user decides to check the physical look and feel of the selected phone in a real shop, he can exploit various augmented memory functions using his internet-enabled PDA. As soon as he grasps a phone from the instrumented shopping shelf, SPECTER generates a comparison table of the features for this particular phone and the best-rated phone that the user found during his preparatory internet search. This is a typical instance of mixed-reality shopping, since the tangible experience with physical products is related to virtual shopping experiences through web browsing. When the user decides to buy a cell phone and puts it in the instrumented shopping basket, this event is recorded in the personal journal together with temporal and spatial information. Later at home, the user of SPECTER can review his digital diary and reflect about his shopping behavior, including entries about which products he has found on the web, which products he has checked in the instrumented shop without buying them, and which products were compared with each other.

The user of the DFKI Cybershopping mall can also look for audio CDs, in particular for soundtracks. People come in contact with soundtracks through various situations — e.g., in a cinema with the family, while watching a DVD at home, or while browsing an Internet store. Sometimes they have a precise idea of the music in question, and sometimes, they have never heard it. This background serves as the scenario in which a user exploits augmented memories by means of SPECTER in order to learn more about soundtracks that might be of interest.

The left side of Fig. 2 shows the user looking at a RFID-tagged CD, which she has grasped from the instrumented rack. The right side shows a screenshot from her PDA, which she is holding in her left hand to access the augmented memory services. SPECTER's personal journal shows that "The Lion King" has been explicitly evaluated by the user (the journal entry is labeled "Classifying") leading to the highest possible rating, visualized on the PDA screen as two thumbs up. For the "Stallion of the Cimarron" SPECTER notes that the user has grasped this CD and then checked a weblink that is automatically offered on the PDA to provide proactively additional product information (the journal entry is labeled "Looking in detail").

In order to actively acquire information about the soundtracks, our user can first browse the Web pages of an Internet store. SPECTER unobtrusively records these actions and assigns to each CD examined by the user a subjective rating based on the user's attention (for more details, see Fig. 8). While shopping, the system provides a listing of services related to the CDs being considered based on situational preferences. For instance, if she is in her favorite shop and has

Fig. 2. Creating a personal journal in SPECTER

spare time, the system may inform her of special offers on similar CDs. The user may exploit her augmented memories in several additional ways. For instance, if a CD is unknown to the user, the system may provide a list of similar CDs known by means of augmented memories, and thus provide a clue about its content. The other way around, the user can tell SPECTER to provide some examples of CDs she likes to the shop in order for them to suggest similar CDs as yet unfamiliar to her. All these actions contribute to her augmented memories and may therefore later become the subject of reflection and introspection.

Humans have memories filled with their experiences. But as an alternative to acquiring experiences on their own, humans often share memories with others (e.g. actively by telling stories or, more modernly, by blogging, passively by watching movies or reading autobiographies and test reports). Given augmented memories created on the basis of observations in instrumented environments and given several users with such memories based on our SPECTER software, the key research question of our SHAREDLIFE project is: Can we reproduce the natural exchange of memories to some degree to enrich the memories of individuals and support their activities?

Fig. 3 illustrates a first version of the SHAREDLIFE system used in our instrumented CD shop. The user's behavior, his ratings and past choices are captured in his augmented memory (see step 1 in Fig. 3). This personal memory can be used for a combination of reminding and recommendation, which we call "recomindation" (see step 2 in Fig. 3). The system reminds the user that he had listened to the soundtrack of "Toy Story" while he was watching the DVD with friends on the 1st of March 2005 at noon. In addition, it recommends to buy the CD, since the augmented memory includes a very favorable personal rating of this soundtrack. The user can publish parts of this shared memory after entering it in his ubiquitous user model (see step 3 in Fig. 3). He can specify privacy

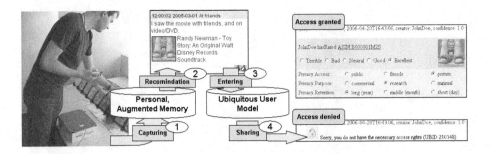

Fig. 3. Sharing augmented personal memories

constraints, so that a selective sharing of the augmented memories becomes possible. For instance, the access to the excellent rating of this soundtrack by the user may be denied for commercial use, but granted for research purposes only (see step 4 in Fig. 3).

4   From Sensor Data to Memories

The first step towards the creation of augmented memories is the automated recording of contextual information as perceived from various types of sensors. In our example scenario, each CD is an RFID-tagged smart object, which allows tracking its presence within the store areas (shelf, basket, cashier). Optionally, these objects may be anthropomorphized in order to facilitate the human-environment communication (cf. [25]). The user's location may be determined using IR, RFID, and/or GPS (see [6]). Biosensors (e.g., electrocardiogram (ECG) electromyogram (EMG), electrodermal activity (EDA), and acceleration (ACC) sensors) provide further information about the user's state, which is applied for choosing an appropriate communication channel and for automatically evaluating events (cf. [5]). Finally, Web services allow the system to acquire rich context information (e.g., the current weather or important events from RSS feeds), which may later on serve as an access key to the memory. In addition, such services are used to realize certain domain-specific features within the user's environment. For instance, SPECTER assists its user with services implemented by [2], such as a similarity search for CDs.

Each of these input sources is linked to a so-called *RDF store* (see [21]). Such a store provides an RDF-based interface to a sensor-specific memory, which is decoupled from the user's augmented memories. Two advantages provide the rationale for this separation:

Efficiency: sensor memories are not bound to SPECTER's RDF-based implementation of the augmented memory. This is of special interest due to the diversity of the perceived data, which may range from raw mass data of biosensors to rich information retrieved from Web services. For the same reason, sensors may implement their own abstraction methods — e.g., a simple mapping is performed in order to translate GPS coordinates into semantically meaningful locations,

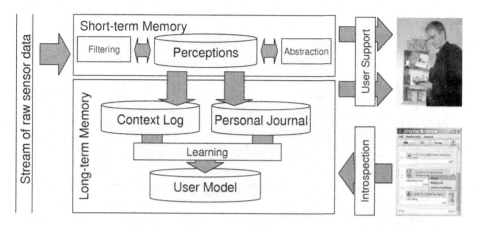

Fig. 4. Building memories from perceptions in SPECTER

whereas input from biosensors is processed on a mobile system by means of dynamic Bayesian networks.

Flexibility: for various reasons, the connection between the augmented memory and some sensors may sometimes be lost (e.g., technical issues, trust issues). In such a case, the user should be allowed to complete the augmented memories at a later time using records from the sensors' memories.

4.1   Modeling Perceptions

Sensors provide the system with *perceptions*; their RDF-encoded content contains simple statements such as "user reaching shelf" or "user holding CD". At any given point of time, the set of all available sensors' latest perceptions defines the *context* of an event in SPECTER.

Information contained in perceptions references an ontology based on the IEEE SUMO and MILO [18]. The user's state is modeled using the general user model ontology (GUMO, cf. [11]), a mid-level ontology which provides applications with a shared vocabulary for expressing statements about users. Furthermore, in order to facilitate the exchange of GUMO statements between different applications, the ontology provides a means of combining such statements with meta statements, e.g., about privacy, trust, and expiry issues. GUMO statements reference a variety of dimensions that describe user properties. Their basic dimensions include `contact information`, `personality`, and `emotional state`. These concepts are the topmost level of a broad range of specialized concepts. The complete ontology can be reviewed and edited online with an ontology browser provided at `http://www.gumo.org/`.

4.2   The Abstraction Pipeline of SPECTER

In order to describe how perceptions are processed and stored we have developed a *memory model* (see Fig. 4). In this model, incoming perceptions are stored in

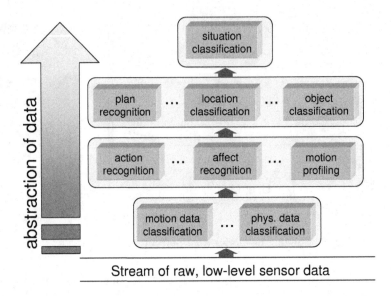

Fig. 5. An abstraction process is performed in order to gain symbolic information from raw sensor data

a short-term memory, which serves two main purposes. First, it is of special relevance for the recognition of situations and thus situated user support. The facts stored in the short-term memory model the user's current context; a BDI planner (JAM, cf. [13]) matches this context against patterns of service bindings specified in the user model (more about this in the next section). The second purpose of the short-term memory is to trigger, based on events and event chains, the construction of episodes in the long-term memory.

Our approach to a long-term memory for intelligent environments is two-fold. First of all, perceptions are stored in a *context log* without further change. This log works primarily as the system's memory. It is linked to the *personal journal*, which consists of entries representing episodes created from one or more perceptions. For instance, the perceptions "user holding A" and "user holding B" can be combined to the episode "user comparing A and B". The creation of journal entries is an abstraction process, which is performed using pre-authored rules expressing commonsense knowledge and domain knowledge.

Fig. 5 illustrates the abstraction pipeline realized in SPECTER. The stream of raw level sensor data is first classified, so that basic motion data can be derived (compare the bottom of Fig. 5). The results of RFID readers can be abstracted to the observation that a certain product has been removed from a particular shopping shelf and put into an instrumented shopping basket. This can be further abstracted to an intended purchase of this product. If the shopper has already put pasta sheets, chopped tomatoes, and beef in his basket then the plan recognition mechanism of SPECTER will generate the hypothesis that the user wants to prepare a lasagne. When an additional plan for preparing tiramisu is recognized and a 12-bottle box of chianti wine is bought, SPECTER may classify the situation on the highest level of abstraction as "shopping for an Italian dinner party" (compare the top of Fig. 5).

The grocery shopping scenario involves complex constraints on cooking for a particular dinner guest, such as availability, food allergies, dietary rules, and religious food preferences. Thus, shopping tips and shared cooking experiences are most welcome and SHAREDLIFE may grant limited access to the augmented memory of friends and family members with cooking expertise.

5   Exploiting Memories

We focused in the previously described shopping scenario on a specific way of exploiting augmented memories. Depending on the user's current context, the system offered recommendations and reminders related to that context with the goal of putting information about past experiences relevant for the current situation into the user's mind. For instance, if the user is inspecting some CD in a shop, the system might come up with cheaper offers previously seen, or with recommendations of similar CDs in this shop. In order to describe such processes we coined the notion *recomindation*; a study with 20 subjects showed that a number of aspects of this paradigm tend to be recognized not only as appropriate and effective in supporting the user's shopping experience, but also as *enjoyable* (cf. [20]).

5.1   Reflection and Introspection

A crucial factor for the quality of such support is information about the relevance of the numerous events recorded over time for a given situation. The relevance is determined by several factors, and one of them is explicit or implicit user feedback. Feedback is represented by ratings, which may be attached to personal journal entries. We experimented with various rating dimensions including *evaluation* (a general quality judgement of objects referenced by the entry), *importance* (how important is the entry with respect to the user's current goals), and *urgency* (how urgent is the entry with respect to the user's goals). Ratings are assigned either explicitly by the user (cf. the left-hand side of Fig. 8) or implicitly based on feedback from biosensors and domain-specific heuristics.

Ratings assigned by the system may differ from the user's perception, e.g., due to noise within the sensor data or inappropriate heuristics. Furthermore, especially in the case of an "untrained" system, the mapping of situations to services might require an adaptation to the user's personal needs as well. The user may address such issues by performing an *introspection* of her augmented memories.

We think of introspection as *"...a process of inward attention or reflection, so as to examine the contents of the mind..."* (cf. [22]). In the case of our approach to augmented personal memories, introspection consists of processes in which the user and/or the system explore the long-term memory in order to learn about the course of events. From the user's point of view this includes the option to explore and to rate journal entries, including those produced in response to

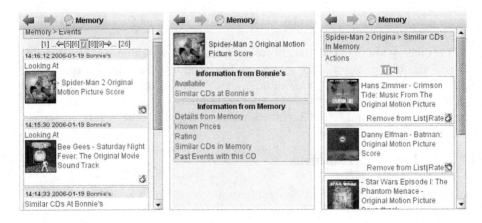

Fig. 6. Views on augmented memories: events, functions, and objects (from the left to the right)

system actions. From the system's point of view introspection is an opportunity to refine, collaboratively with the user, the user model.

Feedback from our test user group indicated that an introspection should be possible in a mobile setting (in order to make use of spare time, e.g., during a train ride) as well as in a desktop setting. This is reflected in SPECTER's user interface to augmented memories, which consists of two major components.

On a mobile device, a *journal browser* lets the user explore and evaluate memories. It provides the user with diverse views on her augmented memories. Fig. 6 illustrates three of these views:

- *Events:* For each event, this view provides information about its context (e.g., location and time), a summary of actions observed by SPECTER, and ratings assigned by system and user. Additional filters may constrain the list of shown events.
- *Functions:* This view lists services related to some object contained in the augmented memory, which can be applied by the user in the current context. Here, the user interface distinguishes environment- and memory-related functions. In our example, the current location is the CD store "Bonnie's", which allows the user to ask for similar CDs on sale ("Similar CDs at Bonnies"). Other functions such as a price comparison ("Known Prices") make use of the augmented memory in order to look up known instances of the CD.
- *Objects:* At many opportunities an object-centered view is applied by the system in order to focus specific sets of information from the memory, such as the outcome of a retrieval process. This view presents a list of objects (here: audio CDs), which can be exploited by the user in diverse ways, for instance, to specify example-based queries to memory and environment.

These views are typically triggered by interactions between user and environment. Complex interactions often involve several views; in our example, the user starts from an event referring the "Spider-Man 2" soundtrack and uses the memory-related function "Similar CDs in Memory" in order to retrieve an object list of similar CDs seen so far.

Other functions of the journal browser are not directly related to the memory, but to the system configuration. These include top-level manipulation of system services (e.g., switching off some support), bookmarking of views, and so-called *reminder points*, which indicate the need for a close review of the current situation.

Such reflection and introspection can be performed on a regular desktop PC using an *introspection environment*. It provides a rich user interface which enables in combination with the planner already mentioned a collaborative introspection of augmented memories (see [3] for a detailed description). If the user is exploring the memory, then the system assists by offering event summaries as well as details and links to the memory and the Web. The latter point to external resources and services, and thus provide another means for retrieving and adding information to the memory.

These activities usually require the user's initiative. The system also proactively checks the memory for situations where clarification might help to improve the user model; examples of such situations are reminder points and user feedback obtained during the execution of some supporting service. When such situations are detected, the user is asked to enter the collaborative process described below.

5.2   Collaborative Critique of Situated User Support

If the user is not confident with the system's suggestions, or wants to set up a new service binding, there is a component explicitly designed for mapping situations to services. This component is described in detail in [4]; at this point we will only summarize its features and focus on its role within the introspection process.

The purpose of this component is to provide a scrutable and easy-to-use interface that allows the user to interact with complex machine-learning processes without the need to deal with the technical subtleties of feature selection or data encoding. The key to our approach is to combine the system's capability to deal with the statistical relevance of a situation's features with the user's ability to name semantically meaningful concepts that can and should be used to describe the characteristics of a situation.

The result of our ongoing effort is a user interface which provides several interaction layers of varying complexity for combining services and situations. Especially relevant for a critique of a situation's features (and thus for configuring the execution of linked services) is the screen shown on the left-hand side of Fig. 7. This shows a list of features, which have been extracted using statistical methods from the memory, as it is presented to the user.

The user may critique this set by deselecting features, or by navigating in their semantic neighborhood using a graphical interface to the underlying ontology. In our example scenario, the user might want to inform the system that recommendations should only be retrieved in certain kinds of shops. A way to achieve this goal is to inspect the list of features and then to refine the shop's features, e.g., by replacing the general shop by a more specific branch.

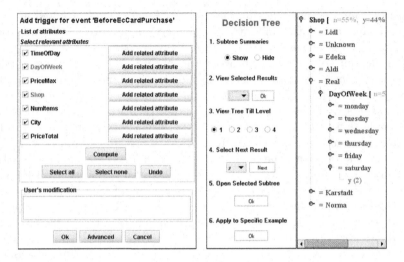

Fig. 7. A user interface for critiquing situational features (left-hand side) and inspecting decision trees generated from these features (right-hand side)

The system applies the adjusted feature set for computing a decision tree. It becomes connected to the short-term memory; from now on it is used to classify the system's observations and thus to decide if the service chosen by the user should be triggered. In order to make this mechanism transparent to the user, the decision tree can be inspected by means of the graphical user interface shown on the right-hand side of Fig. 7. This interface provides various ways of navigating the decision tree, and offers additional information about the relevance of the selected nodes based on the number of positive and negative examples taken from the personal journal.

6   Selective Memory Sharing

So far we focused on personal use of augmented memories. However, there is often the need to communicate personal memories with the environment: for instance, the user may select items from the memory and provide these as examples to the environment in order to personalize services offered there. Of course, these applications may exploit such data for building their own model of the user.

This way of *sharing* personal augmented memories matches the idea of ubiquitous user modeling, which can be described as *"...ongoing modeling and exploitation of user behavior with a variety of systems that share their user models..."* (see [10]). In the following, we will illustrate how by means of a platform for ubiquitous user modeling, namely U2M (cf. www.u2m.org), such sharing mechanisms can be realized.

Within U2M, the concept of *sharing* is split up into *exchanging* and *integrating* statements about users. The former is realized by a user model server that provides a service-based architecture for distributed storage and retrieval of

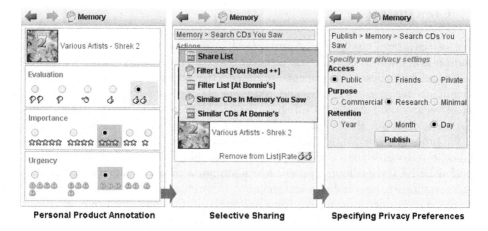

Personal Product Annotation   **Selective Sharing**   **Specifying Privacy Preferences**

Fig. 8. Selective memory sharing with privacy constraints

statements about users. The integration of statements is achieved with the accretion model according to [14] together with a multilevel conflict resolution method described in [10], which also solves the problem of contradictory information.

What statements can be retrieved and how they are integrated depends on several layers of metadata attached to the statements by means of reification. From the outermost to the innermost layer, these are: `administration`, `privacy`, `explanation`, and `situation`. They establish a sequence of access constraints which have to be met in order to obtain the reified statement. The privacy layer in this sequence is of special interest. It implements the following privacy attributes: `key`, `owner`, `access`, `purpose`, and `retention`. The user model server checks these attributes in order to deliver as much information as possible without violating the user's preferences. Combined with the other layers, complex situational access constraints can be established, such as "friends only & at my home & for personal purposes".

Fig. 8 depicts how this technology allows for sharing information extracted from an augmented personal memory. In our example, the user evaluated the CD "Shrek 2" very positively (two thumbs up, left-hand side of Fig. 8). Since no specific situational context is provided for this evaluation, the ratings for importance and urgency have been set by the system to a neutral default value. A context menu allows the user to initiate a sharing process at any time for lists of objects – here: audio CDs – retrieved from the augmented memory. The middle of Fig. 8 shows that the user has selected a list including the CD "Shrek 2". The user may specify privacy-related preferences (here: about access, purpose, and retention) explicitly for the current sharing process (right-hand side of Fig. 8) or rely on U2M's default reasoning which derives privacy preferences from personal defaults set in the ubiquitous user model. Once submitted, U2M makes the data accessible to other users with respect to the user's privacy preferences.

7   Conclusions

One of the grand challenges of current AI research is to create instrumented environments and to capture the physical and communicative interaction of an individual with these environments, to process this multimodal and massive data stream, to recognize, classify, and represent its digital content in a context-sensitive way, and finally to integrate behavior understanding with reasoning and learning about the individual's day by day experiences.

We presented our SHAREDLIFE project, in which we are creating augmented episodic memories that are personal and sharable. We described the experimental DFKI Cybershopping mall, which supports mixed-reality shopping and which augments the usual physical shopping experience with personalized virtual shopping assistance known from some online shops. Our research is aimed at exploiting the networked infrastructure for more personalized shopping assistance like digital shopping list support, automatic comparison shopping, cross- and up-selling, proactive product information, and in-shop navigation. In the technical core of the paper, we described in detail the abstraction pipeline and the first implementation of the memory sharing mechanism of SHAREDLIFE, which is based on a ubiquitous user modeling server.

Our future work on SHAREDLIFE will address the question of how the sharing of augmented memories can contribute to the communication within small, potentially ad-hoc formed groups. We want to provide mechanisms for automated and semi-automated memory sharing. Such mechanisms must not only take into account situated access constraints on privacy and trust (e.g., in order to distinguish between situations of everyday life and emergency cases), but also the structure of the group (e.g., to define experts or opinion leaders).

Acknowledgements. This research was supported by the German Ministry of Education and Research (BMBF) under grant 524-40001-01 IW C03 (project SPECTER) and under grant 01 IW F03 (project SHAREDLIFE). Furthermore, the German Research Foundation (DFG) supported this research in its Collaborative Research Center on Resource-Adaptive Cognitive Processes, SFB 378, Project EM 4, BAIR. Finally, we would like to thank all project members for their valuable contributions.

References

1. K. Aizawa, D. Tancharoen, S. Kawasaki, and T. Yamasaki. Efficient retrieval of life log based on context and content. In *CARPE'04: Proceedings of the the 1st ACM workshop on Continuous archival and retrieval of personal experiences*, pages 22–31, New York, NY, USA, 2004. ACM Press.
2. Amazon Inc. Amazon e-commerce service, April 2005. http://www.amazon.com-/gp/aws/landing.html.
3. S. Baldes, A. Kröner, and M. Bauer. Configuration and introspection of situated user support. In *LWA 2005, Lernen Wissensentdeckung Adaptivität*, pages 3–7, Saarland University, Saarbrücken, Germany, 2005.

4. M. Bauer and S. Baldes. An Ontology-Based Interface for Machine Learning. In John Riedl, Anthony Jameson, Daniel Billsus, and Tessa Lau, editors, *IUI 2005: International Conference on Intelligent User Interfaces*, pages 314–316, New York, 2005. ACM.

5. B. Brandherm, H. Schultheis, M. von Wilamowitz-Moellendorff, T. Schwartz, and M. Schmitz. Using physiological signals in a user-adaptive personal assistant. In *Proceedings of the 11th International Conference on Human-Computer Interaction (HCII-2005)*, Las Vegas, Nevada, USA, 2005.

6. B. Brandherm and T. Schwartz. Geo referenced dynamic Bayesian networks for user positioning on mobile systems. In Thomas Strang and Claudia Linnhoff-Popien, editors, *Proceedings of the International Workshop on Location- and Context-Awareness (LoCA), LNCS 3479*, pages 223–234, Munich, Germany, 2005. Springer-Verlag Berlin Heidelberg.

7. B. Clarkson. *Life Patterns: structure from wearable sensors*. PhD thesis, School of Architecture and Planning, Massachusetts Institute of Technology, Cambridge, MA, 2002.

8. S. Dumais, E. Cutrell, J. Cadiz, G. Jancke, R. Sarin, and D. C. Robbins. Stuff i've seen: a system for personal information retrieval and re-use. In *SIGIR '03: Proceedings of the 26th annual international ACM SIGIR conference on Research and development in informaion retrieval*, pages 72–79. ACM Press, 2003.

9. J. Gemmell, L. Williams, K. Wood, R. Lueder, and G. Bell. Passive Capture and Ensuing Issues for a Personal Lifetime Store. In *Proceedings of The First ACM Workshop on Continuous Archival and Retrieval of Personal Experiences (CARPE '04)*, pages 48–55, New York, USA, 2004.

10. D. Heckmann. *Ubiquitous User Modeling*. PhD thesis, Department of Computer Science, Saarland University, Germany, 2005.

11. D. Heckmann, B. Brandherm, M. Schmitz, Tim Schwartz, and Baroness Margeritta von Wilamowitz-Moellendorf. GUMO - the general user model ontology. In *Proceedings of the 10th International Conference on User Modeling*, pages 428–432, Edinburgh, Scotland, Jun 2005. LNAI 3538: Springer, Berlin Heidelberg.

12. E. Horvitz, S. Dumais, and P. Koch. Learning predictive models of memory landmarks. In *Proceedings of the CogSci 2004: 26th Annual Meeting of the Cognitive Science Society*, Chicago, USA, August 2004.

13. M. J. Huber. JAM: a BDI-theoretic mobile agent architecture. In *AGENTS '99: Proceedings of the third annual conference on Autonomous Agents*, pages 236–243, New York, NY, USA, 1999. ACM Press.

14. J. Kay, B. Kummerfeld, and P. Lauder. Personis: A server for user models. In Paul de Bra, Peter Brusilovsky, and R. Conejo, editors, *Proceedings of the Second International Conference on Adaptive Hypermedia and Adaptive Web-Based Systems*, LNCS 2347, pages 203–212. Springer-Verlag Berlin Heidelberg, May 2002.

15. A. Kröner, D. Heckmann, and W. Wahlster. SPECTER: Building, exploiting, and sharing augmented memories. In Kiyoshi Kogure, editor, *Workshop on Knowledge Sharing for Everyday Life 2006 (KSEL06)*, pages 9–16, Kyoto, Japan, February 2006. ATR Media Information Science Laboratories. Invited talk.

16. N. Kuwahara, H. Noma, K. Kogure, N. Hagita, N. Tetsutani, and H. Iseki. Wearable auto-event-recording of medical nursing. In *Proceedings of the Ninth IFIP TC13 International Conference on Human-Computer Interaction (INTERACT 2003)*, September 2003.

17. M. Lamming and M. Flynn. "Forget-Me-Not": Intimate computing in support of human memory. In *Proceedings of FRIEND21, the 1994 International Symposium on Next Generation Human Interface, Meguro Gajoen, Japan*, Meguro Gajoen, Japan, 1994.

18. I. Niles and A. Pease. Towards a standard upper ontology. In *FOIS '01: Proceedings of the international conference on Formal Ontology in Information Systems*, pages 2–9. ACM Press, 2001.

19. M. Philipose, K. P. Fishkin, M. Perkowitz, D. Patterson, and D. Hähnel. The probabilistic activity toolkit: Towards enabling activity-aware computer interfaces. Technical Report IRS-TR-03-013, Intel Research Seattle, Seattle, WA 98115, USA, 2003.

20. C. Plate, N. Basselin, A. Kröner, M. Schneider, S. Baldes, V. Dimitrova, and A. Jameson. Recomindation: New functions for augmented memories. In V. Wade and H. Ashman, editors, *Adaptive hypermedia and adaptive web-based systems: Proceedings of AH 2006*, Berlin, 2006. Springer.

21. M. Schneider. RDF:Stores – a lightweight approach on managing shared knowledge. In Jianhua Ma, Hai Jin, Laurence T. Yang, and Jeffrey J.-P. Tsai, editors, *Proceedings of the 3rd International Conference on Ubiquitous Intelligence and Computing (UIC 2006)*. Springer, 2006.

22. M. Sharples, D. Hogg, C. Hutchinson, and S. Torrance. *Computers and Thought: A Practical Introduction to Artificial Intelligence*. Bradford Book, 1989.

23. K. Stathis, O. de Bruijn, and S. Macedo. Living memory: agent-based information management for connected local communities. *Interacting with Computers*, 14(6):663–688, 2002.

24. Y. Sumi and K. Mase. Supporting awareness of shared interests and experiences in community. *International Journal of Human-Computer Studies*, 56(1):127–146, January 2002.

25. R. Wasinger and W. Wahlster. The anthropomorphized product shelf: Symmetric multimodal interaction with instrumented environments. In E. Aarts and J. Encarnação, editors, *True Visions: The Emergence of Ambient Intelligence*. Springer, Heidelberg, Berlin, New York, February 2006.

Author Index

Lecture Notes in Artificial Intelligence (LNAI)

Vol. 3918: W.K. Ng, M. Kitsuregawa, J. Li, K. Chang (Eds.), Advances in Knowledge Discovery and Data Mining. XXIV, 879 pages. 2006.

Vol. 3913: O. Boissier, J. Padget, V. Dignum, G. Lindemann, E. Matson, S. Ossowski, J.S. Sichman, J. Vázquez-Salceda (Eds.), Coordination, Organizations, Institutions, and Norms in Multi-Agent Systems. XII, 259 pages. 2006.

Vol. 3910: S.A. Brueckner, G.D.M. Serugendo, D. Hales, F. Zambonelli (Eds.), Engineering Self-Organising Systems. XII, 245 pages. 2006.

Vol. 3904: M. Baldoni, U. Endriss, A. Omicini, P. Torroni (Eds.), Declarative Agent Languages and Technologies III. XII, 245 pages. 2006.

Vol. 3900: F. Toni, P. Torroni (Eds.), Computational Logic in Multi-Agent Systems. XVII, 427 pages. 2006.

Vol. 3899: S. Frintrop, VOCUS: A Visual Attention System for Object Detection and Goal-Directed Search. XIV, 216 pages. 2006.

Vol. 3898: K. Tuyls, P.J. 't Hoen, K. Verbeeck, S. Sen (Eds.), Learning and Adaption in Multi-Agent Systems. X, 217 pages. 2006.

Vol. 3891: J.S. Sichman, L. Antunes (Eds.), Multi-Agent-Based Simulation VI. X, 191 pages. 2006.

Vol. 3890: S.G. Thompson, R. Ghanea-Hercock (Eds.), Defence Applications of Multi-Agent Systems. XII, 141 pages. 2006.

Vol. 3885: V. Torra, Y. Narukawa, A. Valls, J. Domingo-Ferrer (Eds.), Modeling Decisions for Artificial Intelligence. XII, 374 pages. 2006.

Vol. 3881: S. Gibet, N. Courty, J.-F. Kamp (Eds.), Gesture in Human-Computer Interaction and Simulation. XIII, 344 pages. 2006.

Vol. 3874: R. Missaoui, J. Schmidt (Eds.), Formal Concept Analysis. X, 309 pages. 2006.

Vol. 3873: L. Maicher, J. Park (Eds.), Charting the Topic Maps Research and Applications Landscape. VIII, 281 pages. 2006.

Vol. 3864: Y. Cai, J. Abascal (Eds.), Ambient Intelligence in Everyday Life. XII, 323 pages. 2006.

Vol. 3863: M. Kohlhase (Ed.), Mathematical Knowledge Management. XI, 405 pages. 2006.

Vol. 3862: R.H. Bordini, M. Dastani, J. Dix, A.E.F. Seghrouchni (Eds.), Programming Multi-Agent Systems. XIV, 267 pages. 2006.

Vol. 3849: I. Bloch, A. Petrosino, A.G.B. Tettamanzi (Eds.), Fuzzy Logic and Applications. XIV, 438 pages. 2006.

Vol. 3848: J.-F. Boulicaut, L. De Raedt, H. Mannila (Eds.), Constraint-Based Mining and Inductive Databases. X, 401 pages. 2006.

Vol. 3847: K.P. Jantke, A. Lunzer, N. Spyratos, Y. Tanaka (Eds.), Federation over the Web. X, 215 pages. 2006.

Vol. 3835: G. Sutcliffe, A. Voronkov (Eds.), Logic for Programming, Artificial Intelligence, and Reasoning. XIV, 744 pages. 2005.

Vol. 3830: D. Weyns, H. V.D. Parunak, F. Michel (Eds.), Environments for Multi-Agent Systems II. VIII, 291 pages. 2006.

Vol. 3817: M. Faundez-Zanuy, L. Janer, A. Esposito, A. Satue-Villar, J. Roure, V. Espinosa-Duro (Eds.), Nonlinear Analyses and Algorithms for Speech Processing. XII, 380 pages. 2006.

Vol. 3814: M. Maybury, O. Stock, W. Wahlster (Eds.), Intelligent Technologies for Interactive Entertainment. XV, 342 pages. 2005.

Vol. 3809: S. Zhang, R. Jarvis (Eds.), AI 2005: Advances in Artificial Intelligence. XXVII, 1344 pages. 2005.

Vol. 3808: C. Bento, A. Cardoso, G. Dias (Eds.), Progress in Artificial Intelligence. XVIII, 704 pages. 2005.

Vol. 3802: Y. Hao, J. Liu, Y.-P. Wang, Y.-m. Cheung, H. Yin, L. Jiao, J. Ma, Y.-C. Jiao (Eds.), Computational Intelligence and Security, Part II. XLII, 1166 pages. 2005.

Vol. 3801: Y. Hao, J. Liu, Y.-P. Wang, Y.-m. Cheung, H. Yin, L. Jiao, J. Ma, Y.-C. Jiao (Eds.), Computational Intelligence and Security, Part I. XLI, 1122 pages. 2005.

Vol. 3789: A. Gelbukh, Á. de Albornoz, H. Terashima-Marín (Eds.), MICAI 2005: Advances in Artificial Intelligence. XXVI, 1198 pages. 2005.

Vol. 3782: K.-D. Althoff, A. Dengel, R. Bergmann, M. Nick, T.R. Roth-Berghofer (Eds.), Professional Knowledge Management. XXIII, 739 pages. 2005.

Vol. 3763: H. Hong, D. Wang (Eds.), Automated Deduction in Geometry. X, 213 pages. 2006.

Vol. 3755: G.J. Williams, S.J. Simoff (Eds.), Data Mining. XI, 331 pages. 2006.

Vol. 3735: A. Hoffmann, H. Motoda, T. Scheffer (Eds.), Discovery Science. XVI, 400 pages. 2005.

Vol. 3734: S. Jain, H.U. Simon, E. Tomita (Eds.), Algorithmic Learning Theory. XII, 490 pages. 2005.

Vol. 3721: A.M. Jorge, L. Torgo, P.B. Brazdil, R. Camacho, J. Gama (Eds.), Knowledge Discovery in Databases: PKDD 2005. XXIII, 719 pages. 2005.

Vol. 3720: J. Gama, R. Camacho, P.B. Brazdil, A.M. Jorge, L. Torgo (Eds.), Machine Learning: ECML 2005. XXIII, 769 pages. 2005.

Vol. 3717: B. Gramlich (Ed.), Frontiers of Combining Systems. X, 321 pages. 2005.

Vol. 3702: B. Beckert (Ed.), Automated Reasoning with Analytic Tableaux and Related Methods. XIII, 343 pages. 2005.

Vol. 3698: U. Furbach (Ed.), KI 2005: Advances in Artificial Intelligence. XIII, 409 pages. 2005.

Vol. 3690: M. Pěchouček, P. Petta, L.Z. Varga (Eds.), Multi-Agent Systems and Applications IV. XVII, 667 pages. 2005.

Vol. 3684: R. Khosla, R.J. Howlett, L.C. Jain (Eds.), Knowledge-Based Intelligent Information and Engineering Systems, Part IV. LXXIX, 933 pages. 2005.

Vol. 3683: R. Khosla, R.J. Howlett, L.C. Jain (Eds.), Knowledge-Based Intelligent Information and Engineering Systems, Part III. LXXX, 1397 pages. 2005.

Vol. 3682: R. Khosla, R.J. Howlett, L.C. Jain (Eds.), Knowledge-Based Intelligent Information and Engineering Systems, Part II. LXXIX, 1371 pages. 2005.